T0250061

FOOD SAFETY CHEMISTRY

TOXICANT OCCURRENCE, ANALYSIS AND MITIGATION

EDITED BY

LIANGLI (LUCY) YU
SHUO WANG
BAO-GUO SUN

CRC Press
Taylor & Francis Group
Boca Raton London New York

CRC Press is an imprint of the
Taylor & Francis Group, an **informa** business

CRC Press
Taylor & Francis Group
6000 Broken Sound Parkway NW, Suite 300
Boca Raton, FL 33487-2742

First issued in paperback 2017

© 2015 by Taylor & Francis Group, LLC
CRC Press is an imprint of Taylor & Francis Group, an Informa business

No claim to original U.S. Government works

ISBN-13: 978-1-4665-9794-5 (hbk)
ISBN-13: 978-1-138-03381-8 (pbk)

This book contains information obtained from authentic and highly regarded sources. Reasonable efforts have been made to publish reliable data and information, but the author and publisher cannot assume responsibility for the validity of all materials or the consequences of their use. The authors and publishers have attempted to trace the copyright holders of all material reproduced in this publication and apologize to copyright holders if permission to publish in this form has not been obtained. If any copyright material has not been acknowledged please write and let us know so we may rectify in any future reprint.

Except as permitted under U.S. Copyright Law, no part of this book may be reprinted, reproduced, transmitted, or utilized in any form by any electronic, mechanical, or other means, now known or hereafter invented, including photocopying, microfilming, and recording, or in any information storage or retrieval system, without written permission from the publishers.

For permission to photocopy or use material electronically from this work, please access www.copyright.com (http://www.copyright.com/) or contact the Copyright Clearance Center, Inc. (CCC), 222 Rosewood Drive, Danvers, MA 01923, 978-750-8400. CCC is a not-for-profit organization that provides licenses and registration for a variety of users. For organizations that have been granted a photocopy license by the CCC, a separate system of payment has been arranged.

Trademark Notice: Product or corporate names may be trademarks or registered trademarks, and are used only for identification and explanation without intent to infringe.

Library of Congress Cataloging-in-Publication Data

Food safety chemistry: toxicant occurrence, analysis and mitigation / edited by Liangli (Lucy) Yu, Shuo Wang, Bao-Guo Sun.
    pages cm
  Includes bibliographical references and index.
  ISBN 978-1-4665-9794-5 (hardback)
  1. Food adulteration and inspection. 2. Food contamination. 3. Food--Analysis. 4. Food--Microbiology. I. Yu, Liangli. II. Wang, Shuo, 1969- III. Sun, Baoguo, 1961-

TX531.F56824 2014
363.19'264--dc23
                                              2014034196

Visit the Taylor & Francis Web site at
http://www.taylorandfrancis.com

and the CRC Press Web site at
http://www.crcpress.com

Contents

Editors

Dr. Liangli (Lucy) Yu is a professor in the Department of Nutrition and Food Science at the University of Maryland, College Park, Maryland; she also served as a "Zhi-Yuan" chair professor in the School of Agriculture and Biology at Shanghai Jiao Tong University. Her research focuses on food safety chemistry, and the chemistry and biochemistry of nutraceuticals and functional foods. Dr. Yu is a leading food chemist in nontargeted detection of food chemical hazards, food adulteration detection, the chemical mechanisms involved in toxicant formation during food processing, and molecular and cellular food toxicology. In addition, Dr. Yu's group has contributed significantly to food antioxidant, food oxidation and antioxidation, and nutraceuticals and functional food chemistry.

Dr. Yu earned her PhD in food science/food chemistry from Purdue University. She also earned an MS in medicinal chemistry and a BS in pharmaceutical chemistry from the China Pharmaceutical University, Nanjing, China. Dr. Yu has more than 140 research articles published/accepted in refereed scientific journals, along with 13 book chapters, and many invited talks and conference presentations. She has also (co)edited a textbook on organic chemistry, a monograph on wheat antioxidants, and a monograph *Cereals and Pulses: Nutraceutical Properties and Health Benefits.* Dr. Yu is a Fellow of the Institute of Food Technologists and a Fellow of the American Chemical Society, Agricultural and Food Chemistry Division. She received several scholarly awards including the 2006 Young Scientist Award from the American Chemical Society, Agricultural and Food Chemistry Division for her contribution in research and development of nutraceuticals and functional foods, and the 2008 Young Scientist Research Award from the American Oil Chemist Society for her contribution in value-added specialty edible seed oils flours, and 2012 Distinguished Alumina Award from the College of Agriculture, Purdue University. She is a member of the Food Ingredients Expert Committee and the Food Ingredient Intentional Adulterants Expert Panel for the Food Chemical Codex of US Pharmacopeia. She serves as an associate editor for the *Journal of Agricultural and Food Chemistry,* served as the associate editor for the *LWT-Food Science and Technology,* and *Journal of Food Science,* and is an editorial board member for *Food Chemistry* and *LWT-Food Science and Technology.* Her research has been reported in many news reports worldwide including CNN, ABC, NBC, and BBC.

Dr. Baoguo Sun has been professor of applied chemistry in Beijing Technology and Business University since 1997. His main research interest focuses on flavors and savory flavorings. He created a characteristic molecular structural unit model for sulfur-containing meat flavors and developed a series of manufacturing technologies for many important meat flavors in China. He also developed the preparation concept for Chinese characteristic meat flavorings, that is, "the identical origin of aroma with raw material," and crucial techniques for preparing meat flavorings based on meat, bone, and fat. He has many great achievements in sulfur-containing flavors and savory flavorings and was awarded second prize for the National Technologic Invention from State Council of the People's Republic of China and two second prizes for the National Scientific and Technologic Progress. He has been granted 15 invention patents and published 9 academic books and over 300 articles. He was honored as an Academician of Chinese Academy of Engineering in 2009. In addition, he is also dedicated to various administration affairs. He is vice president of the Beijing Technology and Business University and serves in many nongovernmental organizations concurrently, including vice president of the China National Light Industry Council, vice president of the Chinese Institute of Food Science and Technology, vice executive director of the China Association of Fragrance Flavor and Cosmetic Industries, vice president of the China Food Additives and Ingredients Association, and vice director of the China Fermentation Industry Association, among others.

Dr. Shuo Wang is a "Cheung Kong Scholar" chair professor and president of Tianjin University of Science and Technology, Tianjin, China. His research focuses on fundamental theory and technology on rapid analysis of trace-level hazardous chemicals in foods and on the formation of hazardous chemicals and their control techniques in food processing. Dr. Wang earned his PhD and MS in agricultural chemistry from the University of Sydney in 1999 and 1994, respectively. He also earned a BS in environmental biology in 1991 from Nankai University, Tianjin, China. Dr. Wang has had more than 270 publications (over 150 in internationally peer-reviewed scientific journals), along with 6 book/book chapters, 9 Chinese patents, 35 invited talks, and a number of conference presentations. Dr. Wang has received dozens of scholarly awards including second prize from the National Science and Technology Progress Award by the Chinese Government (2012), first prize from the Tianjin Science and Technology Progress Award by Tianjin Municipal Government (2010), second prize from the Science and Technology Progress Award by the National Education Ministry (2009), second prize from the Tianjin Natural Science Award by Tianjin Municipal Government (2009), second prize from the Science and Technology Progress Award by Tianjin Municipal Government (2005, 2008). He was also awarded the National Natural Science Funds for Distinguished Young Scholars of China, Distinguished Professor by Tianjin Municipal Government, among others. He is an editorial board member for the *Journal of Agricultural and Food Chemistry* (American Chemical Society), *LWT-Food Science and Technology* (Elsevier), and several Chinese peer-reviewed journals such as *Food Science, Food Research and Development, Food and Machinery, Food Safety, Journal of Chinese Institute of Food Science and Technology,* and *Food Industry and Technology.*

Contributors

Casimir C. Akoh
Department of Food Science and
   Technology
University of Georgia
Athens, Georgia

Didem Peren Aykas
Food Science and Technology
   Department
The Ohio State University
Columbus, Ohio

Xinyu Chen
Department of Food Science and
   Nutrition
Zhejiang University
Hangzhou, People's Republic of China

Ka-Wing Cheng
School of Biological Sciences
Stony Brook University
Stony Brook, New York

Zeyuan Deng
State Key Lab of Food Science and
   Technology
Nanchang University
Nanchang, Jiangxi, People's Republic
   of China

Boyan Gao
Institute of Food and Nutraceutical
   Science
Shanghai Jiao Tong University
Shanghai, People's Republic of China

and

Department of Nutrition and Food
   Science
University of Maryland
College Park, Maryland

Chi-Tang Ho
Department of Food Science
Rutgers University
New Brunswick, New Jersey

Haiqiu Huang
Department of Nutrition and Food
   Science
University of Maryland
College Park, Maryland

and

Institute of Food and Nutraceutical
   Science
Shanghai Jiao Tong University,
Shanghai, People's Republic of
   China

Ram Chandra Reddy Jala
Centre for Lipid Research
CSIR—Indian Institute of Chemical
   Technology
Hyderabad, India

Hongyan Li
State Key Lab of Food Science and
   Technology
Nanchang University
Nanchang, Jiangxi, People's Republic
   of China

Jing Li
State Key Lab of Food Science and
   Technology
Nanchang University
Nanchang, Jiangxi, People's Republic
   of China

Xihong Li
Key Laboratory of Food Nutrition and
   Safety
Tianjin University of Science and
   Technology
Tianjin, People's Republic of
   China

Jeffrey C. Moore
U.S. Pharmacopeial Convention
Rockville, Maryland

Magdi M. Mossoba
Center for Food Safety and Applied
   Nutrition
Food and Drug Administration
College Park, Maryland

Huan Rao
State Key Lab of Food Science and
   Technology
Nanchang University
Nanchang, Jiangxi, People's Republic
   of China

Luis Rodriguez-Saona
Food Science and Technology
   Department
The Ohio State University
Columbus, Ohio

Xi Shao
Department of Food Science
Rutgers University
New Brunswick, New Jersey

Takayuki Shibamoto
Department of Environmental
   Toxicology
University of California
Davis, California

Baoguo Sun
School of Food and Chemical
   Engineering
Beijing Technology and Business
   University
Beijing, People's Republic of
   China

Yao Tang
Key Laboratory of Food Nutrition and
   Safety
Tianjin University of Science and
   Technology
Tianjin, People's Republic of China

and

Guelph Food Research Centre
Agriculture and Agri-Food Canada
Guelph, Ontario, Canada

Zi Teng
Department of Nutrition and Food
   Science
University of Maryland
College Park, Maryland

Rong Tsao
Guelph Food Research Centre
Agriculture and Agri-Food
   Canada
Guelph, Ontario, Canada

Yen-Chen Tung
Department of Food Science
Rutgers University
New Brunswick, New Jersey

Pierina Visciano
Facoltà di Bioscienze e Tecnologie
   Agro-Alimentari e Ambientali
University of Teramo
Teramo, Italy

Jing Wang
School of Food and Chemical
  Engineering
Beijing Technology and Business
  University
Beijing, People's Republic of China

Junping Wang
College of Food Engineering and
  Biotechnology
Tianjin University of Science and
  Technology
Tian Jing, People's Republic of China

Ming-Fu Wang
School of Biological Sciences
The University of Hong Kong
Hong Kong

Qin Wang
Department of Nutrition and Food
  Science
University of Maryland
College Park, Maryland

Shuo Wang
College of Food Engineering and
  Biotechnology
Tianjin University of Science and
  Technology
Tian Jing, People's Republic of China

Gang Xie
College of Food Engineering and
  Biotechnology
Tianjin University of Science and
  Technology
Tian Jing, People's Republic of China

and

Academy of State Administration of
  Grain
Beijing, People's Republic of China

Xuebing Xu
Wilmar Global R&D Center
Aarhus University
Aarhus, Denmark

Liangli (Lucy) Yu
Institute of Food and Nutraceutical
  Science
Shanghai Jiao Tong University
Shanghai, People's Republic of
  China

and

Department of Nutrition and Food
  Science
University of Maryland
College Park, Maryland

Bing Zhang
Guelph Food Research Centre
Agriculture and Agri-Food Canada
Guelph, Ontario, Canada

and

State Key Lab of Food Science and
  Technology
Nanchang University
Nanchang, Jiangxi, People's Republic
  of China

Xiaowei Zhang
Institute of Food and Nutraceutical
  Science
Shanghai Jiao Tong University
Shanghai, People's Republic of
  China

Xin-Chen Zhang
School of Biological Sciences
The University of Hong Kong
Hong Kong

Yu Zhang
Department of Food Science and
  Nutrition
and
Fuli Institute of Food Science
Zhejiang University
Hangzhou, People's Republic of
  China

1 Food Safety Chemistry
An Overview

Haiqiu Huang and Liangli (Lucy) Yu

CONTENTS

1.1 INTRODUCTION

Food safety is of paramount importance to both consumers and the food industry. According to the Food and Agriculture Organization (FAO), food safety is defined as "absence or acceptable and safe levels of contaminants, adulterants, naturally occurring toxins or any other substance that may make food injurious to health on an acute or chronic basis."[1] In 2008, a perspective article published in the *New England Journal of Medicine* indicated that "globalization and international agribusiness allow problems with food supply to spread around the planet all too quickly,"[2] which indicates that food safety has become a critical concern in human health and life quality worldwide regardless of the geological location or the degree of economic development.

Until now, major attention on food safety has been given to food microbiology-related issues resulting from unexpected and sudden microorganism contamination and outbreak. However, food microbiology-related issues are only one part of food safety problems. With the rapid development of the food industry and the majority of food products being processed, chemical/chemistry-related food safety and security issues have become more prominent and critical. Recently, a number of food safety incidents related to substances such as trans-fats, melamine, acrylamide, and 3-MCPD (3-monochloropropane-1,2-diol) esters have come to our attention. These incidents are characterized by long lag phases before they emerge, compared to those of microbiology-related issues. Therefore, a new field of study that emphasizes on the chemistry nature of food safety issues is necessary to tackle these specific problems. Food safety chemistry aims at defining and understanding existing and emerging food safety issues from a chemistry perspective, and to cultivate future research and development of effective regulation and prevention strategies.

Many factors, including the pre-/post-harvest treatments and storage, composition and food formulation, processing conditions, and product storage, may alter the level and profile of foodborne contaminants in the final food products, and ultimately have an impact on human health and wellness. Additionally, food ingredients may be a source of contamination. With the development of the food industry, most food products are manufactured from various ingredients rather than only raw materials. Contaminants may be generated during processing and storage of these ingredients, and carried over to the final products.

Understanding the chemical and biochemical mechanisms involved in the formation of these food contaminants is essential for developing practical approaches to control and reduce their levels in food ingredients and products. Meanwhile, there is a gap in the systematic review and discussion of these foodborne chemical contaminants, the mechanisms involved in their formation during food processing and storage, their toxic effects and the biological mechanisms behind them, the updated analytical techniques for food quality control, other research efforts on these chemicals, and regulatory-related concerns and suggestions.

1.2 CHEMISTRY AND FOOD SAFETY

Food production can be considered a complex chemical reaction in which certain amounts and types of substrates are introduced, thereby creating the required reaction. However, such systems are extremely complicated and difficult to control, and sometimes lead to undesirable reactions resulting in unexpected or unwanted contaminants. These contaminants can affect the quality of the food produced and also lead to potential adverse effects on human health. Thus, food safety chemistry research aims to maximize the favored reactions leading to food products and also minimize the generation of contaminants.

With the development of food processing and storage technologies, a wide range of novel techniques have been employed in an attempt to improve food quality and stability. However, the additional steps of food processing and storage may generate a new category of food contamination. This emerging group of contaminants is yet to be extensively and thoroughly studied and discussed, which will be the focus of this book. The major chemical food contaminants will be discussed for their toxic effects and the biological mechanisms behind their toxicity, their absorption and distribution in animals, factors influencing their levels in foods and food ingredients, the chemical and biochemical mechanisms involved in their formation, possible approaches to reducing their levels, and the analytical methods for their detection and regulation status if applicable.

A major part of the effort in understanding and controlling the chemistry of food production has been exerted to ensure food quality and safety. Chemistry has been a powerful tool in revealing and understanding the mechanism of formation of undesirable food components and their effects on humans, in which organic chemistry, analytical chemistry, and biochemistry play an important part. The progress in chemical analysis methods greatly boosts the ability and accuracy in detecting food composition as well as contaminants. The employment of liquid and gas chromatography, mass spectrometry, various types of spectroscopy, and a wide variety of biochemistry methods, including gene-expression assays, protein assays, and metabolic analyses, make it possible for food researchers and the food industry to take a closer look at food production and pinpoint the potential safety issues involved. In this book, specific detection and analysis methods for each category of food contaminants are discussed in their respective chapters.

1.3   INTENTIONAL AND ACCIDENTAL FOOD CONTAMINATION

Food contamination refers to the presence of harmful chemicals and microorganisms, which may cause consumer illness. Unlike most microbiological contaminants, the impact of chemical contaminants on consumer health and well-being is only apparent after a period of prolonged exposure, and often at low levels.

Chemical contamination in food can be categorized into two groups, intentional contamination and accidental contamination. Intentional contamination refers to the inclusion of potential toxic or harmful chemical components in food when the manufacturer is aware of the risk; however, accidental contamination occurs when the manufacturer is unaware of the potential risk or the risk of the chemical(s) in question is(are) not yet discovered or reported.

Intentional contamination mostly exists in the food product in its final form, does not go through reactions, or is unaffected during food production. The melamine incident is a case of intentional contamination. In 2008, an incident involving milk and infant formula being adulterated with melamine was reported. In an attempt to increase the level of protein in milk products, melamine was intentionally added to raw milk or milk product, which led to kidney stones and other renal failure in consumers, especially among young children. The melamine incident aroused great attention to intentional contaminants in food; the chemistry and risk of melamine in food are reviewed in Chapter 14.

Accidental contamination mostly results from unknown or unattended reactions during food processing or new processing technologies, and usually appears in the form of a by-product of the intended food product. Trans-fats, acrylamide, and 3-MCPD and its fatty acid esters are all examples of accidental contamination. A major source of trans-fat is hydrogenation of unsaturated fat, which was originally a procedure to improve food quality and stability. The occurrence and formation of trans-fat as well as detection methods and health implications are reviewed and discussed in Chapter 8. Acrylamide is found in many cooked starchy foods, and is believed to be related to Maillard reaction; Chapter 2 is dedicated to the chemistry and safety issues behind the formation of acrylamide. Maillard reaction, one of the most important chemical reactions in food production and processing, has been shown to be a risk factor by generating toxic compounds. Maillard reaction products and their safety chemistry are reviewed in Chapters 3, 4, and 5. 3-MCPD, which has been proven to be a toxic substance, is found in various foodstuffs, such as acid-hydrolyzed vegetable proteins, bread, meat, and beer. The formation mechanisms and potential risk of 3-MCPD are discussed in Chapter 6. 3-MCPD fatty acid esters have been detected in recent years and appear to result from the refining process of fats and oils. Oil is one of the most important and universal food ingredients in food production. The formation of 3-MCPD esters and their potential safety issues are covered in Chapter 7. The mechanisms of the formation of trans-fats, acrylamide, and 3-MCPD fatty acid esters, and their adverse effects in humans are under active study. In addition, though limited processing is involved in sea food and meat production, storage may turn out to be another critical step in reducing and controlling contamination. Contaminants in seafood and meat are covered in Chapters 10 through 12. Food additive is a category of highly processed food ingredients, and the risk of contamination in food additives is reviewed in Chapter 13. It can be expected that, with increasing numbers of new food products and processing technologies coming into the market, research and control technologies for food contamination during food formulation, storage, and processing will be in greater demand in the future.

1.4   BRIEF SUMMARY OF THE FOLLOWING CHAPTERS

Unsafe food causes many acute and lifelong diseases, and food safety issues constitute a growing public health concern; these factors make the importance of ensuring food safety self-evident. The major objectives of this book are to provide a timely summary of current research findings related to food chemical contaminants, and to serve as a long-lasting guide for scientists in the fields of food science, food safety, food ingredients, postharvest treatment and processing, food product research and development, crop and soil science, human nutrition, and to eventually promote food quality and human health. In the following chapters, the current knowledge of food contaminants and related food safety issues will be reviewed and discussed, including appearances of foodborne chemical contaminants, their mechanism of formation, their potential adverse effects in human health, and the updated analytical techniques for food safety analysis and quality control.

REFERENCES

1. FAO. The Importance of Food Quality and Safety for Developing Countries. http://www.fao.org/trade/docs/LDC-foodqual_en.htm (accessed Jan. 28, 2014).
2. Ingelfinger, J. R., Melamine and the global implications of food contamination. *New England Journal of Medicine* 2008, 359(26), 2745–2748.

2 Chemistry and Safety of Acrylamide

Yu Zhang and Xinyu Chen

CONTENTS

2.1   INTRODUCTION

In April 2002, researchers at Stockholm University in Sweden found, for the first time, a considerable amount of potential carcinogen acrylamide in fried or baked potatoes and cereals, which prompted investigation of the acrylamide levels in various foods by the Swedish National Food Administration (SNFA) (SNFA, 2002). The findings have attracted wide attention and evoked an international health alarm. Also, analytical results were quickly validated by British Food Standards Agency (BFSA) and several government agencies from Norway, the United States, Australia, New Zealand, and Canada. Meanwhile, many international organizations and research institutions performed research on the analytical approaches, formation mechanism, mitigation recipes, toxicology, and risk assessment of acrylamide in foods. In March 2005, the Joint Expert Committee on Food Additives (JECFA) of the World Health Organization (WHO) and the Food and Agriculture Organization (FAO) together announced that certain foods processed or cooked at high temperature, especially Western-style snacks, might contain considerable levels of acrylamide and harm human health to a certain extent (INFOSAN, 2005). Later, in May 2010, European Food Safety Authority (EFSA) published the monitoring report about the acrylamide levels in various foods based on the analytical results of 3461 samples from 22 European Union (UN) members and Norway (EFSA, 2010). The acrylamide hazard in diet has attracted continuous concerns worldwide in recent decades.

Acrylamide, a colorless and odorless crystalline powder with a melting point of 84.5°C, is soluble in water, acetone, and ethanol. The α,β-unsaturated double bond in acrylamide may easily react with nucleophiles via the Michael addition. Long-term exposure to acrylamide may cause damage to the nervous system in both humans and animals (Huang et al., 2011). Acrylamide is also regarded as a potentially genetic and reproductive toxin with mutagenic and carcinogenic properties according to both *in vitro* and *in vivo* studies (Lineback et al., 2012; Zhang et al., 2009). On the basis of the animal studies, the International Agency for Research on Cancer (IARC) has classified acrylamide as the Group 2A substance, which is probably carcinogenic to humans (IARC, 1994).

2.2   ANALYTICAL CHEMISTRY OF ACRYLAMIDE

Some analytical methods were initially established to determine the acrylamide contents in several agricultural products, such as mushrooms, field crops, and sugars prior to the finding of acrylamide in heat-processed foods in 2002 (Bologna et al., 1999; Castle, 1993; Tekel et al., 1989). However, these methods could not cope with the analysis of acrylamide in heat-processed foods at trace levels. The mass spectrometry (MS)-based chromatographic methods were then developed for the quantification of acrylamide. In summary, gas chromatography coupled with mass spectrometry (GC–MS) and liquid chromatography coupled with tandem mass spectrometry (LC–MS/MS) seem to be the most useful and authoritative methods for acrylamide analysis (Keramat et al., 2011; Zhang et al., 2005). In recent years, the improvement of MS-based techniques and validation of non-MS-based quantitative methods were especially concerned. Besides the repeatability, sensitivity, and precision, the easy-to-use and rapid analysis methods were also considered.

2.2.1 General Knowledge of Analytical Methods

Figure 2.1 shows the general extraction and clean-up steps of acrylamide before sample injection for the GC–MS or LC–MS/MS analysis. Water at room temperature has been used to extract acrylamide from various sample matrices in most published analytical methods because acrylamide is a good hydrophilic small molecule (Wenzl et al., 2003). Besides, methanol can also be used to extract acrylamide considering further rapid concentration (Tateo and Bononi, 2003). Young et al. (2004) suggested that acrylamide could be extracted by a saturated aqueous solution of sodium chloride so that the emulsification process can be significantly reduced and the high recovery of acrylamide can be obtained. Accelerated solvent extraction (ASE) (Friedman, 2003) has also been tested for extracting acrylamide from solid and semi-solid samples. ASE uses conventional liquid solvents at elevated temperature and pressure to increase the efficiency of the extraction process. Increased temperature accelerates the extraction kinetics, while elevated pressure keeps the solvent below its boiling point, thus enabling safe and rapid extractions.

To control the recoveries and keep track of possible losses during the whole sample pretreatment, an internal standard is usually added into the sample matrix after homogenization. Most of the published studies used $^{13}C_3$-acrylamide produced by Cambridge Isotope Laboratories (Andover, MA, USA) as the internal standard. Besides, $^{13}C_1$-acrylamide, D_3-acrylamide, N,N-dimethylacrylamide, methacrylamide, and propionamide also have been used as acrylamide internal standards (Zhang et al., 2009).

Some research studies about acrylamide analysis included a defatting step before or in combination with the extraction procedure, which was achieved by extraction with hexane, petroleum ether, or cyclohexane. The step may be necessary for high-fat foods. Moreover, some protein-rich sample matrices need to have a deproteinating step, which was achieved by adding a high level of methanol, acetonitrile, or saline solution to the aqueous extract. Delatour et al. (2004) achieved the protein precipitation step within 1 min by the addition of potassium hexacyanoferrate (II) trihydrate (Carrez I) and zinc sulfate heptahydrate (Carrez II) under continuous swirling. Whether the extraction step of acrylamide needs to have a defatting or deproteinating step depends on the sample matrices (Keramat et al., 2011; Oracz et al., 2011).

Before analysis, clean-up procedures comprising the combination of several solid-phase extractions (SPE) may also be necessary. A previous study used a combination of three different cartridges: Oasis MAX (mixed-mode anion exchange), Oasis MCX (mixed-mode cation exchange), and ENVI-Carb (graphitized carbon) (Becalski et al., 2003). Also, such combination can be achieved by using a single SPE column consisting of an inhouse-prepared mixture of C18, strong cation (SCX), and anion exchange (SAX) sorbents (Bortolomeazzi et al., 2012). Oasis HLB cartridge, the main

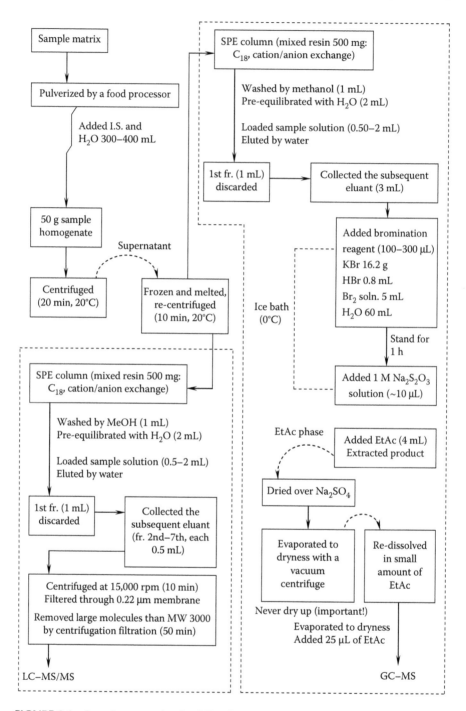

FIGURE 2.1 Sample preparation for GC–MS or LC–MS/MS determination of acrylamide. I.S., internal standard; fr., fraction; EtAc, ethyl acetate; MW, molecular weight. (Modified from Zhang, Y. et al. 2005. *J. Chromatogr. A* 1075, 1–21.)

supporter of which is a hydrophilic–lipophilic balance and a water-wettable reversed-phase sorbent for all compounds and all of general SPE needs, is usually used. Many research studies reported the application of this cartridge before the chromatographic analysis of acrylamide (Roach et al., 2003; Zhang et al., 2005). Oasis MCX cartridge, the main supporter of which is a mixed-mode cation-exchange reversed-phase sorbent for bases with a high selectivity for basic compounds, can also be used. Good accuracy and high recovery of chromatographic method validation during the acrylamide analysis can be achieved when effective clean-up procedures are well designed.

2.2.2   GC-BASED ANALYTICAL METHODS

Although acrylamide can be determined without derivatization, the molecule is normally brominated to generate a derivative that has improved GC properties when using GC–MS. The acrylamide derivative is identified by its retention time and by the ratio of characteristic MS ions. During the routine analysis of acrylamide by GC–MS, the derivatization step may be used before the clean-up step.

In recently reported derivative methods, conversion of acrylamide into 2,3-dibromo-propionamide (2,3-DBPA) is performed, which involves addition of potassium bromide (KBr), hydrobromic acid (HBr), and a saturated Br_2 solution to the derivatization reaction system (Becalski et al., 2003; Zhu et al., 2008). The excess bromine is removed by addition of sodium thiosulfate until the solution becomes colorless so that the derivative reaction is terminated. Under the above conditions, the yield of 2,3-DBPA becomes invariant when the reaction time is more than 1 h. This derivative is less polar compared to the original compound, and is therefore easily soluble in nonpolar organic solvents such as ethyl acetate and hexane. However, Andrawes et al. (1987) have shown that, under certain conditions, 2,3-DBPA can be converted into a more stable derivative 2-bromopropenamide (2-BPA) on the inlet of the GC or directly on the capillary column. Because this decomposition (dehydrobromination) may yield poor repeatability and accuracy, it is preferable to deliberately convert 2,3-DBPA into the stable 2-BPA prior to GC analysis, which can be readily done by adding 10% of triethylamine to the final extract before injection (Andrawes et al., 1987). Regardless of the derivative method used before chromatographic analysis, the purpose of derivatization is to reduce the polarity of acrylamide and to improve the retention time and peak regularization.

The ions monitored for specific identification of the analyte, 2-BPA, include $[C_3H_4NO]^+ = 70$, $[C_3H_4{}^{79}BrNO]^+ = 149$, and $[C_3H_4{}^{81}BrNO]^+ = 151$. The ion $[C_3H_4{}^{79}BrNO]^+$ is regarded as the quantitative ion while the others are used for qualitative observation. The ions monitored for identification of the internal standard, 2-bromo($^{13}C_3$) propenamide, include $[^{13}C_2H_3{}^{81}Br]^+ = 110$ and $[^{13}C_3H_4{}^{81}BrNO]^+ = 154$, whereas $[^{13}C_3H_4{}^{81}BrNO]^+$ is regarded as the quantitative ion. The separation of acrylamide analyte after derivatization was performed on standard GC capillary columns of middle to high polarity with a length of 30 m and an internal diameter of 0.25 mm (standard in GC–MS).

2.2.3   IMPROVEMENT OF GC–MS METHODS

The GC–MS methods with or without derivatization of acrylamide were well reviewed by Castle and Eriksson (2005). The derivatization method was improved and achieved via the reaction between potassium bromate ($KBrO_3$) and potassium bromide (KBr) in an acidic medium (Zhang et al., 2006). The preparation of the strong acid HBr and saturated Br_2 solution is difficult and hazardous, and should be handled carefully. Nevertheless, the use of $KBrO_3$ and KBr combination is relatively more convenient and safe, and the reaction is performed in about 30 min at cold storage temperature with excellent reproducibility. For the chromatographic conditions, polar columns of the Carbowax type are mainly applied to chromatographic separation and chemical ionization MS in selected ion-monitoring mode for analyte detection (Wenzl et al., 2007). During GC–MS analysis, the co-elution of some impurities may interfere with the quantitative results of acrylamide. Biedermann and Grob (2008) found that 3-hydroxy-propionitrile (3-HPN) may be co-eluted with acrylamide causing false high acrylamide values. To eliminate such problems, they selected the Carbowax 1000 column with higher polarity. Results showed that acrylamide was eluted with a clear separation after 3-HPN peak. Alternatively, it is also possible to get 3-HPN eluted after acrylamide using a high molecular weight Carbowax combined with adequate tuning of the separation conditions (Biedermann and Grob, 2008). Overall, current robust GC–MS method with optimized derivatization and sample pretreatment procedures could greatly improve selectivity, sensitivity, and repeatability. The method could also be applied to the quantification of acrylamide in some new sample matrices, such as malt (Mikulíková and Sobotová, 2007) and cigarette mainstream smoke (Diekmann et al., 2008).

2.2.4 LC-Based Analytical Methods

Besides GC–MS, LC–MS/MS techniques have also been employed for the routine analysis or exposure survey of acrylamide because this chromatographic technique has high sensitivity and requires no derivatization step.

LC–MS/MS has a high selectivity when working in multiple reaction monitoring (MRM) mode, in which the transition from a precursor ion to its product ion is monitored. For the acrylamide monomer detection using MRM, injected samples from the liquid chromatograph system enter the ionization source at atmospheric pressure. These ions are sampled through a series of orifices and ion optics into the first quadrupole where they are filtered according to a mass-to-charge ratio (m/z) of 72 for acrylamide. The mass separated ions $[CH_2 = CHCONH_2]^+$ then pass into the ion tunnel collision cell, with an axial field, where they either undergo collision-induced decomposition (CID) or pass unhindered to the second quadrupole. The fragment ions of acrylamide after collision using argon as collision gas, that is, $[CH_2 = CHCO]^+$ for m/z 55 and $[CH_2 = CHCNH]^+$ for m/z 54, are then mass analyzed by the second quadrupole. Finally, the selected and transmitted ions are detected by a conversion dynode, phosphor, and photomultiplier detection system. The output signal is amplified, digitized, and presented to the data system (e.g., MassLynx NT). The transition 72 > 55 is always selected as the quantitative ion for the determination of acrylamide because it shows a high relative abundance. Other transitions, such as 72 > 54, 72 > 44, and 72 > 27, have been used for the analyte confirmation of acrylamide (Zhang et al., 2005). For the detection of the isotopically labeled acrylamide, which is used as internal standard, the monitored transitions are 75 > 58 for $[^2H_3]$ and $[^{13}C_3]$ acrylamides and 73 > 56 for $[^{13}C_1]$ acrylamide.

Although the high selectivity offered by MS/MS shows much more clear analyte elution than chromatograms based on other chromatographic techniques, interference can still inevitably occur. Peaks showing identical retention time to acrylamide and deuterated acrylamide have been observed. However, similar to the experiences of private and official food control laboratories, problems have been encountered during the analysis of complex matrices due to interference compounds appearing in the characteristic acrylamide transitions (for either the internal standard or the analyte). A promising approach is to extract the analyte into a polar organic solvent, such as ethyl acetate. Sanders et al. (2002) used ethyl acetate to extract acrylamide from the aqueous phase (removing interference constituents such as salt, sugars, starches, amino acids, etc.). The ethyl acetate extract could then be concentrated and analyzed by LC–MS/MS. Becalski et al. (2003) also reported the existence of an early eluting compound that interferes when transition of 72 > 55 was used for the detection of acrylamide. Increasing the column length from 100 to 150 mm and applying isolute multimode cartridges during sample preparation eliminated the interference well. Jezussek and Schieberle (2003) developed a new LC–MS method for the quantification of acrylamide based on a stable isotope dilution assay and derivatization with 2-mercaptobenzoic acid. In stable isotope dilution assays, the differentiation between acrylamide and its internal standard was performed by recording the respective molecular masses and/or mass fragments. Due to the low-molecular weight of acrylamide, and also due to its resulting low mass fragment ions, the background interference may impede the analysis. Better confirmation of the analyte and selectivity can be achieved with a two-stage mass spectrometer, that is, double quadrupole, by monitoring more than one characteristic mass transition. However, triple quadrupole mass spectrometers are quite expensive and the less-expensive single-stage LC mass spectrometers are not sensitive enough for the direct analysis of acrylamide. Such isotope dilution assay involving derivatization of acrylamide with 2-mercaptobenzoic acid prior to MS analysis could be useful for not only improving the sensitivity of acrylamide analysis, but also reducing analytical cost by achieving the transition from two-stage mass spectrometer to single-stage mass spectrometer.

2.2.5 IMPROVEMENT OF LC–MS/MS METHODS

The LC–MS/MS analysis of acrylamide was developed based on the method published by Rosén and Hellenäs (2002), and has been largely improved by several research groups (Keramat et al., 2011; Wenzl et al., 2003; Zhang et al., 2005). First, the retention of acrylamide in the column greatly affects the separation efficiency and quantitative precision. Rosén et al. (2007) comparatively investigated the effects of solid phase on the chromatographic retention of acrylamide. During SPE, a hydroxylated polystyrene-divinylbenzene copolymer phase (ENV+) gave the strongest retention. During HPLC, the best retention was achieved with a phase comprising porous graphitic carbon (Hypercarb) using water as the mobile phase. Besides, the solid phase of the selected column should be suitable for the separation and retention of strong polar compounds including acrylamide. The MS determination was also improved. The effect of different atmospheric-pressure interfaces on the LC–MS/MS determination of acrylamide was further demonstrated. Marín et al. (2006)

recommended the Ion Sabre atmospheric-pressure chemical ionization (APCI) as the interface and obtained the highest sensitivity for acrylamide (LOD 0.03 μg/L) and eliminated the matrix effects compared to electrospray ionization (ESI) and APCI. In different foods and model systems, the formation of acrylamide has been shown to correlate with preprocessing levels of asparagine, fructose, glucose, or the products of asparagine and reducing sugars. Therefore, the change of asparagine and sugar contents is greatly related to the acrylamide level. A previous study used HPLC and amino acid analysis kit to quantify the contents of sugars and asparagine, respectively (Knol et al., 2005). However, such methods cost considerable time and require the analysis of large amount of samples. Nielsen et al. (2006) developed an LC–MS/MS method for simultaneous analysis of acrylamide (LOD 0.013 mg/kg), asparagine (LOD 1.8 mg/kg), glucose (LOD 96 mg/kg), fructose (LOD 552 mg/kg), and sucrose (LOD 23 mg/kg) in bread. This method is useful for further investigation of acrylamide formation in various food matrices. With the development of chromatographic technique, many interference-free, generic, and robust LC–MS or LC–MS/MS methods have been validated, which can be regarded as possible alternative methods for the quantification of acrylamide (Gökmen and Şenyuva, 2006; Marchettini et al., 2013; Şenyuva and Gökmen, 2006).

2.2.6 New Analytical Methods Other Than GC or LC Technique

Several MS-based methods have been developed in the past decades to determine acrylamide in water, foods, and biological samples. Additional new analytical methods have also been recently validated, including capillary zone electrophoresis (CZE), micellar electrokinetic capillary chromatography (MEKCC), time-of-flight mass spectrometry (TOF–MS), electrochemical detection, biosensors, and enzyme-linked immunosorbent assay (ELISA).

Regarding the capillary electrophoresis (CE) techniques, the microemulsion electrokinetic chromatography (MEEKC) was first proposed for the determination of acrylamide without derivatization, but this method generally may have a relatively high LOD value. An improved MEKCC method was developed for the separation and quantification of acrylamide and has been approved as a reliable method (LOD 0.1 μg/mL) (Zhou et al., 2007). On the basis of the above techniques, better results were obtained for CZE, after derivatization with 2-mercaptobenzoic acid to obtain an ionic compound (LOD 0.07 μg/mL). To further improve the detection limits and to apply the method over a wide range of samples, field-amplified sample injection (FASI) was proposed (LOD 1 μg/L) (Bermudo et al., 2006). Based on the FASI-CE technique, Bermudo et al. (2007) showed the applicability of CE coupled with MS/MS for the analysis of acrylamide in foods and obtained both good linearity and precision. In addition to the FASI-CE method, a relative field-amplified sample stacking (FASS) was also developed and regarded as a simple, rapid, and inexpensive choice (Tezcan and Erim, 2008).

With the rapid development of TOF–MS technique, the LC or GC method combined with a high resolution TOF–MS was used for the analysis of acrylamide and was validated by the Food Analysis Performance Assessment Scheme (FAPAS) (Dunovská et al., 2006). Compared to robust GC–MS or LC–MS/MS, the applicability of TOF–MS for the quantification of acrylamide still needs to be optimized and improved. Electrochemical detection of DNA damage induced by acrylamide and its metabolites is an alternative new technique using the graphene-ionic liquid-Nafion modified pyrolytic graphite electrode (Qiu et al., 2011).

2.2.7 RAPID AND EASY-TO-USE METHODS

For the sample pretreatment, routine procedure prior to instrumental analysis includes sample preparation, addition of internal standard (IS), defatting or removing proteins, extraction, concentration, and clean-up. Such a tedious procedure is not suitable for the analysis of large amounts of samples. Mastovska and Lehotay (2006) optimized a fast and easy procedure in which deuterated acrylamide IS was added to homogenized sample together with hexane, water, acetonitrile, magnesium sulfate, and sodium chloride, followed by a SPE step. For the chromatographic analysis, an improved method, ultra high-performance liquid chromatography (UHPLC) coupled with MS/MS was developed (Zhang et al., 2007). Compared to routine LC–MS/MS, the UHPLC–MS/MS method supplies a rapid quantitative procedure of acrylamide with a run time of only 3 min. Using this improved technique, the simultaneous determination of asparagine, glucose, fructose, and acrylamide was successfully achieved (Zhang et al., 2011). Besides, the UHPLC method may increase the efficiency and resolution because the particles with 1.7 μm size in UHPLC columns allow the chromatographic analysis under much higher pressure and faster flow rate.

MS is a preferable system for acrylamide detection among various detectors. However, the disadvantage of MS detection is that the mass of acrylamide itself or its fragment ions are not specific due to the presence of co-extractives that yield the same magnitude of m/z with acrylamide. The validation of cost-effective, easy-to-use, or widely applicable methods is highly demanded. Several non-MS detection methods have been gradually developed and optimized. For the LC analysis, the ultraviolet (UV) or diode array detection (DAD) at wavelengths of 210 and 225 nm has been successfully used for the detection and quantification of acrylamide (Keramat et al., 2011). For the GC analysis, the electron capture detection (ECD) is widely applied because of its high sensitivity (Zhang et al., 2006; Zhu et al., 2008). Meanwhile, a recently developed headspace/SPME/GC method using a nitrogen–phosphorus detector (NPD) was validated for analyzing acrylamide formed in an aqueous polyacrylamide solution treated by heat or photo-irradiation (Ishizuka et al., 2008). Also, the reversed-phase direct immersion single-drop microextraction technique has been applied to acrylamide analysis by replacing the organic solvent of the extracting phase (nonpolarity) with a micro-droplet of water (polarity) for the solubility improvement of acrylamide during GC analysis (Kaykhaii and Abdi, 2013). Besides the LC and GC techniques, the method of thin-layer chromatography (TLC) with fluorescence detection after derivatization with dansulfinic acid was also developed. This method features high specificity and sensitivity while being a cost-effective method in terms of equipment (no MS/MS) and standards (no isotope labeled ones) (Alpmann and Morlock, 2008).

To achieve rapid screening, high-throughput screening and inexpensive biological methods were selected for the analysis of large amounts of samples, including electrochemical biosensors (Sun et al., 2013) and ELISA (Quan et al., 2011). However, these new methods still need to be optimized due to their vague repeatability and precision, and their analytical results have to be preferably confirmed by other robust methods.

2.2.8 PROFICIENCY TEST

The need of a certified matrix reference for acrylamide in a food matrix is emphasized by competent authorities as a tool to improve comparability, ensuring accuracy, and traceability of analytical results. Several studies focused on the development of such certified reference materials in various foods for interlaboratory validation of quantitative analysis of acrylamide (Dabrio et al., 2008; Kim et al., 2010; Koch et al., 2009). The production of such a reference includes the material processing, homogeneity, stability assessment, material characterization, and the acrylamide mass fraction value assignment. The validation can be performed when the analysis is integrated within the scope of an authorized proficiency test controlled by the official FAPAS for accreditation (Owen et al., 2005). The decision of such a test will be made by FAPAS after comparing the quantitative results with the measurement results from individual laboratories and the assignment value. Although such a proficiency test is not a prerequisite for performing the quantification of acrylamide, the analytical accuracy and repeatability will be mandatory if a high score of such a test for the determination of a reference material should be approved by FAPAS.

2.3 FORMATION OF ACRYLAMIDE UNDER MAILLARD REACTION CONDITIONS

2.3.1 MAILLARD REACTION AND ACRYLAMIDE

The Maillard reaction plays an important role in the appearance and taste of various foods. Despite all the work that has been conducted and the great progress that has been achieved, the mechanism of the Maillard reaction is still a controversial issue. Such a controversial topic focuses on the formation of advanced glycation end products (AGEs) and melanoidins.

The two contributions published in *Nature* clarified the correspondence between the Maillard reaction and acrylamide. Original findings from Mottram et al. (2002) indicated that the Maillard reaction involving asparagine could produce acrylamide, and might elucidate the elevated concentrations of acrylamide in certain plant-derived foods after cooking. They also proposed the pathways for the formation of acrylamide after Strecker degradation of the amino acids asparagine and methionine in the presence of dicarbonyl products from the Maillard reaction. Furthermore, a new insight from Stadler et al. (2002) showed that N-glycoside formation could be favored in food-processing matrices that combine conditions of high temperature and water loss. When such condensation occurs between reducing sugars and certain amino acids, a direct pathway is opened up to the potential generation of acrylamide.

2.3.2 FUNDAMENTAL MECHANISTIC STUDIES

Shortly after the discovery of acrylamide in heat-treated foods (SNFA, 2002), numerous research groups in academic schools, industry, and official laboratories commenced studies on its possible sources and corresponding mechanisms. Several hypotheses on formation pathways were discussed at the very early stages of investigations, focusing initially on vegetable oils or lipids, since the problem mainly encompassed carbohydrate-rich foods that were fried or baked. In general, some critical and direct precursors contributing to the formation of acrylamide were demonstrated as 3-aminopropionamide, decarboxylated Schiff base (Zyzak et al., 2003), decarboxylated Amadori product (Yaylayan et al., 2003), acrylic acid (Becalski et al., 2003), and acrolein (Yasuhara et al., 2003). At the early stages of mechanistic study, researchers focused on heat-processing parameters affecting the formation of acrylamide. Heating equal molar amounts of asparagine and glucose at 180°C for 30 min resulted in the formation of 368 μmol of acrylamide per mol of asparagine (Stadler et al., 2002). No acrylamide was found in unheated or boiled foods. However, Ezeji et al. (2003) found that acrylamide is formed during the boiling or autoclaving of starch. Some amino acids producing low amounts of acrylamide include alanine, arginine, aspartic acid, cysteine, glutamine, methionine, threonine, and valine. Overall, the formation of acrylamide depends on multiple factors including heating temperature, heating time, pH, types of precursors, and molar ratios of amino acid to reducing sugar.

Several researchers have shown that the formation of acrylamide in foods is related to the Maillard reaction and especially the amino acid asparagine via water or food matrix model systems (Mottram et al., 2002; Stadler et al., 2002). The link of acrylamide to asparagine, which directly provides the backbone chain of acrylamide molecule, was established by labeling experiments. In detail, mass spectral studies showed that the three carbon and the nitrogen atoms of acrylamide are all derived from asparagine. Combined with the isotope substitution technique, key reaction intermediates including Schiff base, decarboxylated Schiff base, and 3-aminopropionamide were confirmed (Zyzak et al., 2003). The mechanism of acrylamide formation from a decarboxylated Amadori product of asparagine is shown in Figure 2.2 (Zyzak et al., 2003). However, lipid oxidation has been suggested as a minor pathway, with acrylic acid as a direct precursor formed via acrolein by oxidative degradation of lipids (Gertz and Klostermann, 2002).

In summary, from Figure 2.2, it can be suggested that acrylamide may be generated via the following pathways: (a) directly from azomethineylide I (Zyzak et al., 2003); (b) β-elimination reaction from the Maillard intermediate, that is, decarboxylated Amadori product (Yaylayan et al., 2003); and (c) loss of ammonia from 3-aminopropionamide derived from azomethineylide II, which has been shown to preferentially proceed under aqueous conditions in the absence of sugars (Granvogl et al., 2004). Besides the main reaction precursors (reducing sugars and amino acids) and key intermediates, it is suggested that lipids can also play an important role in acrylamide formation according to acrolein pathway. In detail, lipids heated at high temperature can lead to the formation of acrolein, which can further react via oxidation to generate acrylic acid or by formation of an intermediate acrylic radical (Umano and Shibamoto, 1987). Both the intermediates could then induce acrylamide formation in the presence of a nitrogen source under favorable reaction conditions (Yasuhara et al., 2003). Another factor under consideration is the general background of food systems such as fat-rich or protein-rich matrix (Claeys et al., 2005).

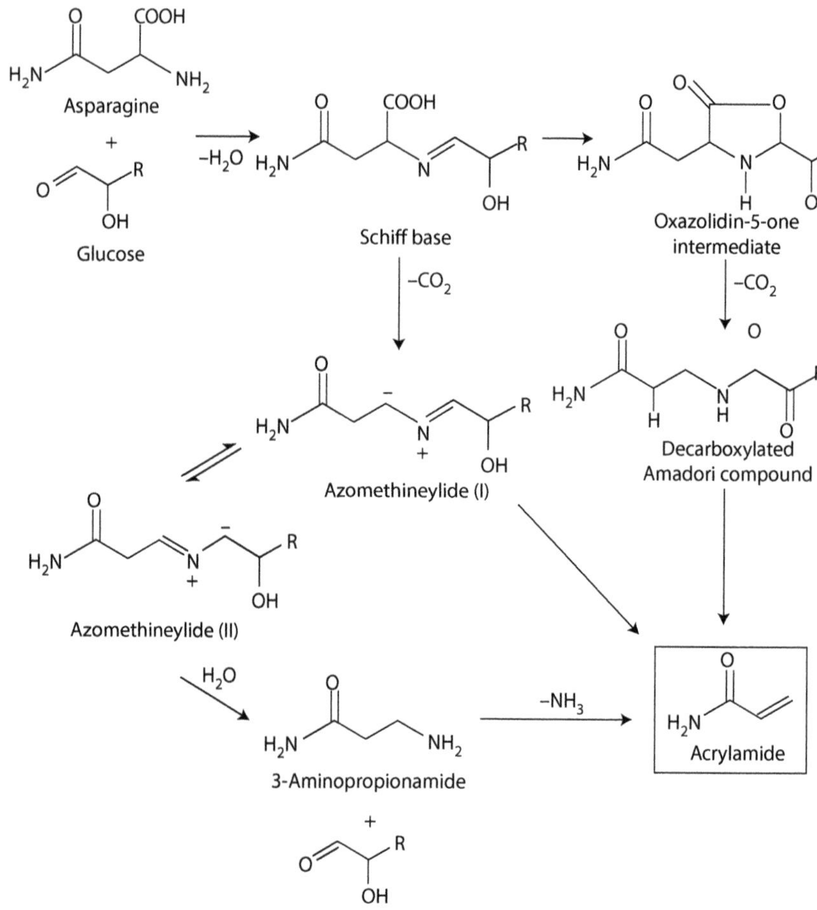

FIGURE 2.2 The mechanism of acrylamide formation.

2.3.3 NEW AND FURTHER UNDERSTANDING ON FORMATION MECHANISM

Several key intermediates in the asparagine pathway were only predicted (Yaylayan et al., 2003; Zyzak et al., 2003). Therefore, the chemical interactions leading to acrylamide remain largely hypothetical. To fill in the gaps under currently proposed routes, Stadler et al. (2004) further demonstrated the mechanism of acrylamide formation in food by comparing the two major hypothetical pathways, that is, via (i) the Strecker aldehyde route and (ii) glycoconjugates of asparagine. They have synthesized appropriate model intermediates and employed them in model systems to obtain a deeper insight into the key precursors and the salient reaction steps. The synthesis of Amadori compounds and N-glycosides of amino acids was further performed. First, Maillard intermediates related to the synthesis of Amadori compounds were prepared and some key intermediates were identified by NMR, which included 2,3:4,5-di-O-isopropylidene-β-D-fructopyranose **1**,2,3:4,5-(di-O--isopropylidene-1-O-trifluoromethanesulfonyl)-β-D-fructopyranose **2**,tert-butyl N-(2,3:4,5-di-O-isopropylidene-1-deoxy-D-fructos-1-yl)-L-asparaginate **3**, sodium N-(2,3:4,5-di-O-isopropylidene-1-deoxy-D-fructos-1-yl)-L-asparaginate **4**, N-(1-deoxy-D-fructos-1-yl)-L-asparagine **5**, N-(2,3:4,5-di-O-isopropylidene-1-deoxy-D-fructos-1-yl)-benzylamine **6**, N-(2,3:4,5-di-O-isopropylidene-1-deoxy-D-fructos-1-yl)-2-phenylethylamine **7**, and N-(1-deoxy-D-fructos-1-yl)-benzylamine **8**. Second, N-glycosides of amino acids were prepared by adapting the general procedure for obtaining N-glycosides through condensation of reducing sugars and amino acids in anhydrous methanol under alkaline conditions. The primary N-glucosides included potassium N-(D-Glucos-1-yl)-L-asparaginate **9** and potassium N-(D-Fructos-2-yl)-L-asparaginate **10**.

The analytes **3–10** are suitable compounds to study the reaction mechanisms leading to acrylamide under food-processing conditions. The experimental results were consistent with the reaction mechanism based on (i) a Strecker-type degradation of the Schiff base leading to azomethineylides, followed by (ii) a β-elimination reaction of the decarboxylated Amadori compound to afford acrylamide. Meanwhile, α-hydroxycarbonyls were much more effective than α-dicarbonyls in converting asparagine into acrylamide (Stadler et al., 2004; Yaylayan, 2009).

Decarboxylated Schiff base (azomethineylide) and decarboxylated Amadori product are both key intermediates contributing to acrylamide formation (Figure 2.2). To identify their relative importance, both intermediates were synthesized and their relative abilities to generate acrylamide under dry and wet heating conditions were investigated (Locas and Yaylayan, 2008). Under both conditions, the decarboxylated Schiff base [N-(D-glucos-1-yl)-3′-aminopropionamide] had the higher intrinsic ability to be converted into acrylamide. In the dry model system, the increase was almost fourfold higher than that for the corresponding decarboxylated Amadori product or 3′-aminopropionamide. However, in the wet system, the increase was twofold higher relative to that for decarboxylated Amadori product but more than 20-fold higher relative to that for 3′-aminopropionamide. Recently, the decarboxylation of asparagine in the presence of alkanals, alkenals, and alkadienals among other lipid derivatives has been investigated to understand the reaction pathways by which some lipid oxidation products are able to convert asparagine into acrylamide. The mechanism for the decarboxylation of asparagine in the presence of alkadienals based on the deuteration results obtained when asparagine/2,4-decadienal model systems were heated in the presence of deuterated water was proposed (Hidalgo et al., 2010).

2.3.4 General Factors Contributing to the Formation of Acrylamide

Many acrylamide-correlative studies focused on the investigation of factors altering acrylamide formation. These factors included heating time, heating temperature, pH, the concentration of precursors and their compositions, heat-processing procedures, and so on. Potato materials and corresponding products such as French fries, potato chips, and crisps are regarded as the most important food matrices for research on acrylamide formation because of high amounts of asparagine in potatoes. Recently, besides the above-mentioned factors, other multiple factors such as potato cultivar, farming systems, field site, fertilization, pesticide/herbicide application, time of harvest, storage time, and temperature may impact the final levels of both amino acid and reducing sugar precursors, which mostly contribute to the formation of acrylamide. The importance of cultivar selection in controlling the acrylamide content has been emphasized in several studies (Amrein et al., 2003; Marchettini et al., 2013). However, sugar control is an alternative way to the change of precursors in food materials, and therefore ultimately leads to the alteration of acrylamide levels in heat-processing products.

2.3.5 Agronomic Factors during Cultivation of Raw Materials

The agronomic factors mainly include fertilization methods, harvest season, and climatic conditions (Halford et al., 2012a). A reverse correlation between the amount of fertilizer applied in potato cultivation and acrylamide content in the edible products has been revealed, since reducing sugar contents were elevated while crude protein and free amino acids decreased when less nitrogen fertilizer was given (De Wilde et al., 2006). Gerendás et al. (2007) observed the same phenomenon, and confirmed that highest acrylamide contents and its precursors were observed in food products made from the tubers grown with high nitrogen and inadequate potassium supply. Furthermore, the influence of sulfur fertilization should be considered. When wheat was grown under conditions of severe sulfate depletion or sulfur deficiency during fertilization, dramatic increases in the concentration of free asparagine were found in the flour and subsequently enhancement of acrylamide content during baking was observed. Independent of fertilization, harvest year, and climatic conditions for altering asparagine and/or reducing sugar levels in crops turned out to be another factor influencing acrylamide formation during baking. Favorable light and temperature conditions during the cultivation period for wheat enhanced amino acid and protein contents, thus promoting the formation of acrylamide during baking (Knutsen et al., 2009).

2.3.6 Variety and Storage Conditions of Raw Materials

Acrylamide formation in crisps can be reduced by using potato varieties with low levels of both asparagine and reducing sugars. Mass transport of precursors during heating is suggested to be important for acrylamide formation in potato crisps (ÅlViklund et al., 2008). A previous study demonstrated that the acrylamide level in potato chips made from tubers stored at low temperature was much higher than that stored at high temperature (Burch et al., 2008). It seems that potato storage at lower temperatures should be avoided for the control of precursors. However, storage at a very high temperature can reduce the preservation period and sensory attributes of raw materials. To cope with this problem, Paleologos and Kontominas (2007) observed a minimum value of acrylamide concentration when the breaded chicken products were stored under refrigeration with a modified atmosphere mixture of 60% CO_2 plus 40% N_2. Overall, the effect of variety and storage conditions of raw materials on acrylamide level in final food products is ascribed to the variation of amino acids and reducing sugars (Halford et al., 2012b).

2.3.7 Change or Modification of Heat-Processing Methods

Blanching, instead of frying or soaking, could significantly affect the acrylamide level in the final food products (Wicklund et al., 2006). However, such ideas do not seem very practical because of the negative impacts on the connatural and sensory charac-teristics of the processed foods. To effectively control the formation of acrylamide, some tips regarding heat-processing methods were recommended such as low-tem-perature vacuum frying, short-time heating, and avoiding the use of palmolein as for modification of processing (Granda and Moreira, 2005; Mulla et al., 2011). Recently, the microwave heating method has been highlighted. Microwave heating as a fast and convenient heat-processing method is widely applied. Microwave heating provides a favorable medium for the occurrence of acrylamide and probably affects the formation and kinetics of acrylamide distinguishingly due to its extraordinary heating style. Yuan et al. (2007) demonstrated that the acrylamide content dramatically increased in potato chips by microwave treatment, which was significantly higher than that produced by traditional frying. Zhang et al. (2008a) systematically investigated and compared the formation of acrylamide in asparagine-sugar microwave heating model systems, which indicated that microwave heating is a potential source for the formation of large amounts of acrylamide. However, microwave frying can also be an alternative for the generation of low levels of acrylamide if the frying time is significantly reduced.

2.4 MITIGATION OF ACRYLAMIDE

2.4.1 Mitigation of Acrylamide

According to current research progress, possible mitigation approaches may include: (i) mitigating the formation and transformation of some key intermediates via modi-fying reaction conditions; (ii) controlling reaction conditions in order to preferably form some other small molecules during final stage of the reaction; (iii) inhibiting the key procedures in the Maillard reaction, such as the formation of Schiff base, Strecker degradation, N-glycoside pathway, and β-elimination reaction of decarbox-ylated Amadori products.

A mechanistic study demonstrated that three other final products including maleimide, succinimide, and niacinamide may be generated besides the formation of acrylamide during asparagine pathway (Yaylayan et al., 2003). Thus, acrylamide can be controlled via inducing the reaction toward the formation of these three products. First, the mitigation of acrylamide can be achieved via the control of decarboxylation of asparagine. Evidently, intramolecular cyclization to form 3-aminosuccinimide inter-mediate is much faster compared with the decarboxylation reaction due to entropy fac-tor. Thus, the formation of maleimide is preferably achieved. Second, the mitigation of acrylamide can be achieved via the control of the Schiff form of N-glycosylasparagine, which can stabilize itself through intramolecular cyclization initiated by the carbox-ylate anion. This may reduce the formation of oxazolidin-5-one. The Schiff form of N-glycosylasparagine can generate Amadori products through Amadori rearrange-ment, which prevent the Schiff base from generating decarboxylated Amadori prod-ucts through intramolecular cyclization and decarboxylation. Subsequently, Amadori products can generate the succinimide Amadori intermediate and further lead to the formation of succinimide. Third, the preferable formation of niacinamide can competi-tively reduce the generation of acrylamide. The formation mechanism of maleimide, succinimide, and niacinamide during asparagine pathway is shown in Figure 2.3.

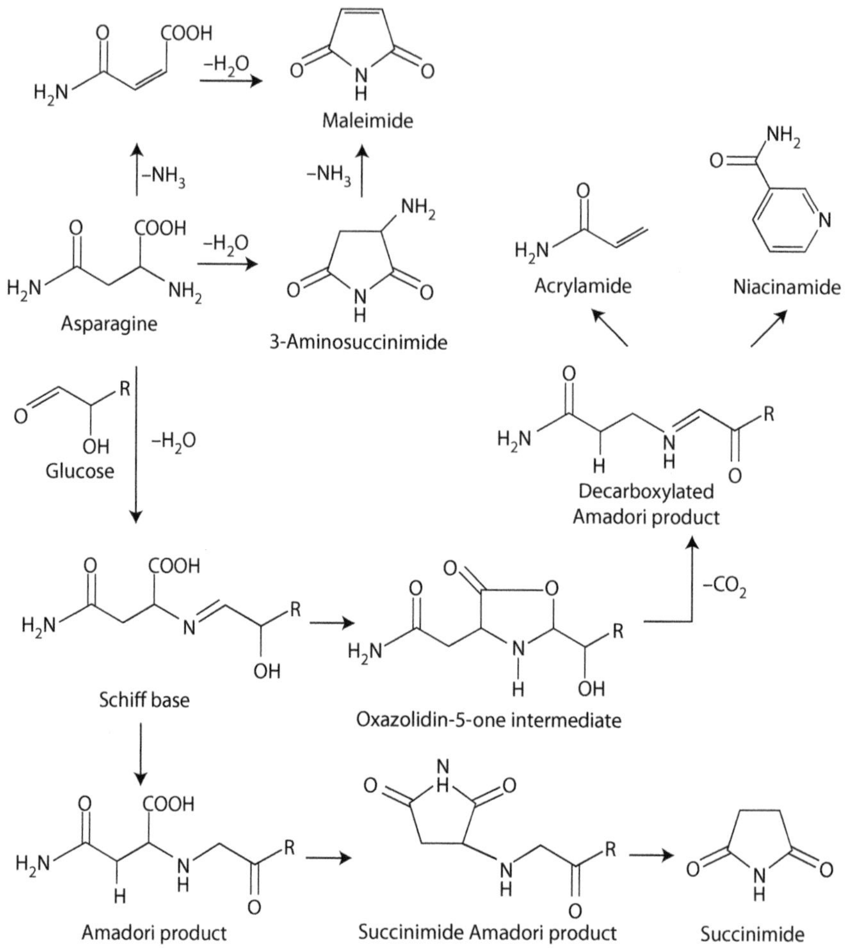

FIGURE 2.3 Mechanistic pathway of reduction of acrylamide and formation of maleimide, succinimide, and niacinamide.

2.4.2 Current Progress on Acknowledged Mitigation Recipes

Recently, some published review articles reported the reduction of acrylamide in potatoes, cereal products, coffee, and so on. The acknowledged mitigation recipes mainly include the modification of raw materials and optimization of heat-processing parameters, such as change of precursors in raw materials, optimization of heat-processing methods and parameters, modification of cultivar techniques, storage temperature of materials, fermentation, and use of additives (Vinci et al., 2012). Some of the representative findings about the mitigation of acrylamide are shown in Table 2.1.

2.4.2.1 Control of Asparagine and Reducing Sugar Contents

A suitable limit for the reducing sugars during prefabricating is a simple and efficient approach to reduce the exposure to acrylamide from the predominant source

TABLE 2.1
Some of Representative Findings about General Mitigation Recipes for the Formation of Acrylamide in Foods

Sample Matrix	Mitigation Method	Main Findings	Reference
French fries	Lactic acid fermentation (45 min and 120 min) and blanching	Inhibitory rates: 79% and 94%	Baardseth et al. (2006)
Rye bread	Fermentation with *Aspergillus niger* glucoamylase	Inhibitory rate: 59.4%	Bartkiene et al. (2013)
French fries	Microwave precooking	Inhibitory rate: 36–60%	Erdoğdu et al. (2007)
Bread	Long-time fermentation	Inhibitory rates: 87% (whole wheat bread) and 77% (rye bran bread)	Fredriksson et al. (2004)
Potatoes	Controlled atmosphere storage and low-dose irradiation	Reduction of reducing sugars in food materials	Geduct et al. (2007)
Potato crisps	Immersion with low-dose acetic acid	Maximal inhibitory rate: 60%	Kita et al. (2004)
Model system	Use of 1%, 5%, and 10% of NaCl solutions	Inhibitory rate: 32%, 36%, and 40%	Kolek et al. (2006)
Cookies	Use of antioxidant of bamboo leaves, sodium isoascorbate, tea polyphenols, vitamin E, and TBHQ	Inhibitory rate: 43.0–71.2%	Li et al. (2012)
French fries	Immersion with water and citric acid solution, blanching	Different inhibitory effects	Pedreschi et al. (2007)
Potato crisps	Use of asparaginase and blanching	Inhibitory rate: 90%	Pedreschi et al. (2011)
French fries	Use of 15 kinds of vitamins	Inhibitory rate: 51% (by niacin) and 34% (by pyridoxal)	Zeng et al. (2009)

for many consumers (Fiselier and Grob, 2005). The acrylamide level in potato chips made from tubers stored at low temperature (<10°C) was much higher than that stored at high temperature (>10°C) based on the consideration of storage conditions (Chuda et al., 2003). Thus, an appropriate storage temperature for the raw materials is required. Meanwhile, the proper fertilization of nitrogen and potassium was in favor of the control of asparagine contents in foodstuffs and the reduction of high amounts of acrylamide during heat processing (Halford et al., 2007). Therefore, the following tips should be considered when the control of precursor contents is used for the limitation of acrylamide concentrations: (i) choice of plant cultivars low in either asparagine or reducing sugars; (ii) avoidance of storage at too low temperatures; (iii) optimization of the blanching process to extract the maximum amount of sugar and asparagine; and (iv) avoidance of adding reducing sugars in further treatments.

2.4.2.2 Optimization of Heat-Processing Parameters or Methods

The control of important processing parameters, which include heating temperature, heating time, and oil type, could be regarded as the most direct way to mitigate acrylamide. The evidence indicated that there is not a simple linear increase between the acrylamide level and the heating temperature, but the acrylamide concentrations are noticed to increase linearly with heating time. The selection of appropriate heating temperature and avoidance of too long heating time are in favor of controlling the formation of large amounts of acrylamide. For the effect of heating temperature, recent finding indicated that acrylamide levels in French fries can be mitigated by half if the final stage of the frying process employs a lower oil temperature (Palazoğlu and Gökmen, 2008). For the effect of oil type, the origin of the deep-frying vegetable oils did not seem to affect the acrylamide formation in potatoes during frying (Mestdagh et al., 2008). However, the formation of acrylamide in food was much higher when using palmolein or frying oils containing silicone (Gertz and Klostermann, 2002).

On the basis of the above contributions, current studies specially focus on the optimization of pretreatments and heat-processing methods. For the modification of pretreatments, additional immersion and blanching procedures before frying could significantly mitigate the acrylamide levels (Mestdagh et al., 2008; Pedreschi et al., 2007). Such mitigation effect may be partly ascribed to the loss of precursors in original materials during pretreatments. For the improvements of heat-processing methods, the mitigation of acrylamide may be achieved by low-temperature heating such as low-temperature vacuum frying. Besides, Erdoğdu et al. (2007) found that when the potato strips were subjected to frying after a microwave precooking step, acrylamide content in the whole potato strip was reduced by 36%, 41%, and 60% for frying at 150°C, 170°C, and 190°C, respectively, in comparison to the control. Therefore, the vacuum frying and microwave precooking techniques are highlighted as potentially effective methods for the mitigation of acrylamide. However, the original sensory attributes of foods should be preferably concerned and guaranteed when using these improved methods during practical heat processing. Besides the above-mentioned reducing methods, the elimination of acrylamide could also be achieved via supercritical fluid extraction (Banchero et al., 2013), lactic acid fermentation (Bartkiene et al., 2013), and use of asparaginase in combination with blanching (Pedreschi et al., 2011).

2.4.3 New and Improved Mitigation Recipes

Based on the maintenance of original sensory attributes and heat-processing procedures, the use of exogenous additives is highlighted as another effective mitigation recipe in recent studies. These exogenous additives may include conventional food additives, food formula, and plant extracts. Such mitigation recipe should comply with the following conditions: (i) the addition level should be properly controlled according to corresponding criteria of food or chemical additives; (ii) the selected additives should be regarded as no toxicity demonstrated by toxicity test from previous publications; and (iii) the additives applied to the food systems cannot affect the connatural and sensory characteristics of the processed foods. During these years, many additives have been found capable of reducing acrylamide formation in the Maillard reaction, as well as the addition levels and methods have also been improved. Meanwhile, the irradiation and genetic modification techniques have been gradually developed for the mitigation of acrylamide (Fan and Mastovska, 2006; Halford et al., 2007).

2.4.3.1 Addition of pH Modifier

The pH modification is an effective way to reduce acrylamide formation in food and model systems. The acrylamide level can be effectively mitigated via lowering pH of the food system. Such change may attribute to protonating the α-amino group of asparagine, which subsequently cannot engage in nucleophilic addition reactions with carbonyl groups (Hamzalioğlu and Gökmen, 2012). Citric acid is a representative compound for the modification of pH, which is approved as an effective acrylamide formation inhibitor (Low et al., 2006). The correlation between pH decrease and acrylamide reduction may not be associated depending on the food products, due to multiple factors or different starting pH values of the products. Although acetic acid can also be used for the mitigation of acrylamide, the sensory attributes of the food matrix may be unacceptable at some circumstances because of the volatile odor of acetic acid. Therefore, the use of acetic acid is not recommended.

2.4.3.2 Addition of Proteins or Amino Acids

In vitro incubation of acrylamide together with glutathione (GSH) at pH 8 significantly reduced the amount of acrylamide monomers to 81%. A higher availability of cysteines (molar ratio 1:10; acrylamide/GSH) led to 48% reduction of acrylamide because acrylamide is most likely bound covalently to glutathione via Michael addition of cysteine residues to the terminal double bond (Schabacker et al., 2004). The degradation mechanism of acrylamide *in vitro* by GSH is mainly via the decomposition of glycine fragment of GSH. The glycine moiety can be eliminated more readily by the cleavage of peptide bonding in the presence of acrylamide. The glycine in the solution is degraded via Strecker degradation to aldehyde, ammonia, and carbon dioxide. These degradation products, such as formaldehyde, react with acrylamide, leading to decomposition of acrylamide to small molecular fragments. Acrylamide can be transformed to acrylate which is susceptible to consecutive decarboxylation and further total oxidation by the catalytic processes of degraded products of glycine. Besides, L-cysteine-containing sulfhydryl group could effectively reduce the acrylamide formation in fried potato crisps (Ou et al., 2008).

2.4.3.3 Addition of Hydrogen Carbonates

The mitigation effect on the formation of acrylamide by hydrogen carbonates is complex. Studies on elimination and net formation of acrylamide caused by sodium hydrogen carbonate ($NaHCO_3$) showed that three different addition levels had a similar effect on acrylamide elimination. Potassium hydrogen carbonate ($KHCO_3$) had an inhibitory effect similar to the sodium compound except that net formation appeared to be slightly higher. Conversely, ammonium hydrogen carbonate (NH_4HCO_3) could dramatically enhance the acrylamide concentration (Amrein et al., 2006). The inter-pretation of great differences among three bicarbonates is also complicated by the change in the temperature curve known to accompany bicarbonate addition.

2.4.3.4   Addition of Mono- or Divalent Cations

The mitigation effect of mono- or divalent cations on the generation of acrylamide has also been investigated. A previous study proved considerable inhibiting effects of sodium chloride (NaCl) on the formation of acrylamide in mixtures of aspara-gine–glucose model systems or potato-based foods (Kolek et al., 2006; Pedreschi et al., 2010). Compared to the systems without NaCl addition, the acrylamide con-tent was lowered roughly by 32%, 36%, and 40% when using 1%, 5%, and 10% of NaCl aqueous solution, respectively. Similarly, the high mitigation efficiency of cal-cium chloride ($CaCl_2$) on the acrylamide formation in fried potato crisps was also demonstrated (Ou et al., 2008). Added divalent cations, such as Ca^{2+}, were found to prevent acrylamide formation completely, whereas monovalent cations, such as Na^+, almost halved the acrylamide formed in the model system (Gökmen and Şenyuva, 2007). A previous study hypothesized that ionic associations involving the ions and charged groups on asparagine and related intermediates were likely to be involved (Lindsay and Jiang, 2005). However, information about how the presence of cations would affect the formation of acrylamide and typical Maillard reaction products is still unclear.

2.4.3.5   Addition of Antioxidants

Many tests have been performed and both positive and negative effects on acryl-amide reduction have been demonstrated by using different antioxidants. The miti-gation effect of pure phenolic compounds isolated from natural antioxidant extracts on the acrylamide formation was demonstrated (Bassama et al., 2010). Biedermann et al. (2002) also found a relatively weak reduction of the acrylamide formation by addition of ascorbic acid to a potato model. Similar results were obtained by Levine and Smith (2005) using ascorbate as the additive.

The above observations indicated that the antioxidant capacity of antioxidants and the antioxidant activity of model systems or food matrices are important for mitigation of acrylamide. Summa et al. (2006) found a direct correlation between acrylamide levels and the antioxidant activity in model cookies. Under the guide of close relationship between the mitigation of acrylamide and antioxidant activity, some new antioxidants were recently found to effectively reduce the acrylamide lev-els in various matrices. The virgin olive oil (VOO) having the highest concentration of ortho-diphenolic compounds was able to efficiently inhibit acrylamide forma-tion in crisps prepared under mild-to-moderate frying conditions (Napolitano et al., 2008). The antioxidant of bamboo leaves (AOB) containing flavonoids, phenolic acids, and lactones was recently approved to effectively mitigate the potential risk of acrylamide in potato-based foods, fried chicken wings, oriental-fried bread sticks, and cookies (Li et al., 2012; Zhang and Zhang, 2008). To elucidate the mitigation mechanism of antioxidants, previous studies investigated the effect of AOB and the extract of green tea (EGT) on the kinetics of acrylamide formation in the aspara-gine–glucose model system under microwave and low-moisture heating conditions.

Both studies indicated that addition of AOB/EGT could reduce the formation of acrylamide during the formation-predominant kinetic stage but could not reduce the acrylamide content during the elimination-predominant kinetic stage of acrylamide (Zhang and Zhang, 2008; Zhang et al., 2008b).

To understand the mechanism behind the mitigation effect of antioxidants, key functional groups which play important roles in these antioxidant compounds need to be clarified. A previous study observed a positive correlation between the carbonyl value of the selected antioxidant additives and acrylamide formation in a model system (Ou et al., 2010). Phenolic compounds bearing carbonyl group may compete with the carbonyl group of reducing sugars in Maillard reactions and impact the kinetics of acrylamide formation and elimination (Hamzalioğlu and Gökmen, 2012).

2.4.3.6 Mitigation of Acrylamide and Sensory Attributes of Products

The use of mitigation recipes during food processing also requires the guarantee of sensory attributes. Current recipes, especially the pH modification, may also have an impact on the sensory quality because low pH inhibits the Maillard reaction and contributes to the generation of undesirable flavors. Acidification processing may result in a sour product taste. However, this effect depends on the soaking or blanching treatment, and the type and concentration of the acid applied. Addition of sulfur-containing amino acids may also generate unpleasant off-flavors upon heating. The use of $CaCl_2$ may improve product texture, but oppositely can cause a bitter after-taste. To cope with the impact on the sensory attributes, the compromise between the mitigation effect of acrylamide and the sensory impact has to be considered. Anese et al. (2009) found that lactic acid fermentation in the presence of glycine led to the most effective decrease of acrylamide formation (up to 70%) and simultaneously kept original browning, flavor, sourness, and crispness of the deep-fried potatoes. A quantitative descriptive analysis can be used for the sensory evaluation according to ISO 6658 (2005). The descriptors may contain browning, sourness, flavor, texture, and crispness.

2.5 TOXICITY AND RISK ASSESSMENT OF ACRYLAMIDE

2.5.1 Toxicity of Acrylamide

Acute toxicity tests showed that the LD_{50} value was 150–180 mg/kg for rats, mice, guinea pigs, and rabbits, suggesting that acrylamide is a moderately toxic substance (Ministry of Health, 2005). In addition, acrylamide has been recognized as a neurotoxin. A long-term exposure test showed that it could cause damage to the nervous system in both humans and animals (mainly peripheral nerves) (LoPachin and Gavin, 2012). Mammalian cell culture in vitro showed possible reproductive toxicity (Dearfield et al., 1988; Wang et al., 2010) and genotoxicity (Mei et al., 2010) of acrylamide. Reproductive and developmental toxicity studies in rats reached its no observed adverse effect model (NOAEL) of 2 mg/kg bw/day. Long-term studies in rats and mice have shown that the incidence for thyroid, adrenal, breast, and reproductive system cancers occurs in a dose-dependent manner with acrylamide intake (Ehlers et al., 2013). Olesen et al. (2008) conducted a prospective cohort study, showing that acrylamide exposure and incidence of breast cancer have a certain relevance in Danish women. However, meta analysis indicated that the current available studies consistently suggested a lack of an increased risk of most types of cancer from exposure to acrylamide (Pelucchi et al., 2011). Acrylamide has strong tissue permeability, and can penetrate through unbroken skin and mucous membranes, or can pass through lungs and digestive tract into the human body. The oral ingestion is considered to be the fastest and most complete way for acrylamide absorption. Drinking water is the primary source of human acrylamide intake. In many countries, acrylamide content of drinking water cannot exceed 0.25 µg/L (Ministry of Health, 2005). Acrylamide released from food packaging materials is also a possible source for human acrylamide intake. The U.S. Food and Drug Administration (US FDA) has limited the amount of monomer in polyacrylamide for use in paper or cardboard in contact with food to 0.2% (2 g/kg).

On May 18, 2010, the European Food Safety Authority released the acrylamide levels in food monitoring report in 2008, which was based on the analytical results for about 3461 food samples from the 22 EU Member States and Norway. The food types included French fries, baked potato chips, home-cooked potato products, bread, breakfast cereals, biscuits, roasted coffee, canned baby food, processed cereal baby food, and other food products. The report showed that in the 22 food categories, the highest average level of acrylamide was found in "coffee substitutes," including grain-based (such as barley or chicory) similar to coffee drinks, at 1124 µg/kg; the lowest mean level of 23 µg/kg was seen in the unspecified bread products (EFSA, 2010).

2.5.2 RISK ASSESSMENT OF ACRYLAMIDE INTAKE

Animal test results showed that NOAEL of acrylamide causing neuropathy is 0.2 mg/kg bw. On the basis of average human intake level (1 µg/kg bw/day) and the high intake level (4 µg/kg bw/day), the margin of exposure (MOE) of populations in average intake and high intake of acrylamide causing neuropathy are 200 and 50, respectively. Furthermore, NOAEL causing reproductive toxicity of acrylamide is 2 mg/kg bw. Thus, the MOEs in average intake and high intake of acrylamide causing genotoxicity are 2000 and 500, respectively. JECFA considers that the risk of such side effects can be ignored, but the high-intake population may not be excluded for the possibility of causing neuropathic changes. It needs to be kept in mind that the available research data, including that from epidemiological, animal, and human studies, are insufficient to draw a definite conclusion. Carcinogenic studies in animals indicated that the benchmark dose lower confidence limit (BMDL) of acrylamide-induced breast cancer was 0.3 mg/kg bw/day. Based on the average intake level (1 µg/kg bw/day) and high intake level (4 µg/kg bw/day) of acrylamide in humans, the MOE of populations in average intake and high intake of acrylamide causing carcinogenicity should be 300 and 75, respectively (JECFA, 2006).

Taking all research results into account, it is recommended to take reasonable measures to reduce acrylamide levels in foods such as avoiding prolonged or high-temperature cooking starchy foods (Ministry of Health, 2005).

2.6   CONCLUSION AND PROSPECTIVES

The state-of-the-art trends of the analytical chemistry, formation, mitigation, and toxicology of acrylamide have been examined by many research groups around the world. For the analytical approaches, besides the pretreatment and instrumental improvement of GC–MS and LC–MS/MS, the recommended fast pretreatment, UHPLC–MS/MS analysis, HPLC–DAD, GC–ECD, TLC, and recent optimized genetic and ELISA techniques significantly accelerate the analytical speed and efficiency for acrylamide in various categories of foods. For the formation of acrylamide, Strecker degradation and N-glycoside pathway during the Maillard reaction are further studied while the importance of decarboxylated Schiff base and decarboxylated Amadori product has been highlighted on a mechanistic basis. For the mitigation recipes of acrylamide, maleimide, succinimide, and niacinamide may be generated besides the formation of acrylamide during asparagine pathway. Thus, acrylamide can be controlled via inducing the reaction toward the formation of these three products. Besides acknowledged recipes including control of precursors and heat-processing parameters, the use of exogenous additives, such as citric acid, proteins and amino acids, hydrogen carbonates, mono- or divalent cations, and antioxidant, is an effective method for the mitigation of acrylamide. Animal test results show that NOAEL of acrylamide causing neuropathy is 0.2 mg/kg bw; on the basis of human average intake (1 μg/kg bw/day) and high intake (4 μg/kg bw/day) of acrylamide, the MOEs of populations causing neuropathy are 200 and 50, respectively. The potential harm to human health should be concerned. It is recommended to take measures to reduce the levels of acrylamide in foods and its health risks in humans.

For the outlook of future research, an acknowledged fact that acrylamide is formed in the Maillard reaction has been adequately demonstrated. With the development of various mitigation recipes of acrylamide, the reduction mechanism needs to be further demonstrated. First, mechanistic study needs to clarify the key functional groups of the additives which play an important role in the mitigation process. Second, further study needs to demonstrate the action positions of these functional groups on the reactions in the asparagine pathway. The preferable approach is to investigate the formation variation of intermediates, such as Schiff base in the reaction between asparagine and fructose, the transform action from glucose to fructose via isomerization, and the formation of acrylamide from intermediates on a kinetic basis. Meanwhile, further study should optimize the compromise recipe between high reduction effect and reasonable sensory attributes. Nevertheless, the edible safety of processing products after addition of mitigation agents needs to be simultaneously concerned. Future toxicological studies *in vivo* should focus on the analytical chemistry, metabolism, toxicokinetics, and mitigation pathways of biomarkers of acrylamide and glycidamide metabolites, including their DNA adducts, hemoglobin adducts, and mercapturic acid conjugates.

REFERENCES

Alpmann, A., Morlock, G. 2008. Rapid and sensitive determination of acrylamide in drinking water by planar chromatography and fluorescence detection after derivatization with dansulfinic acid. *J. Sep. Sci.* 31, 71–77.

Ål Viklund, G., Olsson, K. M., Sjöholm, I. M., Skog, K. I. 2008. Variety and storage conditions affect the precursor content and amount of acrylamide in potato crisps. *J. Sci. Food Agric.* 88, 305–312.

Amrein, T. M., Andres, L., Manzardo, G. G. G., Amadò, R. 2006. Investigations on the promoting effect of ammonium hydrogen carbonate on the formation of acrylamide in model systems. *J. Agric. Food Chem.* 54, 10253–10261.

Amrein, T. M., Bachmann, S., Noti, A., Biedermann, M., Barbosa, M. F., Biedermann-Brem, S., Grob, K., Keiser, A., Realini, P., Escher, F., Amadó, R. 2003. Potential of acrylamide formation, sugars, and free asparagine in potatoes: A comparison of cultivars and farming systems. *J. Agric. Food Chem.* 51, 5556–5560.

Andrawes, F., Greenhouse, S., Draney, D. 1987. Chemistry of acrylamide bromination for trace analysis by gas chromatography and gas chromatography-mass spectrometry. *J. Chromatogr.* 399, 269–275.

Anese, M., Bortolomeazzi, R., Manzocco, L., Manzano, M., Giusto, C., Nicoli, M. C. 2009. Effect of chemical and biological dipping on acrylamide formation and sensory properties in deep-frying potatoes. *Food Res. Int.* 42, 142–147.

Baardseth, P., Blom, H., Skrede, G., Mydland, L. T., Skrede, A., Slinde, E. 2006. Lactic acid fermentation reduces acrylamide formation and other Maillard reactions in French fries. *J. Food Sci.* 71, C28–C33.

Banchero, M., Pellegrino, G., Manna, L. 2013. Supercritical fluid extraction as a potential mitigation strategy for the reduction of acrylamide level in coffee. *J. Food Eng.* 115, 292–297.

Bartkiene, E., Jakobsone, I., Juodeikiene, G., Vidmantiene, D., Pugajeva, I., Bartkevics, V. 2013. Study on the reduction of acrylamide in mixed rye bread by fermentation with bacteriocin-like inhibitory substances producing lactic acid bacteria in combination with *Aspergillus niger* glucoamylase. *Food Control* 30, 35–40.

Bassama, J., Brat, P., Bohuon, P., Boulanger, R., Günata, Z. 2010. Study of acrylamide mitigation in model system: Effect of pure phenolic compounds. *Food Chem.* 123, 558–562.

Becalski, A., Lau, B. P.-Y., Lewis, D., Seaman, S. W. 2003. Acrylamide in foods: Occurrence, sources, and modeling. *J. Agric. Food Chem.* 51, 802–808.

Bermudo, E., Núñez, O., Moyano, E., Puignou, L., Galceran, M. T. 2007. Field amplified sample injection-capillary electrophoresis-tandem mass spectrometry for the analysis of acrylamide in foodstuffs. *J. Chromatogr. A* 1159, 225–232.

Bermudo, E., Núñez, O., Puignou, L., Galceran, M. T. 2006. Analysis of acrylamide in food products by in-line preconcentration capillary zone electrophoresis. *J. Chromatogr. A* 1129, 129–134.

Biedermann, M., Grob, K. 2008. In GC-MS, acrylamide from heated foods may be coeluted with 3-hydroxy propionitrile. *Eur. Food Res. Technol.* 227, 945–948.

Biedermann, M., Noti, A., Biedermann-Brem, S., Mozzetti, V., Grob, K. 2002. Experiments on acrylamide formation and possibilities to decrease the potential of acrylamide formation in potatoes. *Mitt. Lebensmittelunters. Hyg.* 93, 668–687.

Bologna, L. S., Andrawes, F. F., Barvenik, F. W., Lentz, R. D., Sojka, R. E. J. 1999. Analysis of residual acrylamide in field crops. *J. Chromatogr. Sci.* 37, 240–244.

Bortolomeazzi, R., Munari, M., Anese, M., Verardo, G. 2012. Rapid mixed mode solid phase extraction method for the determination of acrylamide in roasted coffee by HPLC-MS/MS. *Food Chem.* 135, 2687–2693.

Burch, R. S., Trzesicka, A., Clarke, M., Elmore, J. S., Briddon, A., Matthews, W., Webber, N. 2008. The effects of low-temperature potato storage and washing and soaking pre-treatments on the acrylamide content of French fries. *J. Sci. Food Agric.* 88, 989–995.

Castle, L. 1993. Determination of acrylamide monomer in mushrooms grown on polyacrylamide gel. *J. Agric. Food Chem.* 41, 1261–1263.

Castle, L., Eriksson, S. 2005. Analytical methods used to measure acrylamide concentrations in foods. *J. AOAC Int.* 88, 274–284.

Chuda, Y., Ono, H., Yada, H., Ohara-Takada, A., Matsuura-Endo, C., Mori, M. 2003. Effects of physiological changes in potato tubers (*Solanum tuberosum* L.) after low temperature storage on the level of acrylamide formed in potato chips. *Biosci. Biotechnol. Biochem.* 67, 1188–1190.

Claeys, W. L., de Vleeschouwer, K., Hendrickx, M. E. 2005. Quantifying the formation of carcinogens during food processing: Acrylamide. *Trends Food Sci. Technol.* 16, 181–193.

Dabrio, M., Sejerøe-Olsen, B., Musser, S., Emteborg, H., Ulberth, F., Emons, H. 2008. Production of a certified reference material for the acrylamide content in toasted bread. *Food Chem.* 110, 504–511.

Dearfield, K. L., Abernathy, C. O., Ottley, M. S., Brantner, J. H., Hayes, P. F. 1988. Acrylamide: Its metabolism, developmental and reproductive effects, genotoxicity, and carcinogenicity. *Mutat. Res.* 195, 45–77.

Delatour, T., Périsset, A., Goldmann, T., Riediker, S., Stadler, R. H. 2004. Improved sample preparation to determine acrylamide in difficult matrixes such as chocolate powder, cocoa, and coffee by liquid chromatography tandem mass spectrometry. *J. Agric. Food Chem.* 52, 4625–4631.

De Wilde, T., de Meulenaer, B., Mestdagh, F., Govaert, Y., Vandeburie, S., Ooghe, W., Fraselle, S., Demeulemeester, K., van Peteghem, C., Calus, A., Degroodt, J.-M., Verhé, R. 2006. Influence of fertilization on acrylamide formation during frying of potatoes harvested in 2003. *J. Agric. Food Chem.* 54, 404–408.

Diekmann, J., Wittig, A., Stalbbert, R. 2008. Gas chromatographic-mass spectrometric analysis of acrylamide and acetamide in cigarette main stream smoke after on-column injection. *J. Chromatogr. Sci.* 46, 659–663.

Dunovská, L., Čajka, T., Hajšlová, J., Holadová, K. 2006. Direct determination of acrylamide in food by gas chromatography-high-resolution time-of-flight mass spectrometry. *Anal. Chim. Acta* 578, 234–240.

Ehlers, A., Lenze, D., Broll, H., Zagon, J., Hummel, M., Lampen, A. 2013. Dose dependent molecular effects of acrylamide and glycidamide in human cancer cell lines and human primary hepatocytes. *Toxicol. Lett.* 217, 111–120.

European Food Safety Authority (EFSA). 2010. Results of acrylamide levels in food from monitoring year 2008. *EFSA J.* 8, 1599.

Erdoğdu, S. B., Palazoğlu, T. K., Gökmen, V., Şenyuva, H. Z., Ekiz, H. İ. 2007. Reduction of acrylamide formation in French fries by microwave pre-cooking of potato strips. *J. Sci. Food Agric.* 87, 133–137.

Ezeji, T. C., Groberg, M., Qureshi, N., Blaschek, H. P. 2003. Continuous production of butanol from starch-based packing peanuts. *Appl. Biochem. Biotechnol.* 106, 375–382.

Fan, X., Mastovska, K. 2006. Effectiveness of ionizing radiation in reducing furan and acrylamide levels in foods. *J. Agric. Food Chem.* 54, 8266–8270.

Fiselier, K., Grob, K. 2005. Legal limit for reducing sugars in prefabricates targeting 50 µg/kg acrylamide in French fries. *Eur. Food Res. Technol.* 220, 451–458.

Fredriksson, H., Tallving, J., Rosén, J., Åman, P. 2004. Fermentation reduces free asparagine in dough and acrylamide content in bread. *Cereal Chem.* 81, 650–653.

Friedman, M. 2003. Chemistry, biochemistry, and safety of acrylamide. A review. *J. Agric. Food Chem.* 51, 4504–4526.

Gerendás, J., Heuser, F., Sattelmacher, B. 2007. Influence of nitrogen and potassium supply on contents of acrylamide precursors in potato tubers and on acrylamide accumulation in French fries. *J. Plant Nutr.* 30, 1499–1516.

Gertz, C., Klostermann, S. 2002. Analysis of acrylamide and mechanisms of its formation in deep-fried foods. *Eur. J. Lipid Sci. Technol.* 104, 762–771.

Gökmen, V., Şenyuva, H. Z. 2007. Acrylamide formation is prevented by divalent cations during the Maillard reaction. *Food Chem.* 103, 196–203.

Gökmen, V., Şenyuva, H. Z. 2006. A generic method for the determination of acrylamide in thermally processed foods. *J. Chromatogr. A* 1120, 194–198.

Granda, C., Moreira, R. G. 2005. Kinetics of acrylamide formation during traditional and vacuum frying of potato chips. *J. Food Process Eng.* 28, 478–493.

Granvogl, M., Jezussek, M., Koehler, P., Schieberle, P. 2004. Quantitation of 3-aminopropionamide in potatoes—A minor but potent precursor in acrylamide formation. *J. Agric. Food Chem.* 52, 4751–4757.

Halford, N. G., Curtis, T. Y., Muttucumaru, N., Postles, J., Elmore, J. S., Mottram, D. S. 2012a. The acrylamide problem: A plant and agronomic science issue. *J. Exp. Bot.* 63, 2841–2851.

Halford, N. G., Muttucumaru, N., Powers, S. J., Gillatt, P. N., Hartley, L., Elmore, J. S., Mottram, D. S. 2012b. Concentrations of free amino acids and sugars in nine potato varieties: Effects of storage and relationship with acrylamide formation. *J. Agric. Food Chem.* 60, 12044–12055.

Halford, N. G., Muttucumaru, N., Curtis, T. Y., Parry, M. A. J. 2007. Genetic and agronomic approaches to decreasing acrylamide precursors in crop plants. *Food Addit. Contam.* 24(S1), 26–36.

Hamzalioğlu, A., Gökmen, V. 2012. Role of bioactive carbonyl compounds on the conversion of asparagine into acrylamide during heating. *Eur. Food Res. Technol.* 235, 1093–1099.

Hidalgo, F. J., Delgado, R. M., Navarro, J. L., Zamora, R. 2010. Asparagine decarboxylation by lipid oxidation products in model systems. *J. Agric. Food Chem.* 58, 10512–10517.

Huang, Y. F., Wu, K. Y., Liou, S. H., Uang, S. N., Chen, C. C., Shih, W. C., Lee, S. C., Huang, C. C. J., Chen, M. L. 2011. Biological monitoring for occupational acrylamide exposure from acrylamide production workers. *Int. Arch. Occup. Environ. Health* 84, 303–313.

International Agency for Research on Cancer (IARC). 1994. *IARC Monographs on the Evaluation of Carcinogenic Risks to Humans. Some Industrial Chemicals.* Lyon, France: IARC, Vol. 60, pp. 389–433.

International Food Safety Authorities Network (INFOSAN). 2005. Information Note No. 2/2005. *Acrylamide in Food Is a Potential Health Hazard.* Geneva, Switzerland: INFOSAN. [Online]: http://www.who.int/foodsafety/fs_mangement/en/No_02_Acrylamide_Mar05_en.pdf.

International Organization for Standardization (ISO) 6658. 2005. *International Standard: Sensory Analysis—Methodology—General Guidance.* [Online]: http://www.iso.org/ iso/iso_catalogue/catalogue_tc/catalogue_detail.htm?csnumber=36226&commid=47858.

Ishizuka, M., Fujioka, K., Shibamoto, T. 2008. Analysis of acrylamide in a complex matrix of polyacrylamide solutions treated by heat and ultraviolet light. *J. Agric. Food Chem.* 56, 6093–6096.

Jezussek, M., Schieberle, P. 2003. A new LC/MS-method for the quantitation of acrylamide based on a stable isotope dilution assay and derivatization with 2-mercaptobenzoic acid. Comparison with two GC/MS methods. *J. Agric. Food Chem.* 51, 7866–7871.

Joint Expert Committee on Food Additives (JECFA). 2006. Evaluation of certain food contaminants. 64th report of the joint FAO/WHO expert committee on food additives. WHO Technical Report Series, No. 930, pp. 8–26, 93 World Health Organization, Geneva, Switzerland. http://whqlibdoc.who.int/trs/WHO_TRS_930_eng.pdf.

Kaykhaii, M., Abdi, A. 2013. Rapid and sensitive determination of acrylamide in potato crisps using reversed-phase direct immersion single drop microextraction-gas chromatography. *Anal. Methods* 5, 1289–1293.

Keramat, J., LeBail, A., Prost, C., Soltanizadeh, N. 2011. Acrylamide in foods: Chemistry and analysis. A review. *Food Bioprocess Technol.* 4, 340–363.

Kim, B., Park, S., Lee, I., Lim, Y., Hwang, E., So, H.-Y. 2010. Development of a certified reference material for the determination of acrylamide in potato chips. *Anal. Bioanal. Chem.* 398, 1035–1042.

Kita, A., Bråthen, E., Knutsen, S. H., Wicklund, T. 2004. Effective ways of decreasing acrylamide content in potato crisps during processing. *J. Agric. Food Chem.* 52, 7011–7016.

Knol, J. J., van Loon, W. A. M., Linssen, J. P. H., Ruck, A.-L., van Boekel, M. A. J. S., Voragen, A. G. J. 2005. Toward a kinetic model for acrylamide formation in a glucose-asparagine reaction system. *J. Agric. Food Chem.* 53, 6133–6139.

Knutsen, S. H., Dimitrijevic, S., Molteberg, E. L., Segtnan, V. H., Kaaber, L., Wicklund, T. 2009. The influence of variety, agronomical factors and storage on the potential for acrylamide formation in potatoes grown in Norway. *LWT-Food Sci. Technol.* 42, 550–556.

Koch, M., Bremser, W., Koeppen, R., Siegel, D., Toepfer, A., Nehls, I. 2009. Development of two certified reference materials for acrylamide determination in foods. *J. Agric. Food Chem.* 57, 8202–8207.

Kolek, E., Šimko, P., Simon, P. 2006. Inhibition of acrylamide formation in asparagine/D-glucose model system by NaCl addition. *Eur. Food Res. Technol.* 224, 283–284.

Levine, R. A., Smith, R. E. 2005. Sources of variability of acrylamide levels in cracker model. *J. Agric. Food Chem.* 53, 4410–4416.

Li, D., Chen, Y. Q., Zhang, Y., Lu, B. Y., Zhang, Y. 2012. Study on mitigation of acrylamide formation in cookies by 5 antioxidants. *J. Food Sci.* 77, C1144–C1149.

Lindsay, R. C., Jiang, S. 2005. Model systems for evaluating factors affecting acrylamide formation in deep fried foods. *Adv. Exp. Med. Biol.* 561, 329–341.

Lineback, D. R., Coughlin, J. R., Stadler, R. H. 2012. Acrylamide in foods: A review of the science and future considerations. *Ann. Rev. Food Sci. Technol.* 3, 15–35.

Locas, C. P., Yaylayan, V. A. 2008. Further insight into thermally and pH-induced generation of acrylamide from glucose/asparagine model systems. *J. Agric. Food Chem.* 56, 6069–6074.

LoPachin, R. M., Gavin, T. 2012. Molecular mechanism of acrylamide neurotoxicity: Lessons learned from organic chemistry. *Environ. Health Perspect.* 120, 1650–1657.

Low, M. Y., Koutsidis, G., Parker, J. K., Elmore, J. S., Dodson, A. T., Mottram, D. S. 2006. Effect of citric acid and glycine addition on acrylamide and flavor in a potato model system. *J. Agric. Food Chem.* 54, 5976–5983.

Marchettini, N., Focardi, S., Guarnieri, M., Guerranti, C., Perra, G. 2013. Determination of acrylamide in local and commercial cultivar of potatoes from biological farm. *Food Chem.* 136, 1426–1428.

Marín, J. M., Pozo, Ó. J., Sancho, J. V., Pitarch, E., López, F. J., Hernández, F. 2006. Study of different atmospheric-pressure interfaces for LC-MS/MS determination of acrylamide in water at sub-ppb levels. *J. Mass Spectrom.* 41, 1041–1048.

Mastovska, K., Lehotay, S. J. 2006. Rapid sample preparation method for LC-MS/MS or GC-MS analysis of acrylamide in various food matrices. *J. Agric. Food Chem.* 54, 7001–7008.

Mei, N., McDaniel, L. P., Dobrovolsky, V. N., Guo, X. Q., Shaddock, J. G., Mittelstaedt, R. A., Azuma, M., Shelton, S. D., McGarrity, L. J., Doerge, D. R., Heflich, R. H. 2010. The genotoxicity of acrylamide and glycidamide in big blue rats. *Toxicol. Sci.* 115, 412–421.

Mestdagh, F., de Wilde, T., Fraselle, S., Govaert, Y., Ooghe, W., Degroodt, J.-M., Verhé, R., van Peteghem, C., de Meulenaer, B. 2008. Optimization of the blanching process to reduce acrylamide in fried potatoes. *LWT-Food Sci. Technol.* 41, 1648–1654.

Mikulíková, R., Sobotová, K. 2007. Determination of acrylamide in malt with GC/MS. *Acta Chim. Slov.* 54, 98–101.

Ministry of Health. 2005. *Risk Assessment of Acrylamide in Foods.* Beijing, China: Ministry of Health. [Online]: http://www.moh.gov.cn/uploadfile/200504/2005413101822861.doc.

Mottram, D. S., Wedzicha, B. L., Dodson, A. T. 2002. Acrylamide is formed in the Maillard reaction. *Nature* 419, 448–449.

Mulla, M. Z., Bharadwaj, V. R., Annapure, U. S., Singhal, R. S. 2011. Effect of formulation and processing parameters on acrylamide formation: A case study on extrusion of blends of potato flour and semolina. *LWT-Food Sci. Technol.* 44, 1643–1648.

Napolitano, A., Morales, F., Sacchi, R., Fogliano, V. 2008. Relationship between virgin olive oil phenolic compounds and acrylamide formation in fried crisps. *J. Agric. Food Chem.* 56, 2034–2040.

Nielsen, N. J., Granby, K., Hedegaard, R. V., Skibsted, L. H. 2006. A liquid chromatography—Tandem mass spectrometry method for simultaneous analysis of acrylamide and the precursors, asparagine and reducing sugars in bread. *Anal. Chim. Acta* 557, 211–220.

Olesen, P. T., Olsen, A., Frandsen, H., Frederiksen, K., Overvad, K., Tjønneland, A. 2008. Acrylamide exposure and incidence of breast cancer among postmenopausal women in the Danish Diet, Cancer and Health study. *Int. J. Cancer* 122, 2094–2100.

Oracz, J., Nebesny, E., Żyżelewicz, D. 2011. New trends in quantification of acrylamide in food products. *Talanta* 86, 23–34.

Ou, S. Y., Shi, J. J., Huang, C. H., Zhang, G. W., Teng, J. W., Jiang, Y., Yang, B. R. 2010. Effect of antioxidants on elimination and formation of acrylamide in model reaction systems. *J. Hazard. Mater.* 182, 863–868.

Ou, S. Y., Lin, Q., Zhang, Y. P., Huang, C. H., Sun, X., Fu, L. 2008. Reduction of acrylamide formation by selected agents in fried potato crisps on industrial scale. *Innov. Food Sci. Emerg. Technol.* 9, 116–121.

Owen, L. M., Castle, L., Kelly, J., Wilson, L., Lloyd, A. S. 2005. Acrylamide analysis: Assessment of results from six rounds of food analysis performance assessment scheme (FAPAS) proficiency testing. *J. AOAC Int.* 88, 285–291.

Palazoğlu, T. K., Gökmen, V. 2008. Reduction of acrylamide level in French fries by employing a temperature program during frying. *J. Agric. Food Chem.* 56, 6162–6166.

Paleologos, E. K., Kontominas, M. G. 2007. Effect of processing and storage conditions on the generation of acrylamide in precooked breaded chicken products. *J. Food Prot.* 70, 466–470.

Pedreschi, F., Granby, K., Risum, J. 2010. Acrylamide mitigation in potato chips by using NaCl. *Food Bioprocess Technol.* 3, 917–921.

Pedreschi, F., Kaack, K., Granby, K., Troncoso, E. 2007. Acrylamide reduction under different pre-treatments in French fries. *J. Food Eng.* 79, 1287–1294.

Pedreschi, F., Mariotti, S., Granby, K., Risum, J. 2011. Acrylamide reduction in potato chips by using commercial asparaginase in combination with conventional blanching. *LWT-Food Sci. Technol.* 44, 1473–1476.

Pelucchi, C., La Vecchia, C., Bosetti, C., Boyle, P., Boffetta, P. 2011. Exposure to acrylamide and human cancer—A review and meta-analysis of epidemiologic studies. *Ann. Oncol.* 22, 1487–1499.

Qiu, Y. Y., Qu, X. J., Dong, J., Ai, S. Y., Han, R. X. 2011. Electrochemical detection of DNA damage induced by acrylamide and its metabolite at the graphene-ionic liquid-Nafion modified pyrolytic graphite electrode. *J. Hazard. Mater.* 190, 480–485.

Quan, Y., Chen, M. L., Zhan, Y. H., Zhang, G. H. 2011. Development of an enhanced chemiluminescence ELISA for the rapid detection of acrylamide in food products. *J. Agric. Food Chem.* 59, 6895–6899.

Roach, J. A. G., Andrzejewski, D., Gay, M. L., Nortrup, D., Musser, S. M. 2003. Rugged LC-MS/MS survey analysis for acrylamide in foods. *J. Agric. Food Chem.* 51, 7547–7554.

Rosén, J., Hellenäs, K.-E. 2002. Analysis of acrylamide in cooked foods by liquid chromatography tandem mass spectrometry. *Analyst* 127, 880–882.

Rosén, J., Nyman, A., Hellenäs, K.-E. 2007. Retention studies of acrylamide for the design of a robust liquid chromatography-tandem mass spectrometry method for food analysis. *J. Chromatogr. A* 1172, 19–24.

Sanders, R. A., Zyzak, D. V., Stojanovic, M., Tallmadge, D. H., Eberhart, B. L., Ewald, D. K. 2002. An LC/MS acrylamide method and its use in investigating the role of asparagine. Presentation at the *Annual AOAC International Meeting*, Los Angeles, CA, September 22–26, 2002.

Schabacker, J., Schwend, T., Wink, M. 2004. Reduction of acrylamide uptake by dietary proteins in Caco-2 gut model. *J. Agric. Food Chem.* 52, 4021–4025.

Şenyuva, H. Z., Gökmen, V. 2006. Interference-free determination of acrylamide in potato and cereal-based foods by a laboratory validated liquid chromatography–mass spectrometry method. *Food Chem.* 97, 539–545.

Stadler, R. H., Blank, I., Varga, N., Robert, F., Hau, J., Guy, P. A., Robert, M.-C., Riediker, S. 2002. Acrylamide from Maillard reaction products. *Nature* 419, 449–450.

Stadler, R. H., Robert, F., Riediker, S., Varga, N., Davidek, T., Devaud, S., Goldmann, T., Hau, J., Blank, I. 2004. In-depth mechanistic study on the formation of acrylamide and other vinylogous compounds by the Maillard reaction. *J. Agric. Food Chem.* 52, 5550–5558.

Summa, C., Wenzl, T., Brohee, M., de la Calle, B., Anklam, E. 2006. Investigation of the correlation of the acrylamide content and the antioxidant activity of model cookies. *J. Agric. Food Chem.* 54, 853–859.

Sun, X. L., Ji, J., Jiang, D. L., Li, X. W., Zhang, Y. Z., Li, Z. J., Wu, Y. N. 2013. Development of a novel electrochemical sensor using pheochromocytoma cells and its assessment of acrylamide cytotoxicity. *Biosens. Bioelectron.* 44, 122–126.

Swedish National Food Administration (SNFA). 2002. *Information about Acrylamide in Food.* Stockholm, Sweden: SNFA. [Online]: http://www.slv.se.

Tateo, F., Bononi, M. 2003. A GC/MS method for the routine determination of acrylamide in food. *Ital. J. Food Sci.* 15, 149–151.

Tekel, J., Farkas, P., Kovác, M. 1989. Determination of acrylamide in sugar by capillary GLC with alkali flame-ionization detection. *Food Addit. Contam.* 6, 377–381.

Tezcan, F., Erim, F. B. 2008. On-line stacking techniques for the nonaqueous capillary electrophoretic determination of acrylamide in processed food. *Anal. Chim. Acta* 617, 196–199.

Umano, K., Shibamoto, T. 1987. Analysis of acrolein from heated cooking oils and beef fat. *J. Agric. Food Chem.* 35, 909–912.

Vinci, R. M., Mestdagh, F., De Meulenaer, B. 2012. Acrylamide formation in fried potato products—Present and future, a critical review on mitigation strategies. *Food Chem.* 133, 1138–1154.

Wang, H., Huang, P., Lie, T., Li, J., Hutz, R. J., Li, K., Shi, F. X. 2010. Reproductive toxicity of acrylamide-treated male rats. *Reprod. Toxicol.* 29, 225–230.

Wenzl, T., de la Calle, B., Anklam, E. 2003. Analytical methods for the determination of acrylamide in food products: A review. *Food Addit. Contam.* 20, 885–902.

Wenzl, T., Lachenmeier, D. W., Gökmen, V. 2007. Analysis of heat-induced contaminants (acrylamide, chloropropanols and furan) in carbohydrate-rich food. *Anal. Bioanal. Chem.* 389, 119–137.

Wicklund, T., Østlie, H., Lothe, O., Knutsen, S. H., Bråthen, E., Kita, A. 2006. Acrylamide in potato crisp-the effect of raw material and processing. *LWT-Food Sci. Technol.* 39, 571–575.

Yasuhara, A., Tanaka, Y., Hengel, M., Shibamoto, T. 2003. Gas chromatographic investigation of acrylamide formation in browning model systems. *J. Agric. Food Chem.* 51, 3999–4003.

Yaylayan, V. A. 2009. Acrylamide formation and its impact on the mechanism of the early Maillard reaction. *J. Food Nutr. Res.* 48, 1–7.

Yaylayan, V. A., Wnorowski, A., Locas, C. P. 2003. Why asparagine needs carbohydrates to generate acrylamide. *J. Agric. Food Chem.* 51, 1753–1757.

Young, M. S., Jenkins, K. M., Mallet, C. R. 2004. Solid-phase extraction and cleanup procedures for determination of acrylamide in fried potato products by liquid chromatography/mass spectrometry. *J. AOAC Int.* 87, 961–964.

Yuan, Y., Chen, F., Zhao, G.-H., Liu, J., Zhang, H.-X., Hu, X.-S. 2007. A comparative study of acrylamide formation induced by microwave and conventional heating methods. *J. Food Sci.* 72, C212–C216.

Zeng, X. H., Cheng, K.-W., Jiang, Y., Lin, Z.-X., Shi, J.-J., Ou, S.-Y., Chen, F., Wang, M. F. 2009. Inhibition of acrylamide formation by vitamins in model reactions and fried potato strips. *Food Chem.* 116, 34–39.

Zhang, Y., Dong, Y., Ren, Y. P., Zhang, Y. 2006. Rapid determination of acrylamide contaminant in conventional fried foods by gas chromatography with electron capture detector. *J. Chromatogr. A* 1116, 209–216.

Zhang, Y., Fang, H. R., Zhang, Y. 2008a. Study on formation of acrylamide in asparagine-sugar microwave heating systems using UPLC-MS/MS analytical method. *Food Chem.* 108, 542–550.

Zhang, Y., Jiao, J. J., Cai, Z. X., Zhang, Y., Ren, Y. P. 2007. An improved method validation for rapid determination of acrylamide in foods by ultra-performance liquid chromatography combined with tandem mass spectrometry. *J. Chromatogr. A* 1142, 194–198.

Zhang, Y., Ren, Y. P., Jiao, J. J., Li, D., Zhang, Y. 2011. Ultra high-performance liquid chromatography-tandem mass spectrometry for the simultaneous analysis of asparagine, sugars, and acrylamide in Maillard reactions. *Anal. Chem.* 83, 3297–3304.

Zhang, Y., Ren, Y. P., Zhang, Y. 2009. New research developments on acrylamide: Analytical chemistry, formation mechanism and mitigation recipes. *Chem. Rev.* 109, 4375–4397.

Zhang, Y., Ying, T. J., Zhang, Y. 2008b. Reduction of acrylamide and its kinetics by addition of antioxidant of bamboo leaves (AOB) and extract of green tea (EGT) in asparagine-glucose microwave heating system. *J. Food Sci.* 73, C60–C66.

Zhang, Y., Zhang, G. Y., Zhang, Y. 2005. Occurrence and analytical methods of acrylamide in heat-treated foods: Review and recent developments. *J. Chromatogr. A* 1075, 1–21.

Zhang, Y., Zhang, Y. 2008. Effect of natural antioxidants on kinetic behavior of acrylamide formation and elimination in low-moisture asparagine-glucose model system. *J. Food Eng.* 85, 105–115.

Zhou, X., Fan, L.-Y., Zhang, W., Cao, C.-X. 2007. Separation and determination of acrylamide in potato chips by micellar electrokinetic capillary chromatography. *Talanta* 71, 1541–1545.

Zhu, Y. H., Li, G. R., Duan, Y. P., Chen, S. Q., Zhang, C., Li, Y. F. 2008. Application of the standard addition method for the determination of acrylamide in heat-processed starchy foods by gas chromatography with electron capture detector. *Food Chem.* 109, 899–908.

Zyzak, D. V., Sanders, R. A., Stojanovic, M., Tallmadge, D. H., Eberhart, B. L., Ewald, D. K., Gruber, D. C., Morsch, T. R., Strothers, M. A., Rizzi, G. P., Villagran, M. D. 2003. Acrylamide formation mechanism in heated foods. *J. Agric. Food Chem.* 51, 4782–4787.

3 Biological and Chemical Activities of Furan

Takayuki Shibamoto

CONTENTS

3.1 INTRODUCTION

It has been known for many years that toxic heterocyclic compounds, such as furan, 5-hydroxymethylfurfural, and 4(5)-methylimidazole, are formed in Maillard reaction systems (Shibamoto, 1983). Accordingly, they are found in large numbers in heat-treated foods and beverages. For example, brewed coffee alone contains over 400 heterocyclic compounds (Flament, 2002). Some of them have been recognized as important flavor chemicals, which give pleasant roasted or toasted flavors to cooked foods. Figure 3.1 shows typical unsubstituted heterocyclic compounds and their physical and sensory properties (Arctander, 1969). Among heterocyclic flavor chemicals shown in this figure, furan may be the only chemical which has both a characteristic desirable flavor and an unacceptable level of toxicity.

Name		Structure	Molecular weight	Boiling point (°C)	Water solubility	Sensory description
Common	UPAC					
Furan	Oxole		68.08	32.0	Insoluble	Peculiar spicy-smoky, slightly cinnamon-like odor
Thiophene	Thiophene		84.14	84.4	Insoluble	Mildly peasant odor, slightly sulfurous odor
Pyrrole	1H-Pyrrole	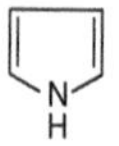	67.09	131.1	Sparingly soluble	Sweet and warm-ethereal, slightly-burnt-nauseating odor
Thiazole	1,3-Thiazole	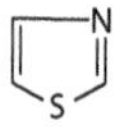	85.13	116.5	Slightly soluble	Characteristic foul odor
Pyridine	Pyridine	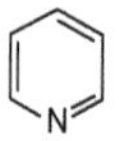	79.10	115.5	Miscible	Pungent, penetrating and diffusive odor, burnt smoky in extreme dilution
Pyrazine	Pyrazine	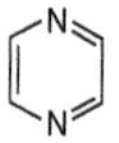	80.09	118.0	Miscible	Pungent, sweet odor, roasted or toasted odor in dilution

FIGURE 3.1 Typical unsubstituted heterocyclic compounds and their physical and sensory properties.

Furan is a five-membered heterocyclic aromatic compound. It possesses four carbon atoms and one oxygen atom as in the structure shown in Figure 3.1. Furan is also called a parent furan or an unsubstituted furan to separate from furan with substituents, such as 5-hydroxymethylfurfural. It is a colorless, flammable, and highly volatile liquid with pungent, bread-like, and caramel-like odor (Arctander, 1969). Its melting point is 85.6°C, and its boiling point 31.4°C, which is the lowest among simple heterocyclic compounds shown in Figure 3.1. Furan (unsubstituted or parent furan) was first prepared in 1870, about 100 years after a substituted furan, 2-furoic acid, was prepared in 1780 (Limpricht, 1870).

After invention of the gas chromatograph, furan has been found in many foods and beverages together with volatile flavor chemicals. A review article published in 1979 reported the presence of furan in coffee, canned beef, milk products, soy protein isolate, rapeseed protein, caramel, and maple syrup (Maga, 1979). Another review article reported that 76 furans, including unsubstituted furan, were found in various Maillard reaction systems consisting of sugars and different amine compounds, including amino acid, protein, or ammonia (Shibamoto, 1983).

Since the International Agency for Research on Cancer (IARC) classified furan as a possible carcinogen (IARC, 1995), its presence in foods and beverages has begun to receive much attention among regulatory agencies. Consequently, the USFDA collected data and reported the presence of furan in numerous foods (FDA, 2004). Therefore, furan is an important compound not only because of its importance as a flavor chemical but also because of concern that it is a toxic chemical found in foods and beverages. Table 3.1 shows selected review articles on furan published previously.

TABLE 3.1
Review Articles on Furan

Title	Contents	No. of Ref. Used	Refs.
Furan in food	Reports from FDA; active metabolite; comparison between furan and acrylamide: list in IARC, mutagenicity, carcinogenicity, target organs, acute toxicity, foods with significantly high contents; and future development.	22	Uneyama et al. (2004)
Precursors, formation, and determination of furan in food	Concentrations in selected food products: baby foods, soups, stew, chili, luncheon meat, and coffee; formation mechanisms from polyunsaturated fatty acid, amino acid, carbohydrate, ascorbic acid; and detection and quantification method.	19	Yaylayan (2006)
A review of the occurrence, formation, and analysis of furan in heat-processed foods	Occurrence in foods: baby and infant foods, coffees, breads, fish, fruits, meats, and vegetables; Toxicology: metabolism and genotoxicity; Analysis: sample preparations, GC, and GC–MS; Formations: from ascorbic acid and fatty acid; dietary exposure; and prospects for mitigation measures.	43	Crews and Castle (2007)
Furan in Kaffee und anderen Lebensmittein	Formation in foods and coffees, analyses, and prospect.	15	Kuballa (2007)
Data requirements for risk assessment of furan in food	Analytical methodology, occurrence, metabolism, toxicology, genotoxicity, and mode of action.	34	Heppner and Schalatter (2007)
Furan in processed food	Occurrence, routes of formation, precursors in foods, mitigation	57	Blank (2008)
Hazards of dietary furan	Analysis of dietary furan, exposure, methods of mitigation, toxicology including mechanism of action.	25	Bolger et al. (2009)
Furan in food—a review	Origin: baby food, fruits, vegetables, coffees, canned products, meats, and bread; formation pathways: from amino acid, carbohydrate, and upon ionizing radiation; metabolism and toxicology; analyses in food; and exposure assessment.	41	Vranova and Ciesarova (2009)
Hazards of dietary furan	Analysis, occurrence, exposure, mitigation/ reduction, metabolism/toxicokinetics, toxicology, and mechanism of action.	25	Bolgerm et al. (2009)
Toxicity and carcinogenicity of furan in human diet	In human diet: baby foods, infant formula, coffees, meat products, soups, cereal products, vegetables, and nutritional/diet drinks; toxicokinetics and metabolism; hepatotoxicity; genotoxicity; carcinogenicity; and human data on toxicity.	85	Bakhiya and Apple (2010)

It should be noted that there are many toxic chemicals, such as 5-hydroxymeth-ylfurfural and 4(5)-methylimidazole formed by the Maillard reaction in foods and beverages, in addition to furan. However, comprehensive reviews on these chemicals have recently been published (Hengel and Shibamoto, 2013; Morales, 2009). Therefore, this chapter focuses on furan alone.

3.2 FORMATION MECHANISMS OF FURAN

3.2.1 Precursors of Furan Formation

It has been known since the mid-twentieth century that the degradation of sugars produces furans and many low-molecular weight carbonyl compounds, such as formaldehyde, acetaldehyde, glyoxal, methylglyoxal, and diacetyl (Hodge, 1967; Newth, 1951; Newth and Wiggins, 1947). These carbonyl compounds subsequently produce heterocyclic flavor compounds with amine compounds via the Maillard reaction (Jiang et al., 2013). For example, 5-hydroxymethyl furfural was formed from D-fructose heated in aqueous solution under acidic conditions (Anet, 1962). When arabinose and mannose were degraded at 500°F, the products furfural and 5-hydroxymethyl furfural, respectively, were formed in the greatest amount (Raeisaenen et al., 2003). One sugar pyrolysis study with isotopic labeling glucose demonstrated that C1 and C2 in glucose split off to form glyoxal upon a direct cyclic Grob fragmentation and then the remaining C3, C4, C5, and C6 formed furan (Paine III et al., 2008).

Lipids also produce carbonyl compounds, which are precursors for the Maillard reaction. The formation of furan derivatives, furanones from beef fat heated with glycine, was reported in the early 1980s (Ohnishi and Shibamoto, 1984). It has been proposed that furan forms from polyunsaturated fatty acids (PUFA) via endoperoxides upon oxidation (Owczarek-Fendor et al., 2011; Yaylayan, 2006).

In addition to sugars and lipids, some chemicals, such as ascorbic acid, are known to produce furan upon heat treatment. For example, about 2 mmol/mol of furan was formed from dry-heating ascorbic acid (Limacher et al., 2007). When ascorbic acid buffered solutions were heated, 0.4 ppb of furan was formed, whereas its formation increased to 13.2 ppb when ascorbic acid was heated with starch under the same conditions (Owczarek-Fendor et al., 2010). The same authors also reported the formation of furan from β-carotene in starch-based systems (Owczarek-Fendor et al., 2011).

3.2.2 Proposed Formation Mechanisms in Foods and Beverages

Figure 3.2 is a summary of typically proposed formation pathways of furan from carbohydrates, lipids and amino acids, as well as ascorbic acid. This figure was constructed using the previously reported pathways (Limacher et al., 2008; Locas and Yaylayan, 2004; Owczarek-Fendor et al., 2011; Paine III et al., 2008; Yaylayan, 2006). Among the many products and intermediates formed prior to the formation of furan, 2-deoxy-aldotetrose seems to be one of the most important intermediates because four possible sources of food ingredients yield furan through this intermediate. Many reactions must be involved in the thermal degradation of food components, such as oxidation, dehydration, hydrolysis, decarboxylation, and condensation to yield furan.

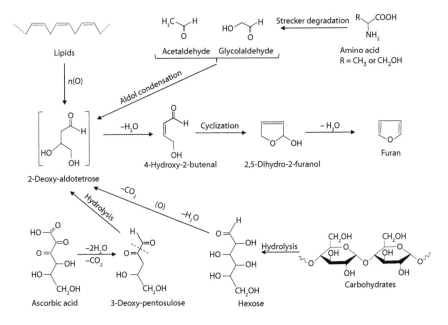

FIGURE 3.2 Summary of typical proposed formation pathways of furan from food components.

3.2.3 FORMATION OF FURAN BY MAILLARD REACTION

Although furan does not contain a nitrogen atom, an amine compound reacts with a sugar to promote the formation of carbonyl compounds (Wang et al., 2009; Yaylayan, 2006), suggesting that an amine compound acts as a catalyst for sugar degradation. An interesting study on furan formation from ^{13}C-labeled sugars has been reported (Yaylayan, 2006). In this study, involving a glucose/serine model system, the major furan formation pathway in glucose degradation involved C3–C4–C5–C6 carbon atoms of glucose, indicating that no carbons from serine were incorporated into the furan. A similar result was obtained when ^{13}C–5-labeled ribose was heated with cysteine (Cerny and Davidek, 2003). These results suggest that amine compounds in a Maillard reaction play an important role in furan formation from sugars. Another study with isotope-labeled glucose in a Maillard reaction system found that alanine and serine influenced furan production, whereas the formation of furan from glucose via fragmentation and recombination did not occur (Van Lancker et al., 2011). However, the presence of alanine, threonine, or serine in Maillard reaction systems promotes furan formation by the recombination of C2 fragment from a hexose (Limacher et al., 2008). In this study, formation of a C2 fragment was determined by identification of acetaldehyde and glycoaldehyde.

The above reports suggest that the most important reaction to produce furan in foods and beverages is the Maillard reaction. Consequently, many studies on furan formation have been conducted using Mallard reaction systems. The Maillard reaction—also called the nonenzymatic browning reaction—was advanced by a French chemist, L. C. Maillard in 1912. He hypothesized that the reaction that accounts for the browning pigments and polymers produced from the reaction occurring between the amino group of an amino acid and the carbonyl group of a sugar upon heat treatment. Later, this concept was expanded to include the reaction that occurs between an amine and a carbonyl compound and was called amino-carbonyl reaction (Fujimaki et al., 1986). Amine sources are not only amino acids but also proteins and phospholipids. Also, carbonyl sources are not only sugars but also carbohydrates and lipids. Therefore, foods contain ideal ingredients for Maillard reaction. A Maillard reaction system consisting of a monosaccharide, such as D-glucose, and an amino acid, such as cysteine, has been commonly used to study the formation of flavor chemicals, including furan, in foods and beverages because the matrix of a Maillard system is much simpler than that of a real food system.

3.2.4 FORMATION OF FURAN BY UV IRRADIATION

In addition to thermal treatment, photochemical formation of furan from sugars is also known to occur. Similar volatile compounds to the ones found in thermal Maillard model systems, including several furan derivatives, were yielded from an L-cysteine/D-glucose Maillard model system upon UV irradiation (Sheldon et al., 1986). Furan formation was observed in aqueous solutions containing sucrose, glucose, or fructose exposed to ionizing radiation (Fan, 2005). When apple cider and its constituent fructose were irradiated by UV, a significant amount of furan was produced (Fan and Geveke, 2007). pH conditions also play some role in furan formation from sugars. For example, the level of furan formed in an aqueous solution of sucrose or ascorbic acid at pH 3 was higher than at pH 7, whereas the level of furan formed from glucose in an aqueous solution at pH 7 was higher than that at pH 3 (Fan, 2005). Over 450 ppb of furan was formed from fructose upon UV irradiation but it formed in much smaller amounts from ascorbic acid (trace) and sucrose (none) (Bule et al., 2010). These results indicate that furan formation involves many factors.

3.3 OCCURRENCE OF FURAN IN FOODS AND BEVERAGES

As mentioned earlier, the presence of furan in foods and beverages has been reported as one of the components of flavor extracts since the early 1960s (Maga, 1979). Furan is one of the components of flavor extracts in numerous flavor and fragrance studies of foods and beverages. When a flavor characteristic of foods and beverages is evaluated, the composition of the flavor chemicals is more important than their absolute amounts. Therefore, reporting GC peak area % of each volatile chemical was generally recognized as an adequate data point in the scientific society in the early research. However, after government agencies reported the carcinogenicity of furan (IARC, 1995; NTP, 1993), its quantitative analysis in foods and beverages became important in assessing food safety. Consequently, many quantitative studies have been conducted on furan since then. Table 3.2 shows the levels of furan reported in different foods and beverages in the review articles as can be seen in Table 3.1.

TABLE 3.2
Amounts of Furan Found in Food and Beverages

Commodity	Amount (µg/kg)	References
Coffee	3–5050	Kuballa et al. (2005); Ho et al. (2005); Crews and Castle (2007); Heppner and Schlatter (2007); Kuballa (2007); La Pera et al. (2009); Vranova and Ciesarova (2009); Guenther et al. (2010); Bicchi et al. (2011); PavesiArisseto et al. (2012).
Vegetables	0.8–58.30	Crews and Castle (2007); Heppner and Schlatter (2007); Kim et al. (2010); Wegener and López-Sánchez (2010).
Fruits	0–37	Crews and Castle (2007); Heppner and Schlatter (2007); Kim et al. (2010); Pavesi Ariseto et al. (2010); Wegener and López-Sánchez (2010).
Baby foods	1–172.7	Bianchi et al. (2006); Crews and Castle (2007); Heppner and Schlatter (2007); Liu et al. (2008); Jestoi et al. (2009); Vranova and Ciesarova (2009); Kim et al. (2010); PavesiArisseto et al. (2010); Lancker et al. (2009); Ruiz et al. (2010).
Meat products	0.9–195.95	Crews and Castle (2007); Heppner and Schlatter (2007); Vranova and Ciesarova (2009); Kim et al. (2010).
Beer	0.8–3.4	Crews and Castle (2007); Heppner and Schlatter (2007).
Cereal products	2–191.3	Crews and Castle (2007); Heppner and Schlatter (2007); Vranova and Ciesarova (2009); Pavesi Arisseto et al. (2012).
Bakery products	2–200	Crews and Castle (2007); Vranova and Ciesarova (2009); Wegener and López-Sánchez (2010).
Miscellaneous	0.5–420	Senyuva and Gökmen (2007); Crews and Castle (2007); Heppner and Schlatter (2007); Vranova and Ciesarova (2009); Van Lancker et al. (2009); Kim et al. (2010).

3.3.1 ANALYTICAL METHODS FOR QUANTIFYING FURAN

Gas chromatography (GC) was invented in the 1950s and became a powerful analytical method for studying volatile flavor chemicals. Moreover, the detection limits of GC were significantly lower than other analytical methods, such as spectrophotometry. Generally, a volatile sample was prepared with an organic solvent extraction and the extract was directly injected into the GC equipped with a frame ionization detector (FID). However, recently, the mainstream analytical method for analyzing volatile flavor chemicals, including furan, has shifted to the use of a gas chromatogram/mass spectrometer (GC/MS). Volatiles in the headspace of a food sample are collected on a solid-phase microextraction system and then transferred to the GC/MS (HS-SPME-GC/MS) (Altaki et al., 2009; Nyman et al., 2008). This HS-SPME-GC/MS method has been used for furan analysis in various food samples, including coffee (Ho et al., 2005; Senyuva and Gökmen, 2007), heated food products (Goldmann et al., 2005; Pérez-Palacios et al., 2012), crackers, corn chips (Nyman et al., 2008), and baby foods (Becalski et al., 2010; Yoshida et al., 2007). In addition to food samples, a unique study analyzing the levels of furan in the plasma of 100 healthy individuals, who consumed a normal diet, was conducted using this method (Lee et al., 2009).

3.3.2 Furan Found in Coffee

The greatest amounts of furan found in foods and beverages are in coffee. The presence of furan in roasted coffee was first reported in 1938 (Johnston and Frey, 1938), before the gas chromatograph was invented. GC analysis of volatile chemicals, including furan, isolated from coffee began to be reported in the early 1960s (Merritt et al., 1963). The quantities of furan in coffee ranged from 3 µg/kg (Heppner and Schlatter, 2007) to 5050 µg/kg (Crews and Castle, 2007). The amount of furan found in coffee powder varies among different reports: 28 ng/g (Hasnip et al., 2006), 3–5050 ng/g (EFSA, 2004), 2279 ± 53 ng/g (Altaki et al., 2007), and 53–587 ng/g (La Pera et al., 2009). Such variability suggests that some aspects of the roasting conditions (temperature, time, moisture content, and pH) play a role in furan formation. One recent study reported that temperature, moisture content, and the presence of ascorbic acids contribute considerably to furan formation in starchy food model systems (Mariotti et al., 2012). In addition to furan, many furan derivatives (furans) have been reported in coffee. The number of furans found in coffee increased steadily from about 10 in the 1950s to over 140 in 2000 (Flament, 2002). Another report indicated that the number of furans found in coffee is more than 800 (Crews and Castle, 2007).

3.3.3 Furan Found in Baby Foods

Since the Food Quality Protection Act (FQPA) was established in 1996, obtaining information on the levels of toxicants, including furan, in foods became extremely important. This was particularly true in the case of foods for babies, where it was urgently necessary to establish adequate food safety laws to protect vulnerable infants from toxicants in foods and beverages. Accordingly, many studies have been conducted to determine the levels of furan in baby products. It has been reported that various baby foods contained 1–172.7 µg/kg furan, as shown in Table 3.2. Furan concentration in baby foods from different countries has also been reported. Furan in 39 different baby food samples ranged from 4 µg/kg (fruit-based samples) to 64.6 µg/kg (rice- and chicken-based samples) in Spain (Ruiz et al., 2010). Furan in 21 different baby food samples obtained in Finnish markets contained 4.7–90.3 µg/kg (Jestoi et al., 2009). Levels of furan in 31 Brazilian baby foods—containing meat, vegetables, cereals, or fruits—had 0–95.5 µg/kg (Pavesi Arisseto et al., 2010). The same study reported that samples containing vegetables and meat showed higher furan levels compared with those containing only fruits. Baby foods purchased in Korea contained furan ranging from 3.43 ng/g (apple or blueberry) to 97.21 ng/g (sweet potato or apple) (Kim et al., 2010). These results suggest that levels of furan in baby foods vary among products prepared with different ingredients.

3.3.4 Furan Found in Miscellaneous Foods and Beverages

As shown in Table 3.2, there are many reports on the presence of furan in various foods and beverages, in addition to coffee and baby foods. Around the time when

GC was invented, qualitative analysis of volatiles, including furan, from various commodities was one of the most popular areas of flavor and fragrance research. The commodities studied included coffee, tea, cocoa (Reymond et al., 1966), and pork liver (Mussinan and Walradt, 1974). For example, furan was reported as one of the 38 components identified in geranium oil (Timmer et al., 1971).

Furan may not be present in fresh fruits and vegetables but once they are processed into some form, such as being canned or jarred, certain amounts of furan is formed. Canned or jarred fruits and vegetables have been shown to contain 0–37 µg/kg and 0.8–58.3 µg/kg furan, respectively (Table 3.2). Canned and jarred meat products contained furan ranging from 0.9 to 196.9 µg/kg (Crews and Castle, 2007; Kim et al., 2010). Furan found in beer, at levels of 0.8–3.4 µg/kg, may be formed from malt during processing (Crews and Castle, 2007; Heppner and Schlatter, 2007). Cereal products and bakery products contained about the same levels of furan (2–200 µg/kg).

In addition to the above foods and beverages, wide ranges of food commodities, such as fish, milk, honey, soups, jams, and sauces, reportedly contain levels of furan ranging from 0 to 125 µg/kg (Heppner and Schlatter, 2007; Kim et al., 2010). These reports indicate that almost all foods and beverages, in particular processed ones, contain certain levels of carcinogenic furan.

3.4   TOXICITY STUDIES ON FURAN

3.4.1   Acute and Subchronic Toxicity

The acute toxicity median lethal dose (LD_{50}) of furan administered by intraperitoneal injection is 5.2 mg/kg for rats and 7.0 mg/kg for mice. The median lethal concentration (LC_{50}) for mice is 0.12 µg/mL (Egle and Gochberg, 1979).

A subchronic oral toxicity study of furan in Fischer-344 rats demonstrated that a furan dose of 0.03 mg/kg bw (body weight) showed a no-observed adverse effect level (NOAEL) for hepatic toxicity (Gill et al., 2010). The acute or subchronic toxicity of furan is relatively low, but its chronic toxicity is more important because it is continually ingested by people through foods and beverages.

3.4.2   Chronic Toxicity: Carcinogenicity

The possible carcinogenicity of furan was reported in the early 1990s (Burka et al., 1991; Wilson et al., 1992). The National Toxicology Program (NTP) in the United States also reported some genotoxicity of furan, including gene mutation, chromosome aberrations, and sister chromatid exchanges (NTP, 1993). Furan was classified as a possible carcinogen in humans (group 2B) by IARC in 1995. Around the same time, liver toxicity of furan was reported by many researchers.

Table 3.3 shows selected studies on furan toxicity. A rapid development of cholangiofibrosis was observed in the liver of rats treated with furan (Elmore and Sirica, 1991). Later studies confirmed that furan induces hepatocellular tumors in rats and mice and bile duct tumors in rats (Bakhiya and Appel, 2010). In addition to liver damage, a recent study reported the histopathological observation of congestion, edema, fibrosis, and tubular damage in the kidneys of growing male rats treated

TABLE 3.3
Toxicology Studies on Furan

Object	Results of Observation	References
Disposition in rat organs	About 80% in urine and expired air, hepatotoxicity may be due to the metabolites.	Burka et al. (1991)
Genotoxicity	Profile of oncogene mutations, tumor development.	Wilson et al. (1992)
Intestinal metaplasia	Increased number of metaplastic intestinal glands in rats and more fibrotic stroma progressed development.	Elmore and Sirica (1992)
Cytolethality	Produced cytolethality and modest GSH depletion with dose dependence.	Carfagna et al. (1993)
Risk assessment	Analysis of 230 commercially jarred ready-to-eat infant food products, evaluation of levels.	Lachenmeier et al. (2009)
Review	Human exposure, applicability in assessing the risk in human diet.	Bakhiya and Appel (2010)
Genotoxic effect	Induced liver toxicity at 250 mg/kg in mouse, significant increase of DNA damage.	Cordelli et al. (2010)
Assessment of *in vivo* genotoxicity	Quantitated chromosome damage in mouse splenocytes.	Leopardi et al. (2010)
Alteration of cell cycle	Significant change in the expression of genes related to cell cycle control and apoptosis.	Chen et al. (2010)
Review	Toxicity, human dietary exposure, results of animal studies.	Moro et al. (2012)
Risk assessment	Estimation of infants exposure by thermally processed jarred food products, margin of exposure.	Minorczyk et al. (2012)
Effect toward thymus	Affected the thymus in growing male rats.	Kockaya et al. (2012)
Mechanism for carcinogenicity	Histopathological evaluation of rat liver evidenced inflammation, single-cell necrosis, and cell proliferation.	Ding et al. (2012)
DNA binding	DNA adduct significantly increased in F344 rats, some DNA damage in liver.	Neuwirth et al. (2012)
Review	Toxicity, human dietary exposure, formation mechanisms, and mitigation.	Mariotti et al. (2013)

with furan (Selmanoğlu et al., 2012). There are no epidemiological studies on furan carcinogenicity, but it has been determined as an animal carcinogen.

3.4.3 MECHANISMS OF CARCINOGENICITY OF FURAN

Many studies have been conducted to elucidate the mechanism of tumor induction by furan. The toxicity and carcinogenicity of furan was proposed to proceed through a process in which furan is oxidized to α,β-unsaturated dialdehyde and *cis*-2-bu-tene-1,4-dial by cytochrome P450. Subsequently, these oxidative metabolites cause adverse effects via forming adducts with protein and DNA nucleophiles (Peterson,

2006). They are also mutagenic in Ames assay strain TA104. Furan metabolites excreted with bile from male F-344 rats orally administered with furan were structurally characterized and it was hypothesized that its metabolite, *cis*-2-butene-1,4-dial, binds to critical target proteins to play a role in toxicity and carcinogenicity (Hamberger et al., 2010). These furan metabolites seem to affect the liver. For example, when the hepatocyte suspensions prepared from F-344 rat livers were treated with cytochrome P450 inhibitor 1-phenylimidazole, the glutathione depletion was delayed (Carfagna et al., 1993).

3.4.4 ASSESSMENT OF SAFETY OF FOODS AND BEVERAGES CONTAINING FURAN

When foods may contain a toxic chemical, such as furan, it must be evaluated, with the most important factor to assess being how much of that toxicant is consumed by an individual. It should also be noted that all substances are poisonous; there is none that is not a poison. Only the dose differentiates a poison from a remedy. This concept was advanced by Paracelsus (1493–1541), who was a medical doctor in Europe, and it remains useful today in evaluating risks in our daily lives by considering examples of well-known substances with low and high toxicity. According to this theory, the amount of furan present in foods and beverages may not cause any adverse effects to humans based on the currently available analytical data. However, some beverages, such as coffee and beer, may cause serious problems because they are consumed consistently and constantly. Among 12 foods and beverages, the average estimated intake of coffee was the worst culprit. Coffee drinkers ingested the highest amount of furan per consumer, in amounts ranging from 2.4 to 115.5 µg/person (Heppner and Schlatter, 2007). The same report indicated that furan consumption through beer is estimated as 1.3–3.4 µg/person.

However, in order to establish satisfactory risk assessment of furan present in foods and beverages, more information is required, including results of studies of chronic but low-dose ingestion and additional analytical data. In the case of furan, the statistical effect is estimated at the 2 mg/kg bw, whereas lower levels than that—albeit levels at which carcinogenicity may appear—must be investigated because volatile chemicals, including furan, formed in foods and beverages by Maillard reaction, are very low yet they are consumed daily by people.

3.5 POSSIBLE MITIGATION OF FURAN IN FOODS AND BEVERAGES

Whenever carcinogens are found in foods, methods by which they are mitigated receive attention. However, it is extremely difficult to reduce the formation of one specific flavor chemical since more than 1000 flavor chemicals are formed simultaneously by Maillard reaction in a heated food. For example, many researchers have been trying to mitigate carcinogenic acrylamide formation in heated foods ever since it was reported in Maillard reaction systems in 2002 but more information is still required before methods of preparing satisfactory low-acrylamide foods can be developed (Friedman, 2003). A recent report also indicated that mitigation of carcinogenic 4(5)-methylimidazole formed by Maillard reaction in a caramel color is extremely difficult because it is formed simultaneously along with many other heterocyclic flavor chemicals (Figure 3.1). If 4(5)-methylimidazole is removed, other important flavor chemicals would be removed as well, and the palatability of foods would be reduced (Hengel and Shibamoto, 2013). These reports indicate that the mitigation of furan formed in heated foods with Maillard reaction is extremely difficult and to date there is virtually no report on satisfactory methods of reducing furan formation in Maillard reaction systems.

3.6 CONCLUSION

Furan has received considerable attention as one of the most important flavor chemicals formed in Maillard reaction systems. Its presence has been reported in various food items, particularly in heat-treated foods and beverages in which Maillard reaction occurs, including coffees and baby foods. Unfortunately, however, furan was determined to be a possible carcinogen for humans by the IARC in 1995. Despite this, the USFDA reported the presence of furan in a wide range of food items in 2004. Consequently, many studies on furan, including analysis, formation mechanisms, and toxicity, have been conducted. Since the HS-SPME-GC/MS method was advanced, the analysis of furan became considerably more efficient. Toxicology studies on the carcinogenicity of furan using experimental animals, such as mice and rats, have been performed and have confirmed its carcinogenicity. However, low-dose studies to assess the safety of the levels of furan present in foods and beverages are still in order.

REFERENCES

Altaki, M.S., Santos, F.J., Galceran, M.T. 2007. Analysis of furan in foods by head-space solid phase microextraction gas chromatography-ion trap mass spectrometry. *J. Chromatogr. A* 1146, 103–109.

Altaki, M.S., Santos, F.J., Galceran, M.T. 2009. Automated headspace solid-phase microextraction versus headspace for the analysis of furan in foods by gas chromatography-mass spectrometry. *Talanta* 78, 1315–1320.

Anet, E.F.I.J. 1962. Formation of furan compounds from sugars. *Chem. Ind.* 262.

Arctander, S. 1969. *Perfume and Flavor Chemicals*. Montclair: Published by the author.

Bakhiya, N., Appel, K.E. 2010. Toxicity and carcinogenicity of furan in human diet. *Arch. Toxicol.* 84, 563–578.

Becalski, A., Hayward, S., Krakalovich, T., Pelletier, L., Roscoe, V., Vavasour, E. 2010. Development of an analytical method and survey of foods for furan, 2-methylfuran and 3-methylfuran with estimated exposure. *Food Addit. Contam. A.* 27, 764–775.

Bianchi, F., Careri, M., Mangia, A., Musci, M. 2006. Development and validation of a solid phase micro-extraction-gas chromatography-mass spectrometry method for the determination of furan in baby-food. *J. Chromatogr. A.* 1102, 268–272.

Bicchi, C., Ruosi, M.R., Cagliero, C., Cordero, C., Liberto, E., Rubiolo, P., Sgorbini, B. 2011. Quantitative analysis of volatiles from solid matrices of vegetable origin by high concentration capacity headspace techniques: Determination of furan in roasted coffee. *J. Chromatogr. A.* 1218, 753–762.

Blank, I. 2008. Furan in processed foods. In *Bioactive Compounds in Foods*, eds. J. Gilbert and H.Z. Senyuva, 292–322. Oxford: Blackwell Pub. Ltd.

Bolgerm, P.M., Tao, S.S.-H., Dinovi, M. 2009. Hazards of dietary furan. In *Process-Induced Food Toxicants*, eds. R.H. Stadler and D.R. Lineback, 117–133. Hoboken: John Wiley and Sons, Inc.

Bule, M.V., Desai, K.M., Parisi, B., Parulekar, S.J., Slade, P., Singhal, R.S., Rodriguez, A. 2010. Furan formation during UV-treatment of fruit juices. *Food Chem.* 122, 937–942.

Burka, L.T., Washburn, K.D., Irwn, R.D. 1991. Disposition of [14C]furan in the male F344 rat. *J. Toxicol. Environ. Health* 34, 245–257.

Carfagna, M.A., Held, S.D., Kedderis, G.L. 1993. Furan-induced cytolethality in isolated rat hepatocytes: Correspondence with in vivo dosimetry. *Toxicol. Appl. Pharm.* 123, 265–273.

Cerny, C., Davidek, T. 2003. Formation of aroma compounds from ribose ad cysteine during the Maillard reaction. *J. Agric. Food Chem.* 51, 2714–2721.

Chen, T., Mally, A., Ozden, S., Chipman, J.K. 2010. Low doses of the carcinogen furan alter cell cycle and apoptosis gene expression in rat liver independent of DNA methylation. *Environ. Health Perspect.* 118, 1597–1602.

Cordelli, E., Leopardi, P., Villani, P., Marcon, F., Macri, C., Cajola, S., Siniscalchi, E., Conti, L., Eleuteri, P., Malchiodi-Albedi, F., Crebelli, R. 2010. Toxic and genotoxic effects of oral administration of furan in mouse liver. *Mutagenesis* 25, 305–314.

Crews, C., Castle, L. 2007. A review of the occurrence, formation and analysis of furan in heat-processed foods. *Trend. Food Sci. Technol.* 18, 365–372.

Ding W., Petibone, D.M., Latendresse, J.R., Pearce, M.G, Muskhelishvili, L., White, G.A., Chang, C.W., Mittelstaedt, R.A., Shaddock, J.G., McDaniel, L.P., Doerge, D.R., Morris, S.M., Bishop, M.E., Manjanatha, M.G., Aldoo, A., Heflich, R.H. 2012. In vivo genotoxicity of furan in F344 rats at cancer bioassay doses. *Toxicol. Appl. Pharmacol.* 261, 164–171.

EFSA J. 2004. Report of the scientific panel on contaminants in the food chain on provisional findings of furan in food. *EFSA J.* 137, 11–20.

Egle, J.L. Jr., Gochberg, B.J. 1979. Respiratory retention and acute toxicity of furan. *Am. Ind. Hyg. Assoc. J.* 40, 310–314.

Elmore, L.W., Sirica, A.E. 1991. Phenotypic characterization of metaplastic intestinal glands and ductular hepatocytes in cholangiofibrotic lesions rapidly induced in the caudate liver lobe of rats treated with furan. *Cancer Res.* 51, 5752–5759.

Elmore, L.W., Sirica, A.E. 1992. Sequential appearance of intestinal mucosal cell types in the right and caudate liver lobes of furan-treated rats. *Hepatology* 16, 1220–1226.

Fan, X. 2005. Formation of furan from carbohydrates and ascorbic acid following exposure to ionizing radiation and thermal processing. *J. Agric. Food Chem.* 53, 7826–7831.

Fan, X., Geveke, D.J. 2007. Furan formation in sugar and apple cider upon ultraviolet treatment. *J. Agric. Food Chem.* 55, 7816–7821.

FDA (US Food and Drug Administration). 2004. Furan in food, thermal treatment; request of data and information. *Fed. Regist.* 69(90), 25911–25913.

Flament, I. 2002. *Coffee Flavor Chemistry*. New York: John Wiley and Sons, Ltd.

Friedman, M. 2003. Chemistry, biochemistry, and safety of acrylamide. A review. *J. Agric. Food Chem.* 51, 4504–4526.

Fujimaki, M., Namiki, M., Kato, H. 1986. Amino-carbonyl reactions in food and biological systems. *Proceedings of the 3rd International Symposium on the Maillard Reaction.* New York: Elsevier.

Gill, S., Bondy, G., Lefebvre, D.E., Becalski, A., Kavanagh, M., Hou, Y., Turcotte, A.M., Barker, M., Weld, M., Vavasor, E., Cooke, G.M. 2010. Subchronic oral toxicity study of furan in Fischer-344 rats. *Toxicol. Pathol.* 38, 619–630.

Goldmann, T., Perisset, A., Scanlan, F., Stadler, R.H. 2005. Rapid determination of furan in heated foodstuffs by isotope dilution solid phase microextraction-gas chromatography-mass spectrometry (SPME-GC/MS). *Analyst* 130, 878–883.

Guenther, H., Hoenicke, K., Biesterveld, S., Gerhard-Rieben, E., Lanz, I. 2010. Furan in coffee: Pilot studies on formation during roasting and losses during production steps and consumer handling. *Food Addit. Contam. A.* 27, 283–290.

Hamberger, C., Kellert, M., Schauer, U.M., Dekant, W., Mally, A. 2010. Hepatobiliary toxicity of furan: Identification of furan metabolites in bile of male f344/n rats. *Drug Metab. Dispos.* 38, 1698–1706.

Hasnip, S., Crews, C., Castle, L. 2006. Some factors affecting the formation of furan in heated foods. *Food Addit. Contam.* 23, 219–227.

Hengel, M., Shibamoto, T. 2013. Carcinogenic 4(5)-methylimidazole found in beverages, sauces, and caramel colors: Chemical properties, analysis, and biological activities. *J. Agric. Food Chem.* 61, 780–78.

Heppner, C.W., Schlatter, J.R. 2007. Data requirements for risk assessment of furan in food. *Food Addit. Contam.* 24(Suppl. 1), 114–121.

Ho, I.P., Yoo, S.J., Tefera, S. 2005. Determination of furan levels in coffee using automated solid-phase microextraction and gas chromatography/mass spectrometry. *JAOAC Int.* 88, 574–576.

Hodge, J. 1967. Origin of flavor in foods nonenzymatic Browning reactions. In *Chemistry and Physiology of Flavors*, eds. H.W. Schultz, E.A. Day, and L.M. Libbey, 465–491. Westport: AVI Pub. Co.

IARC (International Agency for Research on Cancer). 1995. IARC Monographs on the Evaluation for Carcinogenic Risks to Humans, Volume 63: Dry Cleaning, Some Chlorinated Solvents and Other Industrial Chemicals, 394–407. Lyon: IARC.

Jestoi, M., Jarvinen, T., Jarvenpaa, E., Tapanainen, H., Virtanen, S., Peltonen, K. 2009. Furan in the baby-food samples purchased from the Finnish markets—Determination with SPME-GC-MS. *Food Chem.* 117, 522–528.

Jiang, Y., Hengel, M., Pan, C., Seiber, J.N., Shibamoto, T. 2013. Determination of toxic α-dicarbonyl compounds, glyoxal, methylglyoxal, and diacetyl released to the headspace of lipid commodities upon heat treatment. *J. Agric. Food Chem.* 61, 1067–1071.

Johnston, W.R., Frey, C.N. 1938. The volatile constituents of roasted coffee. 60, 1624–1627.

Kim, T.-K., Kim, S., Lee, K.-G. 2010. Analysis of furan in heat-processed foods consumed in Korea using solid phase microextraction-gas chromatography/mass spectrometry (SPME-GC/MS). *Food Chem.* 123, 1328–1333.

Kockaya, E.A., Kilic, A., Karacaoglu, E., Selmanoglu, G. 2012. Does furan affect the thymus in growing male? *Drug Chem. Toxicol.* 35, 316–323.

Kuballa, T. 2007. Furan in Kaffee und anderen Levensmitteln. *J. Verbr. Lebensm.* 2, 429–433.

Kuballa, T., Stier, S., Strichow, N. 2005. Furan in Kattee und Kaffeegetränken. *Deutsche Lebensmittel-Rundschau1* 101, 229–235.

La Pera, L., Liberatore, A., Avellone, G., Fanara, S., Dugo, G., Agozzino, P. 2009. Analysis of furan in coffee of different provenance by head-space solid phase microextraction gas chromatography-mass spectrometry: Effect of brewing procedures. *Food Addit. Contam.* 26, 786–792.

Lachenmeier, D.W., Reusch, H., Kuballa, T. 2009. Risk assessment of furan in commercially jarred baby foods, including insights into its occurrence and formation in freshly home-cooked foods for infants and young children. *Food Addit. Contam. A.* 26, 776–785.

Lee, Y.-K., Jung, S.-W., Lee, S.J., Lee, K.-G. 2009. Analysis of residual furan in human blood using solid phase microextraction-gas chromatography/mass spectrometry (SPME-GC/MS). *Food Sci. Biotechnol.* 18, 379–383.

Leopardi, P., Cordelli, E., Villani, P., Cremona, T.P., Conti, L., De Luca, G., Crebelli, R. 2010. Assessment of in vivo genotoxicity of the rodent carcinogen furan: Evaluation of DNA damage and induction of micronuclei in mouse splenocytes. *Mutagenesis* 25, 57–62.

Limacher, A., Kerier, J., Conde-Petit, B., Blank, I. 2007. Formation of furan and methylfuran from ascorbic acid in model systems and food. *Food Addit. Contam.* 24, 122–135.

Limacher, A., Kerler, J., Davidek, T., Schmalzried, F., Blank, I. 2008. Formation of furan and methylfuran by Maillard-type reactions in model systems and food. *J. Agric. Food Chem.* 56, 3639–3547.

Limpricht, H. 1870. Ueberdastetraphenol C_4H_4O. *Ber Dtsch Chem Ges* 3, 90–91.

Locas, C.P., Yaylayan, V.A. 2004. Origin and mechanistic pathways of formation of the parent furan—A food toxicant. *J. Agric. Food Chem.* 52, 6830–6836.

Liu, P., Xue, Y., Jin, Q., Xu, J., Zhang, Z., Wu, G. 2008. Determination of furan in baby foods using headspace gas chromatography-mass spectrometry. *Sepu* 26, 35–38.

Maga, J.A. 1979. Furans in foods. *CRC Crit. Rev. Food Sci. Nutr.* 11, 355–400.

Mariotti, M.S., Granby, K., Rozowski, J., Pedreschi, F. 2013. Furan: A critical heat induced dietary contaminant. *Food Funct.* 4, 1001–1015.

Mariotti, M., Granby, K., Fromberg, A., Risum, J., Agosin, E., Pedreschi, F. 2012. Furan occurrence in starchy food model systems processed at high temperatures: Effect of ascorbic acid and heating conditions. *J. Agric. Food Chem.* 60, 10162–10169.

Merritt, C., Bazinet, M.L., Sullivan, J.H., Robertson, D.H. 1963. Coffee aroma, mass spectrometric determination of the volatile components from ground coffee. *J. Agric. Food Chem.* 11, 152–155.

Minorczyk, M., Góralczyk, K., Struciński, P., Hernik, A., Czaja, K., Lyczewska, M., Korcz, W., Starski, A., Ludwicki, J.K. 2012. Risk assessment for infants exposed to furan from ready-to-eat thermally processed food products in Poland. *Rocz. Panstw. Zakl. Hig.* 63, 403–410.

Morales, F. 2009. Hydroxymethylfurfural (HMF) and related compounds. In *Process-Induced Food Toxicants*, eds. R.H. Stadler and D.R Lineback, 117–133. Hoboken: John Wiley and Sons, Inc.

Moro, S., Chipman, J.K., Wegener, J.W., Hamberger, C., Dekant, W., Mally, A. 2012. Furan in heat-treated foods: Formation exposure, toxicity, and aspects of risk assessment. *Mol. Nutr. Food Res.* 56, 1179–1211.

Mussinan, C.J., Walradt, J.P. 1974. Volatile constituents of pressure cooked pork liver. *J. Agric. Food Chem.* 22, 827–831.

Neuwirth, C., Mosesso, P., Pepe, G., Fiore, M., Malfatti, M., Turteltaub, K., Dekant, W., Mally, A. 2012. Furan carcinogenicity: DNA binding and genotoxicity of furan in rats in vivo. *Mol. Nutr. Food Res.* 56, 1363–1374.

Newth, F.H. 1951. The formation of furan compounds from hexoses. *Adv. Carbohydr. Chem.* 6, 83–106.

Newth, F.H., Wiggins, L.F. 1947. The conversion of sucrose into furan compounds. Part III. Some amidio-furans. *J. Chem. Soc.* 396–398.

NTP. 1993. Toxicology and carcinogenesis studies of furan (CAS No. 110-00-9) in F344/N rats and B6C3F1 mice (gavage studies), NTP Technical Report No. 402. Research Triangle Park, NC: US Department of Health and Human Services, Public Health Service, National Institute of Health.

Nyman, P.J., Morehouse, K.M., Perfeitti, G.A., Diachenko, G.W., Holcomb, J.R. 2008. Single-laboratory validation of a method for the determination of furan in foods by using headspace gas chromatography/mass spectrometry, part 2-low-moisture snack foods. *JAOAC Int.* 91, 414–421.

Ohnishi, S., Shibamoto, T. 1984. Volatile compounds from heated beef fat and beef fat with glycine. *J. Agric. Food Chem.* 32, 987–992.

Owczarek-Fendor, A., de Meulenaer, B., Scholl, G., Adams, A., van Lancker, F., Yogendrarajah, P., Eppe, G., de Pauw, E., Scippo, M.-L., de Kimpe, N. 2010. Furan formation from vitamin C in a starch-based model system: Influence of the reaction conditions. *Food Chem.* 121, 1163–1170.

Owczarek-Fendor, A., de Meulenaer, B., Scholl, G., Adams, A., van Lancker, F., Eppe, G., de Pauw, E., Scippo, M.-L. 2011. Furan formation from lipids in starch-based model systems, as influenced by interactions with antioxidants and proteins. *J. Agric. Food Chem.* 59, 2368–2376.

Paine, J.B. III, Pithawalla, Y.B., Naworal, J.D. 2008. Carbohydrate pyrolysis mechanisms from isotopic labeling Par 4. The pyrolysis of *D*-glucose: The formation of furans. *J. Anal. Appl. Pyrolysis* 83, 37–63.

Pavesi Arisseto, A., Vicente, F., De Figueitredo Toledo, M.C. 2010. Determination of furan levels in commercial samples of baby food from Brazil and preliminary risk assessment. *Food Addit. Contam.* 27, 101–1059.

Pérez-Palacios, T., Petisca, C., Melo, A., Ferreira, I.M. 2012. Quantification of furanic compounds in coated deep-fried products simulating normal preparation and consumption: Optimization of HS-SPME analytical conditions by response surface methodology. *Food Chem.* 135, 1337–1343.

Peterson, L.A. 2006. Electrophilic intermediates produced by bioactivation of furan. *Drug Metab. Rev.* 38, 615–626.

Raeisaenen, U., Pitkaenen, I., Halttunen, H., Hurtta, M. 2003. Formation of the main degradation compounds from arabinose, xylose, mannose and arabinitol during pyrolysis. *J. Ther. Anal. Calori.* 72, 481–488.

Reymond, D., Mueggler-Chavan, F., Viani, R., Vautaz, L., Egli, R.H. 1966. Gas chromatographic analysis of steam volatile aroma constituents: Application to coffee, tea and cocoa aromas. *J. Gas Chromatogr.* 4, 28–31.

Ruiz, E., Santillana, M.I., Nieto, M.T., Cirugeda, M.E., Sanchez, J.J. 2010. Determination of furan in jarred baby food purchased from the Spanish market by headspace gas chromatography-mass spectrometry (HS-GC-MS). *Food Addit. Contam.* 27, 1208–1214.

Selmanoğlu, G., Karacaoglu, E., Kiliç, A., Koçkaya, E.A., Akay, M.T. 2012. Toxicity of food contaminant furan on liver and kidney of growing male tats. *Environ. Toxicol.* 27, 613–622.

Senyuva, H.Z., Gökmen, V. 2007. Potential of furan formation in hazelnuts during heat treatment. *Food Addit. Contam.* 24, 136–142.

Sheldon, S.A., Russell, G.F, Shibamoto, T. 1986. Photochemical and thermal activation of model Maillard reaction systems. In *Amino-Carbonyl Reactions in Food and Biological Systems. Proceedings of the 3rd International Symposium on the Maillard Reaction*, eds. M. Fujimaki, M. Namiki, and H. Kato, 145–154. New York: Elsevier.

Shibamoto, T. 1983. Heterocyclic compounds in browning and browning/nitrite model systems: Occurrence, formation mechanisms, flavor characteristics and mutagenic activity. In *Instrumental Analysis of Foods*. Vol. I, eds. I.G. Charalambous and G. Inglett, 229–278. New York: Academic Press.

Timmer, R., terHeide, R., de Valois, P.J., Wobben, H.J. 1971. Qualitative analysis of the most volatile natural components of reunion geranium oil (*Pelargonium roseum* Bourbon). *J. Agric. Food Chem.* 19, 1066–1068.

Uneyama, C., Toda, M., Morikawa, K., Yamamoto, M. 2004. Furan in foods. *Shokuhin Eiseigaku Zasshi* 45, J249–J251.

Van Lancker, F., Adams, A., Owczarek-Fendor, A., De Meulenaer, B., De Kimpe, N. 2009. Impact of various food ingredients on the retention of furan in foods. *Mol. Nutr. Food Res.* 53, 1505–1511.

Van Lancker, F. Adams, A., Owczarek-Fendor, A., de Meulenaer, B., de Kimpe, N. 2011. Mechanistic insights into furan formation in Maillard model systems. *J. Agric. Food Chem.* 59, 229–235.

Vranova J., Ciesarova, Z. 2009. Furan in food—A Review. *Czech J. Food Sci.* 27, 1–10.

Wang, Y., Juliani, H.R., Simon, J.E., Ho, C.-T. 2009. Amino acid-dependent formation pathways of 2-acetylfuran and 2,5-dimethyl-4-hydroxy-3[2H]-furanone in the Maillard reaction. *Food Chem.* 115, 233–237.

Wegener, J.-W., López-Sánchez, P. 2010. Furan levels in fruit and vegetables juices, nutrition drinks and bakery products. *Anal. Chim. Acta* 672, 55–60.

Wilson, D.M., Goldsworthy, T.L., Popp, J.A., Butterworth, B.E. 1992. Evaluation of genotoxicity, pathological lesions, and proliferation in livers of rats and mice treated with furan. *Environ. Mol. Mutagen.* 19, 209–222.

Yaylayan, V.A. 2006. Precursors, formation and determination of furan in food. *J. Verbr. Lebensm.* 1, 5–9.

Yoshida, I., Isagawa, S., Kibune, N., Hamano-Nagaoka, M., Maitani, T. 2007. Rapid and improved determination of furan in baby foods and infant formulas by headspace GC/MS. *Shokuhin Eiseigaku Zasshi* 48, 83–89.

4 Maillard Reaction Products II
Reactive Dicarbonyl Compounds and Advanced Glycation End Products

Yen-Chen Tung, Xi Shao, and Chi-Tang Ho

CONTENTS

4.1 INTRODUCTION

Nonenzymatic modification of proteins has been implicated in the pathogenesis of diabetes, atherosclerosis, neurodegenerative diseases, and normal aging. The modification can arise from direct exposure to reactive oxygen, chloride, or nitrogen species, and from reaction with low-molecular-weight reactive carbonyl species (RCS). The accumulation of various RCS derived from carbohydrates as well as their subsequently induced protein modification is proposed to constitute a state of "carbonyl stress." These RCS are responsible for the formation of advanced glycation end products (AGEs), and their roles in the development of various age-related diseases have been increasingly recognized. Carbohydrate-derived RCS include methylglyoxal (MGO), glyoxal (GO), and 3-deoxyglucosone (3-DG). Both RCS and AGEs can be formed endogenously (body) and exogenously (food) (Singh et al., 2001; Wang and Ho, 2012). RCS are reactive intermediates formed from the interaction of a carbonyl group mainly from reducing sugars with an amino group from proteins, peptides, and amino acids known as Maillard reaction. The Maillard reaction plays a very important role in food, and provides flavor, aroma, and color to food during heating, processing, and storage (Wu et al., 2011). The carbonyl groups of RCS interact with proteins to form various AGEs or to crosslink proteins. This process induces carbonyl stress and causes proteins to lose their function, damage tissue, and accelerate cell dysfunction (Lesgardset al., 2011; Wu and Juurlink, 2002). More and more research studies show that RCS and AGEs are connected with diabetes and its complication, as well as age-related disorders, such as Alzheimer's disease and atherosclerosis (Picklo et al., 2002; Ramasamy et al., 2012; Sena et al., 2012). RCS and AGEs can be found in food and biological system, and it is very important to develop standardized and accurate methods to analyze them.

In this chapter, the formation and analysis of RCS and AGEs, and the environmental factors affecting their formation are reviewed. The toxicities of RCS and AGEs and their relation to diabetes, atherosclerosis, and age-related diseases will also be discussed.

4.2 CHEMICAL MECHANISM FOR FORMATION

Formation of MGO, GO, 3-DG (structures shown in Figure 4.1), and AGEs are all related to Maillard reaction. Maillard reaction, or nonenzymatic glycation, is a complex series of reactions among reducing sugars and amino groups of amino acids, peptides, proteins, and DNA (Cho et al., 2007; Nemet et al., 2006). As the first step of Maillard reaction, the proteins in the tissues are modified by reducing sugars (e.g., glucose) through the reaction between a free amino group of proteins and a carbonyl group of the sugars, leading to the formation of fructosamines via a Schiff base by Amadori rearrangement. Then, both Schiff's base and Amadori product further

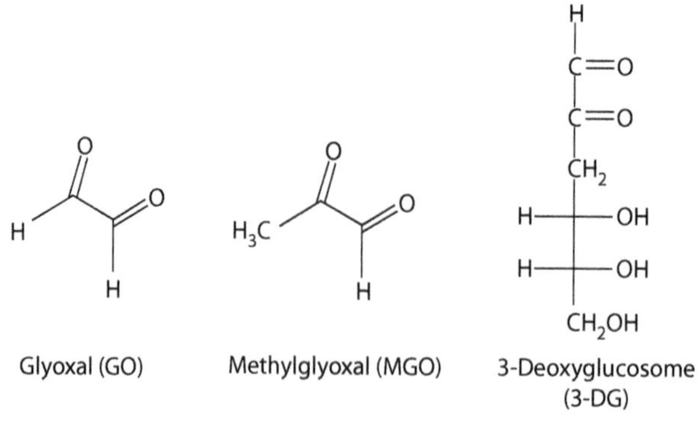

Glyoxal (GO) Methylglyoxal (MGO) 3-Deoxyglucosome (3-DG)

FIGURE 4.1 Chemical structures of MGO, GO, and 3-DG.

undergo a series of reactions through dicarbonyl intermediates (e.g., GO, MGO, and 3-DG) to form AGEs (Singh et al., 2001). GO and MGO, the two major α-dicarbonyl compounds found in humans, are extremely reactive and readily modify lysine, arginine, and cysteine residues on proteins (Nagaraj et al., 2002).

RCS can be produced by exogenous pathways and exist abundantly in our food such as cookies, coffee, cocoa, beverages, and cigarette smoke (Arribas-Lorenzo and Morales, 2010; Nemet et al., 2006; Tan et al., 2008).

Sugar autoxidation, Maillard reaction, lipid oxidation, and microbial fermentation can form MGO (Wang and Ho, 2012). Generally, carbohydrate is a key precursor in the formation of MGO. MGO is formed from fragmentation of sugar by the retro–aldol condensation in alkaline condition. The more glucose or monosaccharide present in foods, the more MGO can be formed (Nursten, 2005).

Endogenously, MGO is primarily formed during glycolysis in cells (Figure 4.2). It may be degraded from triose phosphate intermediate such as dihydroxyacetone phosphate and glyceraldehydes 3-phosphate. This process can be nonenzymatic, or catalyzed by triosephosphateisomerase (Pompliano et al., 1990) and MGO synthase (Price and Knight, 2009). Except being transformed from glucose, fructose, or ethanol, the elevated MGO may be from the metabolism of aminoacetone or acetone from lipolysis catalyzed by semicarbazide-sensitive amine oxidase or acetol monooxygenase (Kalapos, 1999; Nemet et al., 2006). The metabolisms of amino acid, glycine, and threonine can also produce MGO as an intermediate.

AGEs are formed from the reaction of reactive dicarbonyl compounds with proteins by combining to amine groups of proteins. Various chemical structures of AGEs have been identified (Barrenscheen and Braun, 1931; Guha et al., 1988; Nagai et al., 2002; Singh et al., 2001).

The most commonly measured AGE under *in vivo* conditions was MGO-lysine dimer (MOLD), which is formed from the reaction of MGO with two lysine residues (Brunner et al., 1998; Nemet et al., 2006; Turk 2010). Moreover, aminoimidazoline

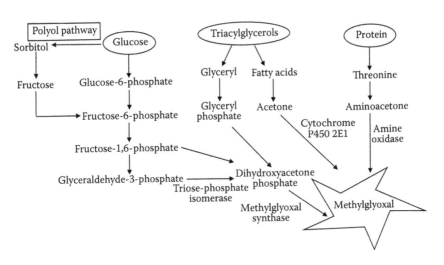

FIGURE 4.2 Pathways for producing MGO *in vivo*.

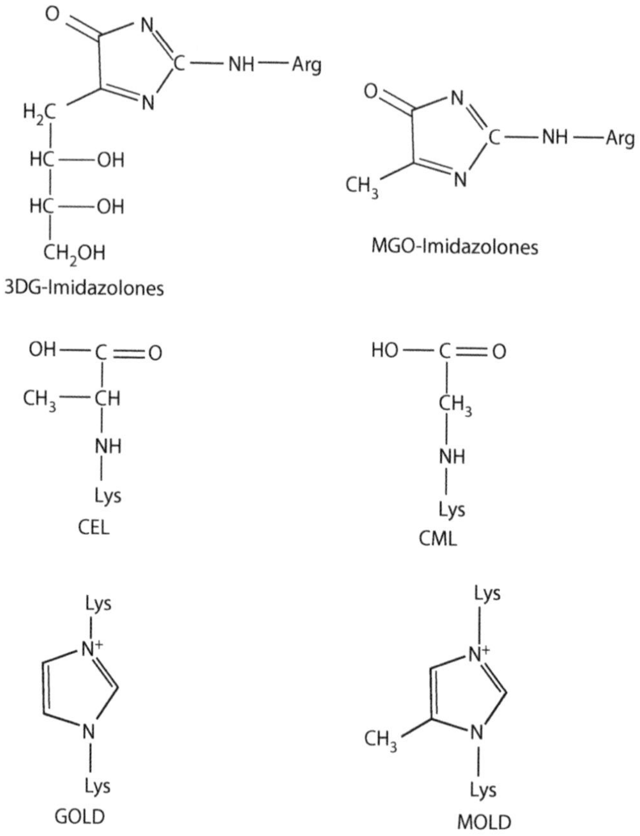

FIGURE 4.3 Chemical structures of AGEs derived from MGO and GO.

imine cross-link was thought to be derived from the reaction of arginine and α-oxoaldimin Schiff base of a lysine residue and MGO (Cho et al., 2007; Lederer and Klaiber, 1999). The reaction of MGO with the guanidine group of arginine can also lead to the formation of AGEs, also known as imidazolone (Ahmed et al., 2002; Paul et al., 1998). CEL derived from MGO is found in bound and free forms *in vivo* (Hakim et al., 2003).

Under physiological conditions, the reactions between GO and protein residues can also lead to the formation of AGEs, such as carboxymethyllysine (CML), glyoxal-lysine dimer (GOLD), imidazolone, or aminoimidazoline imine cross-link. The chemical structures of AGEs derived from both MGO and GO are shown in Figure 4.3.

4.3 FACTOR AFFECTING THE FORMATION OF REACTIVE DICARBONYL COMPOUNDS AND ADVANCED GLYCATION END PRODUCTS DURING MAILLARD REACTION

The formation of MGO, GO, and AGEs are all related with Maillard reaction (Cho et al., 2007; Nemet et al., 2006), and Maillard reaction is affected by several factors, such as types of carbohydrates and amino compounds, time, temperature, pH, water activity, and high pressure. The importance of processing conditions on the outcome of the Maillard reaction has been discussed by Ames (1998).

It is well-known that the optimal water activity for Maillard reaction is in the range of 0.5–0.75. Food-processing methods will also affect MGO, GO, and AGEs formation during Maillard reaction (Arribas-Lorenzo and Morales, 2010; Goldberg et al., 2004; Labuza and Saltmarch, 1981; O'Brien et al., 1989). The level of reactive dicarbonyl compounds, MGO, and GO in foods or model systems can be affected by heat processing (roasting, frying, baking, and pasteurizing), temperature, different carbohydrate, and other nutrient components. The formation of MGO from dextrin is 1.5-fold higher than starch in aqueous solution (Homoki-Farkas et al., 1997). When cookies are baked at 190°C at different times, the formation of MGO and GO could increase with time (Arribas-Lorenzo and Morales, 2010; Daglia et al., 2007). Similarly, the formation of AGEs is also affected by nutrient composition, temperature, water content, pH, cooking method, and time. Food with higher lipid or protein content has higher AGEs level than carbohydrate food. Different cooking methods at different temperatures will cause different levels of AGEs. Therefore, the levels of AGEs have the following order according to processing method: oven frying (230°C) > deep frying (180°C), and broiling (225°C) > roasting (177°C) > boiling (100°C). The AGE formation at dry heat condition is 10–100 times higher than the uncooked food. A higher temperature will lead to higher AGEs. Studies have also shown that temperature and cooking method have more effect than the cooking time (Goldberg et al., 2004; Uribarri et al., 2010). Different reactants can affect the formation of AGEs. Previous studies showed that glucose has slower rate than fructose to form AGEs *in vivo*. Dietary intake of egg white with fructose will have 200-fold of AGEs formed than egg white alone (Goldin et al., 2006; Koschinsky et al., 1997).

4.4 TOXIC EFFECTS

MGO, GO, and other RCS formed from glycoxidation as well as by the subsequent carbonyl modification of proteins defined as "carbonyl stress" (Golej et al., 1998; Turk, 2010), which is a highlighted phenomenon for several diseases such as diabetes and atherosclerosis, influence some aging-related process. Studies have shown that GO, MGO, 3-DG, and AGEs could cause cytotoxicity by forming reactive oxygen species to damage cells (Kalapos, 1999; Shangari and O'Brien, 2004). As an example, it has been demonstrated that direct administration of MGO to Sprague–Dawley (SD) rats significantly increased the levels of renal carboxyethyllysine (CEL), systolic blood pressure, urinary albumin excretion, urinary thiobarbituric acid-reactive substances excretion, and the renal nitrotyrosine expression in the kidney compared to non-MGO-treated rats. These results suggested that high level of MGO under *in vivo* conditions may induce insulin resistance and salt-sensitive hypertension (Guo et al., 2009).

4.4.1 Risk of Diabetes

Diabetes is a heterogeneous disorder that is generally accompanied by multiple complications. It involves the glucose and lipid metabolism in peripheral tissues to the biological activities of insulin and inadequate insulin secretion by pancreatic β cells. Epidemiological and large prospective clinical studies have confirmed that hyperglycemia is the most important factor for the onset and progress of diabetic complications, both in type 1 (insulin-dependent) and type 2 diabetes mellitus (Reichard et al., 1993). Increasing evidence identifies the formation of AGEs as a major pathogenic link between hyperglycemia and diabetes-related complications (Singh et al., 2001).

A growing body of evidence indicates that the increase in reactive carbonyl intermediates is a consequence of hyperglycemia in diabetes. Carbonyl stress leads to increased modification of proteins, followed by oxidant stress and tissue damage (Baynes and Thorpe, 1999). Several studies have shown that diabetic patients have higher levels of MGO and GO in their plasma compared to non-diabetic patients (Khuhawar et al., 2006; Lapolla et al., 2003; Odani et al., 1999). In one of the reports (Khuhawar et al., 2006), the amount of MGO found in diabetic patients was in the range of 16–27 μg/dL, while the amount found in non-diabetic patients was in the range of 3.0–7.0 μg/dL. Diabetic patients were reported to have higher serum concentration of 3-DG than healthy ones (Niwa et al., 1993). Thus, decreasing the levels of GO, MGO, and other RCS may provide a useful approach in preventing the formation of AGEs.

High-dietary AGEs intake has been reported to increase the serum level of MGO compared to low-dietary AGE intake group (Nemetet al., 2006; Uribarri et al., 2010). Excessive glycotoxic intermediates, such as MGO, GO, and 3-DG, can also cause long-term complications of diabetes (Turk, 2010). Glycation can be found in the aging process and is related with increasing glucose concentrations. 3-DG has been proved to participate in aging and diabetic complications. AGEs could accelerate chemical modification of proteins by glucose with high blood sugar that is a pathogenesis of diabetic complications, such as nephropathy, retinopathy, neuropathy, and atherosclerosis (Jakuš and Rietbrock, 2004; Niwa et al., 1993; Turk, 2010).

4.5 ANALYTICAL METHOD FOR RCS

4.5.1 HIGH-PERFORMANCE LIQUID CHROMATOGRAPHY

Since MGO cannot be measured directly, most high-performance liquid chromatography (HPLC) methods are based on MGO derivatization into quinoxaline adducts with diamino derivatives of benzene or naphthalene (Figure 4.4). These quinoxalines can easily be monitored either by UV detector at 300–360 nm or by fluorescent detector with excitation wavelengths at 300–360 nm and emission wavelengths at 380–450 nm (Nemet et al., 2004).

The HPLC method where MGO was derivatized into the corresponding fluorescent pteridine derivative with 6-hydroxy-2,4,5-triaminopyrimidine (TRI) has also been described. The major disadvantage is that the derivatization can form two isomers from MGO and affect the accuracy of quantification (Espinosa-Mansilla et al., 1998).

In addition, the electrospary ionization/liquid chromatography/mass spectrometry (ESI/LC/MS) method has been developed based on the quantification of dicarbonyl

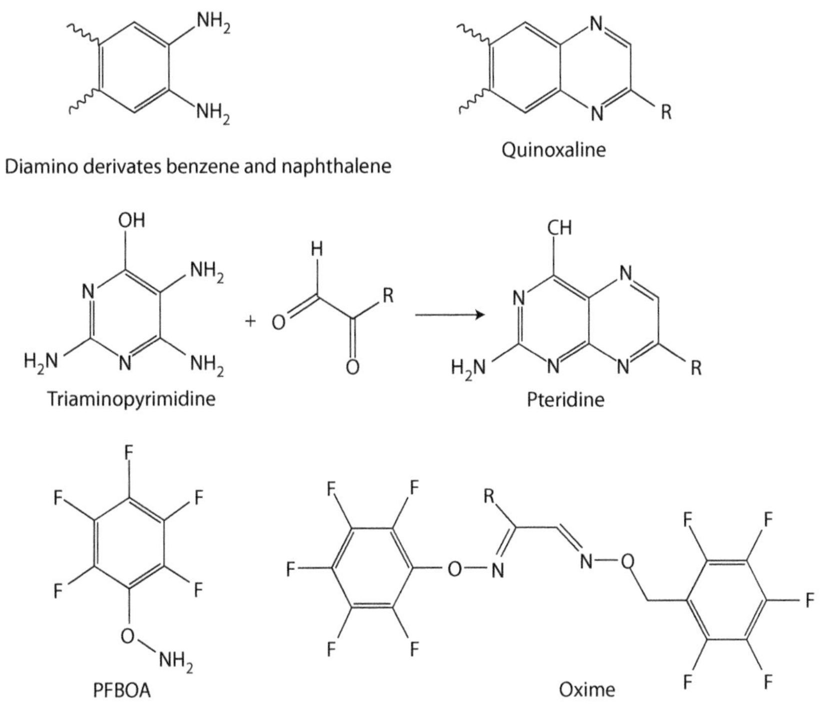

FIGURE 4.4 Derivatization of MGO, GO with diamino derivatives of benzene and naphthalene, triaminopyrimidine, and PFBOA.

adducts with 2,3-diaminonaphthalene. This method offers the advantage of greater specificity for measuring dicarbonyl compounds (Randella et al., 2005). However, it requires time-consuming liquid–liquid extraction or solid-phase extraction (SPE) and considerable quantities of highly pure and usually toxic organic solvents.

Generally, the injection of the sample to the instrumentation would be the last step of MGO quantification. However, the concentration of MGO in samples might be too low to be measured. So, the pre-concentration of the sample is often necessary. Usually, the liquid–liquid extraction and SPE may be used, followed by evaporation of organic solvent or freeze-drying. The HPLC analysis is usually performed on the reverse phase C18 columns and eluted with water solutions of acids or buffers in combination with methanol or acetonitrile under gradient or isocratic conditions.

4.5.2 GAS CHROMATOGRAPHY (GC)

The GC methods were relied on MGO derivatization by 1,2-diaminobenzene with detection either on mass spectrometry/selected ion monitor (MS/SIM) or on specific nitrogen-phosphorus detector (NPD), and by O-(2,3,4,5,6-pentafluorobenzyl)-hydroxylamine hydrochloride (PFBOA) (Figure 4.4) detection either on MS/SIM detector, electron-capture detector, NPD, or flame photometric detector (Lapolla et al., 2003). The development of GC method offers the advantage of less sample preparation and low-price equipment to perform highly specific measurements. The time-consuming extraction process and the toxic organic solvents are considered to be the disadvantages of GC method. The GC analysis is usually performed on fused-silica capillary columns with helium as carrier gas.

4.6 METHODOLOGIES FOR QUANTIFICATION OF AGES

Although there are experimental evidences for AGEs accumulation with age and under certain pathological states such as diabetes, it is difficult to compare the results between laboratories. Moreover, there is now a number of AGEs that are structurally heterogeneous, making it difficult to ensure that the detected AGEs is relevant to the complications observed *in vitro* or *in vivo*. To date, AGEs measurement is confined to investigative laboratories. Currently, there is no commonly accepted or widely used method to detect AGEs, nor any commercially available kits. The lack of internal standards is a critical challenge for required accuracy and reproducibility for each sample run. Currently, the most common methods used for the detection are HPLC (Sell and Monnier, 1990), LC/MS (Nagai et al., 2003), enzyme-linked immunosorbent assay (ELISA) (Makita et al., 1996; Münch et al., 1997), and immunohistochemistry (Soulis et al., 1997). A monoclonal antibody called 6D12, which recognizes CML, is commercially available (Ikeda et al., 1996).

4.6.1 FLUORESCENCE

The fluorescent property of AGEs was found by Monnier (Lapolla et al., 2004), who demonstrated that fluorescent pigments showed similar fluorescence spectra to AGEs. Since then, the measurement of fluorescence intensity after excitation at 370 nm, at emission 440 nm was widely used for characterizing and quantifying AGEs. Fluorescence was expressed as the relative fluorescence intensity in arbitrary units (A.U.) (Hilt et al., 2003). However, the measurement of AGEs by means of fluorometric analysis is limited due to its low specificity.

4.6.2 HPLC METHODS

Due to the structurally heterogeneous properties of AGEs, the HPLC methods are usually used to quantify and identify the amino acid residues from enzymatic digestion of the most severely glycated portion of proteins or conjugated amino acids with reactive dicarbonyl intermediates under *in vitro* condition. Difference in the sample preparation, separation, and detection methods should be considered based on the structural properties of each specific AGE during HPLC method development.

For instance, the determination of CEL in glucose-modified bovine serum albumin (AGE-BSA) by HPLC may be affected by co-elution with CML due to the similar chemical structure of those two AGEs. The HPLC method coupled with a styrene-divinylbenzene co-polymer resin and coupled with sulfonic group cation exchange column followed by a ninhydrin post-column derivatization and detection by fluorescence (e.g., 271 nm/em. 503 nm) was developed for the quantification of CEL contents of modified BSA (Hilt et al., 2003).

4.6.3 Liquid Chromatography-Tandem Mass Spectrometry Methods

Liquid chromatography-tandem mass spectrometry (LC-MS/MS) is currently the method of choice for the analysis of AGEs. For instance, a stable-isotope-dilution LC-MS/MS method has been widely used to analyze CML and CEL, the two major non-enzymatic chemical modifications on tissue proteins, which serve as biomarkers of oxidative stress resulting from sugar and lipid oxidation (Gonzalez-Reche et al., 2006).

Since high AGE concentrations reflect the production of AGE-modified proteins which, being chemically altered, show different biological activity, activating a macrophage response with consequent internalization and digestion (Nagai et al., 2003). AGE-modified proteins can then generate a series of AGE-modified and highly reactive peptides, which react with plasma lipoprotein (low-density lipoprotein, LDL) to form AGE-modified LDL and cross-links with collagen. In order to obtain information on the preferential glycation sites of the protein, the LC-MS/MS methods were also used to perform the structural elucidation (Lapolla et al., 2004).

4.7 POSSIBLE APPROACHES TO CONTROL LEVELS OF AGEs AND RCS *IN VIVO* AND *IN VITRO*

Several animal studies have shown that dietary flavonoids had potential antidiabetes effect. For instance, the significant reductions of serum glucose were observed in epigallocatechingallate (EGCG)-treated SD rats and in male, lean, obese Zucker rats (Hakim et al., 2003). Moreover, the daily treatment of green tea to rats with streptozotocin (STZ)-induced diabetes showed inhibition of diabetic cataracts by lowering plasma and lens glucose levels. The same phenomenon was also noted in a black tea treatment group (Tsuneki et al., 2004). Green tea lowered the blood glucose concentrations in the genetically diabetic db/db mice 2–6 h after administration at 300 mg/kg bw; whereas, no effect was observed in the control mice (Hakim et al., 2003).

Besides flavonoids from tea, naringenin and its glycosides, quercetin, luteolin glycoside, genistein, and diacylated anthocyanins have also shown significant effects on reducing blood glucose levels in STZ-induced diabetic rats (Choi et al., 1991; Lee, 2006; Matsui et al., 2002; Vessalet al., 2003).

The detailed mechanisms for the prevention of diabetic complications by flavonoids are remaining for further studies. Several studies have shown that those effects could partially be due to the inhibition of AGEs formation. Wu and Yen (2005) reported the inhibitory effect of naturally occurring flavonoids on the formation of AGEs. Kaempferol, luteolin, naringenin, quercetin, rutin, and tea catechins were screened in their study, and the results showed that (–)-epicatechin gallate (ECG), EGCG, luteolin, and rutin significantly inhibited the formation of AGEs under the *in vitro* condition. Rutter et al. (2003) also reported that green tea extract was able to delay collagen aging in C57Bl/6 mice by blocking AGEs formation and collagen cross-linking. In addition, the EGCG was able to reduce renal AGEs accumulation and its related protein expression in the kidney cortex based upon STZ-induced diabetic nephropathy model rats (Knekt et al., 2002). An *in vitro* study has demonstrated that green tea polyphenols dose-dependently inhibited AGE-stimulated proliferation and p44/42 mitogen-activated protein kinase expression of rat vascular smooth muscle cells (Ping et al., 2004).

Since RCS are responsible for the formation of AGEs, their roles in the development of various aging-related diseases have been increasingly recognized (Baynes and Thorpe, 2000). Therefore, decreasing the levels of dicarbonyl compounds, and consequently inhibiting the formation of AGEs would be a possible approach to prevent the development of diabetic complications and other age-related diseases. There is thus a prompt need to develop effective strategies to protect from RCS-associated pathogenic conditions, and this will remain one of the major research directions with global intention. A very limited number of chemical agents have been found to suppress or prevent excessive generation and accumulation of cellular RCS, and their therapeutic potential has been recognized only recently.

Several natural botanical extracts and phytochemicals have recently been evaluated for their effects on RCS-induced modification of protein structure. Yet, only a very limited number of natural products have demonstrated RCS-trapping capacity (Hu et al., 2012; Lo et al., 2006, 2011; Lv et al., 2010; Peng et al., 2008; Sang et al., 2007; Shao et al., 2008; Tan et al., 2008, 2011; Wang and Ho, 2012). In an earlier study, it was found that epigallocatechin-3-gallate (EGCG) could rapidly trap both MGO and GO under neutral or alkaline conditions (Lo et al., 2006). Data showed that EGCG was more reactive than lysine and arginine in trapping MGO or GO, indicating that EGCG has the potential to compete with lysine and arginine *in vivo* and, therefore, may prevent the formation of AGEs (Sang et al., 2007). In addition, it was found that EGCG was more reactive at trapping MGO than the pharmaceutical agent, aminoguanidine, which has been shown to inhibit the formation of AGEs by trapping of reactive dicarbonyl compounds *in vivo* (Nilsson 1999).

Three major products have been purified from the reaction between EGCG and MGO at a 3:1 mole ratio. Their structures were identified as two mono-MGO adducts and one di-MGO adduct of EGCG with the MGO conjugated at positions 6 and 8 of the EGCG A-ring (Figure 4.5). The results clearly indicate that the major active site of EGCG is at positions 6 and 8 of the A-ring and the gallate ring does not play an important role in trapping reactive dicarbonyl species. The slightly alkaline pH can increase the nucleophilicity at positions 6 and 8 of the A-ring of EGCG, facilitating the addition of MGO at these two positions to form mono- and di-MGO adducts (Lo et al., 2006; Sang et al., 2007).

Besides EGCG, catechin, epicatechin, theaflavin (Lo et al., 2006), proanthocyanidins (Peng et al., 2008), phloretin, phloridzin (Shao et al., 2008), genistein (Lv et al., 2011), curcumin (Hu et al., 2012), and a stilbeneglycoside from *Polygonum multiforum* Thunb. (Lv et al., 2010) could effectively trap MGO. These compounds may represent a new group of 1,2-dicarbonyl scavenging agents. However, these hypotheses must be proven by *in vitro* and *in vivo* studies with the AGEs being accurately analyzed. In addition, different from traditional views on drugs (most drugs elicit their effects via transient interactions with membrane-spanning receptors that modulate cellular signaling pathways), ideally, the carbonyl scavengers should show minimal activity toward drug receptors, thus minimizing undesirable pharmacological effects. Rather, the administration of carbonyl scavengers should proceed in the

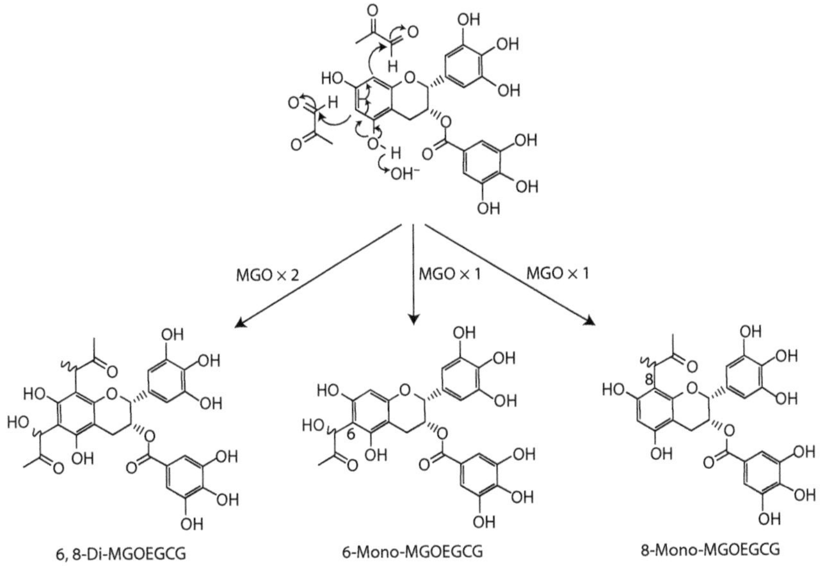

FIGURE 4.5   Adducts of MGO and epigallocatechin-3-gallate (EGCG).

expectation that they rapidly sequester carbonyl species in cells, thus blocking the adduction of macromolecules and any downstream damages. Whether these phenolic compounds can selectively perform this function also requires further study.

REFERENCES

Ahmed, N., Argirov, O.K., Minhas, H.S., Cordeiro, C.A., Thornalley, P.J. 2002. Assay of advanced glycation endproducts (AGEs): Surveying AGEs by chromatographic assay with derivatization by 6-aminoquinolyl-N-hydroxysuccinimidyl-carbamate and application to N-epsilon-carboxymethyl-lysine- and N-epsilon-(1-carboxyethyl)lysine-modified albumin. *Biochem. J.* 364, 1–14.

Ames, J.M. 1998. Applications of the Maillard reaction in the food industry. *Food Chem.* 62, 431–439.

Arribas-Lorenzo, G., Morales, F.J. 2010. Analysis, distribution, and dietary exposure of glyoxal and methylglyoxal in cookies and their relationship with other heat-induced contaminants. *J. Agric. Food Chem.* 58, 2966–2972.

Barrenscheen, H.K., Braun, K. 1931. Farb-und Fallungsreaktionen des Methylglyoxals. *BiocheniischeZeitschrift* 233, 296–304.

Baynes, J.W., Thorpe, S.R. 1999. Role of oxidative stress in diabetic complications: A new perspective on an old paradigm. *Diabetes* 48, 1–9.

Baynes, J.W., Thorpe, S.R. 2000. Glycoxidation and lipoxidation in atherogenesis. *Free Radic. Res.* 28, 1708–1716.

Brunner, N.A., Brinkmann, H., Siebers, B., Hensel, R. 1998. NAD1-dependent glyceraldehyde-3-phosphate dehydrogenase from *Thermoproteus tenax*. *J. Biol. Chem.* 279, 6149–6156.

Cho, S.J., Roman, G., Yeboah, F., Konishi, Y. 2007. The road to advanced glycation end products: A mechanistic perspective. *Curr. Med. Chem.* 14, 1653–1671.

Choi, J.S., Yokozawa, T., Oura, H. 1991. Improvement of hyperglycemia and hyperlipemia in streptozotocin-diabetic rats by a methanolic extract of *Prunus davidiana* stems and its main component, prunin. *Planta Med.* 57, 208–211.

Daglia, M., Papetti, A., Aceti, C., Sordelli, B., Spini, V., Gazzani, G. 2007. Isolation and determination of α-dicarbonyl compounds by RP-HPLC-DAD in green and roasted coffee. *J. Agric. Food Chem.* 55, 8877–8882.

Espinosa-Mansilla, A., Duran-Meras, I., Salinas, F. 1998. High-performance liquid chromatographic-fluorometric determination of glyoxal, methylglyoxal, and diacetyl in urine by prederivatization to pteridinic rings. *Anal. Biochem.* 255, 263–273.

Goldberg, T., Cai, W., Peppa, M., Dardaine, V., Baliga, B.S., Uribarri, J., Vlassara, H. 2004. Advanced glycoxidation end products in commonly consumed foods. *J. Am. Diet Assoc.* 104, 1287–1291.

Goldin, A., Beckman, J.A., Schmidt, A.M., Creager, M.A. 2006. Advanced glycation end products. *Circulation* 114, 597–605.

Golej, J., Hoeger, H., Radner, W., Unfried, G., Lubec, G. 1998. Oral administration of methylglyoxal leads to kidney collagen accumulation in the mouse. *Life Sci.* 63, 801–807.

Gonzalez-Reche, L.M., Kucharczyk, A., Musiol, A.K., Kraus, T. 2006. Determination of N-ε-(carboxymethyl)lysine in exhaled breath condensate using isotope dilution liquid chromatography/electrospray ionization tandem mass spectrometry. *Rapid Commun. Mass Spectrom.* 20, 2747–2752.

Guha, M.K., Vander Jagt, D.L., Creighton, D.J. 1988. Diffusion-dependent rates for the hydrolysis reaction catalyzed by glyoxalase II from rat erythrocytes. *Biochemistry* 27, 8818–8822.

Guo, Q., Mori, T., Jiang, Y., Hu, C., Osaki, Y., Yoneki, Y., Sun, Y., Hosoya, T., Kawamata A., Ogawa, S. 2009. Methylglyoxal contributes to the development of insulin resistance and salt sensitivity in Sprague–Dawley rats. *J. Hypertens.* 27, 1664–1671.

Hakim, I.A., Alsaif, M.A., Alduwaihy, M., Al-Rubeaan, K., Al-Nuaim, A.R., Al-Attas, O.S. 2003. Tea consumption and the prevalence of coronary heart disease in Saudi adults: Results from a Saudi national study. *Prev. Med.* 36, 64–70.

Hilt, P., Schieber, A., Yildirim, C., Arnold, G., Klaiber, I., Conrad, J., Beifuss, U., Carle, R. 2003. Detection of phloridzin in strawberries (*Fragaria* x *ananassa* Duch.) by HPLC-PDA-MS/MS and NMR spectroscopy. *J. Agric. Food Chem.* 51, 2896–2899.

Homoki-Farkas, P., Örsi, F., Kroh, L. 1997. Methylglyoxal determination from different carbohydrates during heat processing. *Food Chem.* 59, 157–163.

Hu, T.Y., Liu C.L., Chyau, C.C., Hu, M.L. 2012. Trapping of methylglyoxal by curcumin in cell-free systems and in human umbilical vein endothelial cells. *J. Agric. Food Chem.* 60, 8190–8196.

Ikeda, K., Higashi, T., Sano, H., Jinnouchi, Y., Yoshida, M., Araki, T., Ueda, S., Horiuchi, S. 1996. N-(epsilon)-(carboxymethyl)lysine protein adduct is a major immunological epitope in proteins modified with advanced glycation end products of the Maillard reaction. *Biochemistry* 35, 8075–8083.

Jakuš, V., Rietbrock, N. 2004. Advanced glycation end-products and the progress of diabetic vascular complications. *Physiol. Res.* 53, 131–142.

Kalapos, M.P. 1999. Methylglyoxal in living organisms: Chemistry, biochemistry, toxicology and biological implications. *Toxicol. Lett.* 110, 145–175.

Khuhawar, M., Kandhro, A., Khand, F. 2006. Liquid chromatographic determination of glyoxal and methylglyoxal from serum of diabetic patients using meso-stilbene diamine as derivatizing reagent. *Anal. Lett.* 39, 2205–2215.

Knekt, P., Kumpulainen, J., Järvinen, R., Rissanen, H., Heliövaara, M., Reunanen, A., Hakulinen, T., Aromaa, A. 2002. Flavonoid intake and risk of chronic diseases. *Am. J. Clin. Nutr.* 76, 560–568.

Koschinsky, T., He, C.J., Mitsuhashi, T., Bucala, R., Liu, C., Buenting, C., Heitmann, K., Vlassara, H. 1997. Orally absorbed reactive glycation products (glycotoxins): An environmental risk factor in diabetic nephropathy. *Proc. Natl. Acad. Sci. U.S.A.* 94, 6474–6479.

Labuza, T., Saltmarch, M. 1981. The nonenzymatic browning reaction as affected by water in foods. In *Water Activity: Influences Food Quality*, edited by Rockland, L.B., Stewart, G.F., Academic Press, New York, NY. pp. 605–650.

Lapolla, A., Fedele, D., Reitano, R., Aricò, N. C., Seraglia, R.,Traldi, P., Marotta, E., Tonani, R. 2004. Enzymatic digestion and mass spectrometry in the study of advanced glycation end products/peptides. *J. Am. Soc. Mass Spectrom.* 15, 496–509.

Lapolla, A., Flamini, R., Vedova, A.D., Senesi, A., Reitano, R., Fedele, D., Basso, E., Seraglia R., Traldi P. 2003. Glyoxal and methylglyoxal levels in diabetic patients: Quantitative determination by a new GC/MS method. *Clin. Chem. Lab. Med.* 41, 1166–1173.

Lederer, M.O., Klaiber, R.G. 1999. Cross-linking of proteins by Maillard processes: Characterization and detection of lysine-arginine cross-links derived from glyoxal and methylglyoxal. *Bioorg. Med. Chem.* 7, 2499–2507.

Lee, J.S. 2006. Effects of soy protein and genistein on blood glucose, antioxidant enzyme activities, and lipid profile in streptozotocin-induced diabetic rats. *Life Sci.* 79, 1578–1584.

Lesgards, J.F., Gauthier, C., Iovanna, J., Vidal, N., Dolla, A., Stocker, P. 2011. Effect of reactive oxygen and carbonyl species on crucial cellular antioxidant enzymes. *Chem. Biol. Interact.* 190, 28–34.

Lo, C.Y., Li, S., Tan, D., Pan, M.H., Sang, S., Ho, C.T. 2006. Trapping reactions of reactive carbonyl species with tea polyphenols in simulated physiological conditions. *Mol. Nutr. Food Res.* 50, 1118–28.

Lo, C.Y., Hsiao, W.T., Chen, X.Y. 2011. Efficiency of trapping methylglyoxal by phenols and phenolic acid. *J. Food Sci.* 76, H90–H96.

Lv, L., Shao, X., Wang, L., Huang, D., Ho, C.T., Sang, S. 2010. Stilbene glucoside from *Polygonum multiforum* Thunb.: A novel natural inhibitor of advanced glycation end product formation by trapping of methylglyoxal. *J. Agric. Food Chem.* 58, 2239–2245.

Lv, L., Shao, X., Chen, H., Ho, C.T., Sang, S. 2011. Genistein inhibits advanced glycation end products formation by trapping of methylglyoxal. *Chem. Res. Toxicol.* 24, 579–586.

Makita, Z., Yanagisawa, K., Kuwajima, S., Bucala, R., Vlassara, H., Koike,T. 1996. The role of advanced glycosylation end-products in the pathogenesis of atherosclerosis. *Nephrol. Dial. Transplant* 11, 31–33.

Matsui, T., Ebuchi, S., Kobayashi, M., Fukui, K., Sugita, K., Terahara, N., Matsumoto, K. 2002. Anti-hyperglycemic effect of diacylated anthocyanin derived from Ipomoea batatas cultivar Ayamurasaki can be achieved through the alpha-glucosidase inhibitory action. *J. Agric. Food Chem.* 50, 7244–7248.

Münch, G., Keis, R., Weßels, A., Riederer, P., Bahner, U., Heidland, A., Niwa, T., Lemke, H.D., Schinzel, R. 1997. Determination of advanced glycation end products in serum by fluorescence spectroscopy and competitive ELISA. *Eur. J. Clin. Chem. Clin. Biochem.* 35, 669–677.

Nagai, R., Araki, T., Hayashi, C.M., Hayase, F., Horiuchi, S. 2003. Identification of *N*-ε-(carboxymethyl)lysine, one of the methylglyoxal-derived AGE structures, in glucose-modified protein: Mechanism for protein modification by reactive aldehydes. *J. Chromatogr. B Analyt. Technol. Biomed. Life Sci.* 788, 75–84.

Nagai, R., Unno, Y., Hayashi, M.C., Masuda, S., Hayase, F., Kinae, N., Horiuchi, S. 2002. Peroxynitrite induces formation of *N*-ε-(carboxymethyl) lysine by the cleavage of Amadori product and generation of glucosone and glyoxal from glucose novel pathways for protein modification by peroxynitrite. *Diabetes* 51, 2833–2839.

Nagaraj, R.H., Sarkar, P., Mally, A., Biemel, K.M., Lederer, M.O., Padayatti, P.S. 2002. Effect of pyridoxamine on chemical modification of proteins by carbonyls in diabetic rats: Characterization of a major product from the reaction of pyridoxamine and methylglyoxal. *Arch. Biochem. Biophys.* 402, 110–119.

Nemet, I., Varga-Defterdarovi, L., Turk, Z. 2004. Preparation and quantification of methylglyoxal in human plasma using reverse-phase high-performance liquid chromatography. *Clin. Biochem.* 37, 875–881.

Nemet, I., Varga-Defterdarovi, L., Turk, Z. 2006. Methylgloxal in food and living organisms. *Mol. Nutr. Food Res.* 50, 1105–1117.

Nilsson, B.O. 1999. Biological effects of aminoguanidine: An update. *Inflamm. Res.* 48, 509–515.

Niwa, T., Takeda, N., Yoshizumi, H., Tatematsu, A., Ohara, M., Tomiyama, S., Niimura, K. 1993. Presence of 3-deoxyglucosone, a potent protein crosslinking intermediate of Maillard reaction, in diabetic serum. *Biochem. Biophys. Res. Commun.* 196, 837–843.

Nursten, H. 2005. *The Maillard Reaction: Chemistry, Biochemistry and Implications*, Royal Society of Chemistry, Cambridge, UK.

O'Brien, J., Morrissey, P., Ames, J. 1989. Nutritional and toxicological aspects of the Maillard browning reaction in foods. *Crit. Rev. Food Sci. Nutr.* 28, 211–248.

Odani, H., Shinzato, T., Matsumoto, Y., Usami, J., Maeda, K. 1999. Increase in three α, β-dicarbonyl compound levels in human uremic plasma: Specific in vivo determination of intermediates in advanced Maillard reaction. *Biochem. Biophys. Res. Commun.* 256, 89–93.

Paul, R.G., Avery, N.C., Slatter, D.A., Sims, T.J., Bailey, A.J. 1998. Isolation and characterization of advanced glycation end products derived from the in vitro reaction of ribose and collagen. *Biochem. J.* 330, 1241–1248.

Peng, X., Cheng, K.W., Ma, J., Chen, B., Ho, C.T., Chen, F., Wang, M. 2008. Cinnamon bark proanthocyanidins as reactive carbonyl scavengers to prevent the formation of advanced glycation endproducts. *J. Agric. Food Chem.* 56, 1907–1911.

Picklo, Sr. M.J., Montine, T.J., Amarnath, V., Neely, M.D. 2002. Carbonyl toxicology and Alzheimer's disease. *Toxicol. Appl. Pharmacol.* 184, 187–197.

Ping, O., Peng, W.L., Xu, D.L., Lai, W.Y., Xu, A.L. 2004. Green tea polyphenols inhibit advanced glycation end product-induced rat vascular smooth muscle cell proliferation. *Di Yi Jun Yi Da XueXueBao* 24, 247–51.

Pompliano, D.L., Peyman, A., Knowles, J.R. 1990. Stabilization of a reaction intermediate as a catalytic device: Definition of the functional role of the flexible loop in triosephosphate isomerase. *Biochemistry* 29, 3186–3194.

Price, C.L., Knight, S.C. 2009. Methylglyoxal: Possible link between hyperglycaemia and immune suppression? *Trends Endocrinol. Metab.* 20, 312–317.

Ramasamy, R., Yan, S.F., Schmidt A.M. 2012. The diverse ligand repertoire of the receptor for advanced glycation endproducts & pathways to the complications of diabetes. *Vascul. Pharmacol.* 57, 160–167.

Randella, E.W., Vasdevb, S., Gillb, V. 2005. Measurement of methylglyoxal in rat tissues by electrospray ionization mass spectrometry and liquid chromatography. *J. Pharmacol. Toxicol. Methods* 51, 153–157.

Reichard, P., Nilsson B.Y., Rosenqvist U. 1993. The effect of long-term intensified insulin treatment on the development of microvascular complications of diabetes mellitus. *N. Engl. J. Med.* 329, 304–309.

Rutter, K., Sell, D.R., Fraser, N., Obrenovich, M., Zito, M., Starke-Reed, P., Monnier, V.M. 2003. Green tea extract suppresses the age-related increase in collagen crosslinking and fluorescent products in C57BL/6 mice. *Int. J. Vitam. Nutr. Res.* 73, 453–460.

Sang, S., Shao, X., Bai, N., Lo, C.Y., Yang, C.S., Ho, C.T. 2007. Tea polyphenol (−)-epigallocatechin-3-gallate: A new trapping agent of reactive dicarbonyl species. *Chem. Res. Toxicol.* 20, 1862–1870.

Sell, D.R., Monnier V.M. 1990. End stage renal disease and diabetes catalyse the formation of a pentose-derived cross-link form aging human collagen. *J. Clin. Invest.* 85, 380–384.

Sena, C. M., Matafome, P., Crisostomo, J., Rodrigues, L., Fernandes, R., Pereira P., Seica, R.M. 2012. Methylglyoxal promotes oxidative stress and endothelial dysfunction. *Pharmacol. Res.* 65, 497–506.

Shangari, N., O'Brien, P.J. 2004. The cytotoxic mechanism of glyoxal involves oxidative stress. *Biochem. Pharmacol.* 68, 1433–1442.

Shao, X., Bai, N., He, K., Ho, C.T., Yang, C.S., Sang, S. 2008. Apple polyphenols, phloretin and phloridzin: New trapping agents of reactive dicarbonyl species. *Chem. Res. Toxicol.* 21, 2042–2050.

Singh, R., Barden, A., Mori, T., Beilin, L. 2001. Advanced glycation end-products: A review. *Diabetologia* 44, 129–146.

Soulis, T., Thallas, V., Youssef, S., Gilbert, R.E., McWilliam, B.G., Murray-McIntosh, R.P., Cooper, M.E. 1997. Advanced glycation end-products and their receptors co-localise in rat organs susceptible to diabetic microvascular injury. *Diabetologia* 40, 619–628.

Tan, D., Wang, Y., Lo, C.Y., Sang, S., Ho, C.T. 2008. Methylglyoxal: Its presence in beverages and potential scavengers. *Ann. N. Y. Acad. Sci.* 1126, 72–75.

Tsuneki, H., Ishizuka, M., Terasawa, M., Wu, J.B., Sasaoka, T., Kimura, I. 2004. Effect of green tea on blood glucose levels and serum proteomic patterns in diabetic (db/db) mice and on glucose metabolism in healthy humans. *BMC Pharmacol.* 4, 18.

Turk, Z. 2010. Glycotoxines, carbonyl stress and relevance to diabetes and its complications. *Physiol. Res.* 59, 147–156.

Uribarri, J., Woodruff, S., Goodman, S., Cai, W., Chen, X., Pyzik, R., Yong, A., Striker, G.E., Vlassara, H. 2010. Advanced glycation end products in foods and a practical guide to their reduction in the diet. *J. Am. Diet. Assoc.* 110, 911–916.

Vessal, M., Hemmati, M.,Vasei, M. 2003. Antidiabetic effects of quercetin in streptozocin-induced diabetic rats. *Comp Biochem. Physiol. C Toxicol. Pharmacol.* 135, 357–364.

Wang, Y., Ho, C.T. 2012. Flavour chemistry of methylglyoxal and glyoxal. *Chem. Soc. Rev.* 41, 4140–4149.

Wu, C.H.,Yen, G.C. 2005. Inhibitory effect of naturally occurring flavonoids on the formation of advanced glycation endproducts. *J. Agric. Food Chem.* 53, 3167–3173.

Wu, C.H., Huang, S.M., Lin, J.A.,Yen, G.C. 2011. Inhibition of advanced glycation endproduct formation by foodstuffs. *Food Funct.* 2, 224–234.

Wu, L., Juurlink, B.H. 2002. Increased methylglyoxal and oxidative stress in hypertensive rat vascular smooth muscle cells. *Hypertension* 39, 809–814.

5 Chemistry and Safety of Maillard Reaction Products III

Heterocyclic Amines and Amino Acid Derivatives

Ka-Wing Cheng, Xin-Chen Zhang, and Ming-Fu Wang

CONTENTS

5.1 INTRODUCTION

Genetic predisposition and environmental factors are both implicated in the etiology of cancer. It is generally accepted that diet is an important contributor to environmental exposure to genotoxicants, which increases the risk of developing cancers. In the 1970s, Japanese scientists first identified potent mutagenic activity in the charred surface of broiled beef and fish (Sugimura et al., 1977). These mutagens were later isolated and characterized as heterocyclic amines (HAs). HAs are polycyclic aromatic compounds that are formed in muscle foods (i.e., meat and fish), and their formation is substantially enhanced by high temperature and prolonged heating, such as while frying, grilling, and roasting. Due to their potent genotoxicity, HAs are a major health concern.

Since their initial discovery, more than 25 HAs have been identified. 2-Amino-1-methyl-6-phenylimidazo(4,5-b)pyridine (PhIP) is the most abundant mutagenic HA found in foods. Many HAs demonstrated strong mutagenicity based on the Ames *Salmonella* mutagenicity test (Felton and Knize, 1990). PhIP, 2-amino-3,4-dimethylimidazo[4,5-f]quinoline (MeIQ), 2-amino-3,8-dimethyl-imidazo[4,5-f] quinoxaline (MeIQx), 2-amino-9H-dipyrido[2,3-b]indole (AαC), 2-amino-3-methyl-9H-dipyrido[2,3-b]indole (MeAαC), 3-amino-1,4-dimethyl-5H-pyrido[4,3-b]indole (Trp-P-1), 3-amino-1-methyl-5H-pyrido[4,3-b]indole (Trp-P-2), and 2-amino-dipyrido[1,2-a:39,29-d]imidazole (Glu-P-2) have been classified by the International Agency for Research on Cancer as possible, and one of them, 2-amino-3-methyl-imidazo[4,5-f]quinoline (IQ) as a probable human carcinogen.

Extensive studies have been conducted to better understand the chemistry of HAs and their relation with human health. In this chapter, the chemistry of HAs, their adverse biological properties, the approaches to reduce occurrence of HAs, and the methods for qualitative and quantitative analyses of HAs will be reviewed and discussed.

5.2 CHEMICAL MECHANISMS INVOLVED IN THE FORMATION OF HETEROCYCLIC AMINES

Different classes of HAs can be formed from proteinaceous food depending on the heating temperature. Those formed in temperatures above 300°C are protein pyrolysates and characterized as 2-amino-pyridine-mutagens or amino-carbolines, while those formed below 300°C are 2-amino-imidazole-type mutagens or aminoimidazole-azaarenes. Based on polarity (Figure 5.1), HAs can also be divided into polar, which are mainly the IQ-, IQx-, as well as the imidazopyridine-type, and nonpolar, which have a common pyridoindole or dipyridoimidazole moiety (Murkovic, 2004). Most HAs have an exocyclic amino group that renders them highly mutagenic. Exceptions are β-carbolines, such as harman and norharman, which are nonmutagenic according to the Ames/*Salmonella* mutagenicity test (Sugimura et al., 1982).

The formation of amino-carbolines depends on amino acids or protein as major precursors. Since creatin(in)e is not required for their formation, they may be produced in foods of animal and plant origins (Jagerstad et al., 1998). Given the drastic thermal environment (>300°C) in which they are formed, it is hypothesized that

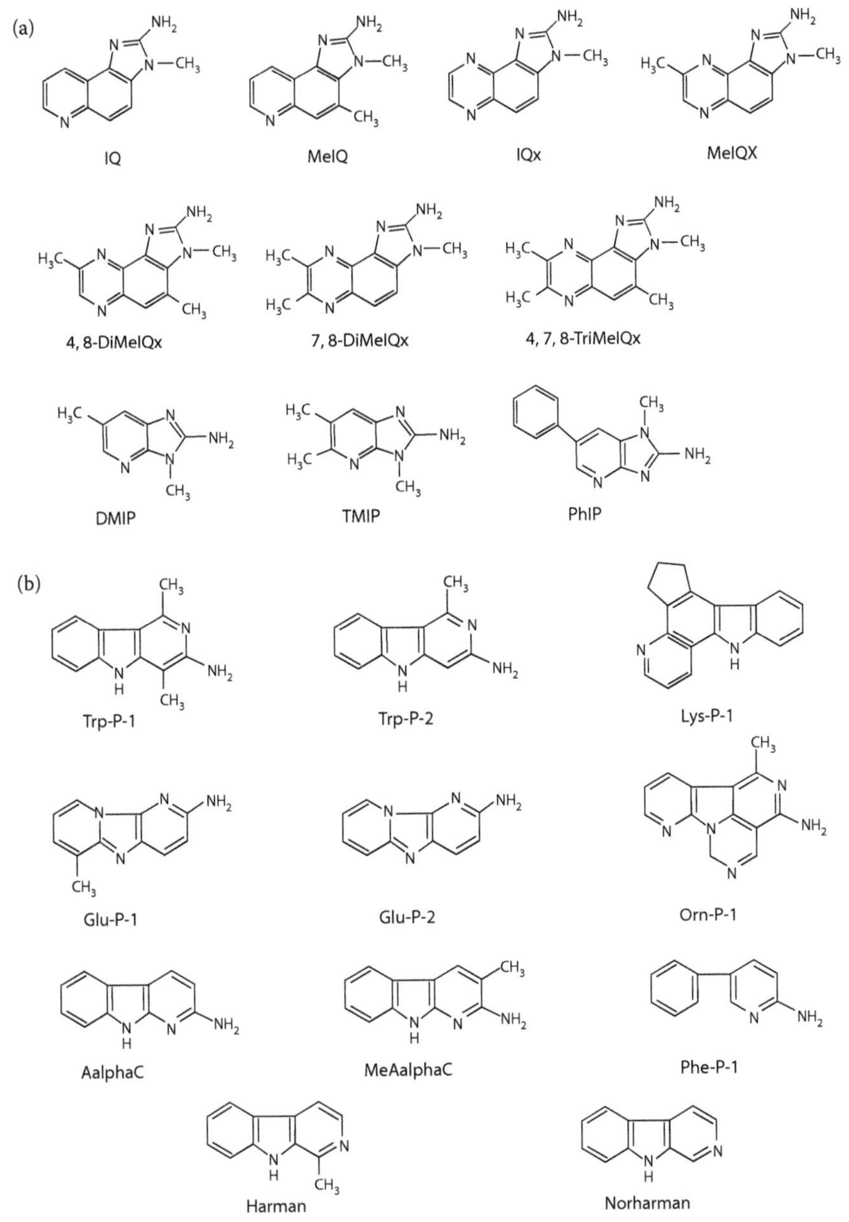

FIGURE 5.1 Chemical structures of (a) polar and (b) nonpolar heterocyclic amines.

amino-carbolines are generated via free radical reactions, which produce many reactive fragments (Skog et al., 2000). These fragments may then condense to form new structures.

The formation pathways of aminoimidazole-azaarenes, including IQ, IQx, and PhIP, are shown in Figure 5.2. These HAs are suggested to have been formed from

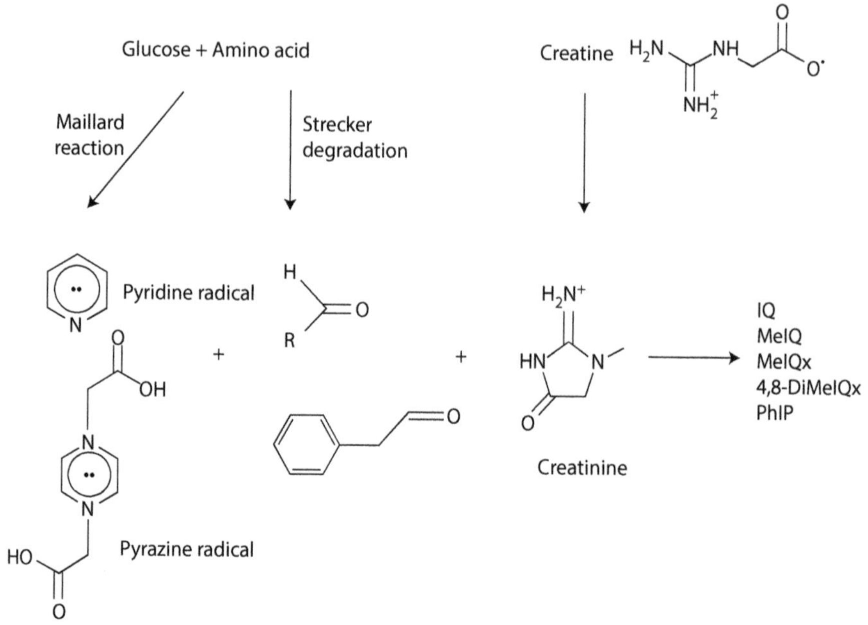

FIGURE 5.2 Suggested pathways for the formation of aminoimidazole-azaarenes.

creatin(in)e, sugars, free amino acids, and some dipeptides through the Maillard reaction and Strecker degradation upon mild heating (<300°C) (Skog et al., 2000). Creatinine forms the amino-imidazo part of IQ and IQx via cyclization and water elimination, while Strecker degradation products, such as pyridines or pyrazines, formed in the Maillard reaction between amino acids and hexoses contributed to the remaining part of the molecule, probably via aldol condensation (Jagerstad et al., 1998). Free radical-mediated pathways have also been proposed. Studies have suggested the involvement of pyridine and pyrazine radicals (Milic et al., 1993) in HAs formation.

5.3 FACTORS AFFECTING THE FORMATION OF HETEROCYCLIC AMINES DURING MAILLARD REACTION

The yield of HAs is governed by various factors, including the type and concentration of the precursors, the heating conditions (temperature and time), and the content and type of fats. It is known that creatinine or creatine, sugars, and amino acids are principal precursors of HAs in food (Skog, 1993). Different types of precursors, in particular amino acids, can give rise to fundamentally different HAs. It is therefore expected that different HAs may predominate in different types of meats and fish under similar heating conditions (Pais et al., 1999). Addition of extra creatinine to the surface of meat prior to frying increased the yield of HAs, indicating the important role of creatinine (Weisburger, 1994). Some studies suggested sugars as a limiting factor, since their levels decreased rapidly to undetectable levels after only 2.5 min of heating (Shin et al., 2002). However, HAs may also be formed in the absence of sugar (Felton and Knize, 1990). It appears that formation of HAs is strongly dependent on the reaction conditions and the presence of different precursors.

Heating time and temperature have a significant impact on the formation of HAs, and the amount of HAs formed increases with increasing temperature and time (Skog et al., 1997). The effect of temperature is especially important. For example, in a simple chemical model system with creatinine, phenylalanine, and glucose as major constituents, yield of PhIP was found to increase when the temperature was raised from 180°C to 225°C (Skog and Jagerstad, 1991), while at low temperatures, this HA has rarely been produced in significant quantities (Johansson, 1995). Others also reported strong enhancement in the formation of many HAs upon an increase in temperature from 190°C to 230°C (Abdulkarim and Smith, 1998). Prolonged heating also increases the formation of HAs (Jagerstad et al., 1998), albeit to a lesser extent compared to an increase in heating temperature.

Recent studies have also provided evidence that the type and amount of fat may play a role in HA formation. The amount of HAs was found, in most cases, to be higher in meat than in fish (Skog et al., 1998). One possible reason may be the presence of high levels of unsaturated fatty acids in fish that trap radical intermediates involved in the HA formation, with themselves being converted into more stable radicals. However, studies examining the effect of fats on HA formation have reported inconsistent effects. Further studies are needed to determine the exact effect of fats on HA formation.

5.4   TOXIC EFFECTS OF HETEROCYCLIC AMINES

HAs have been found to possess various detrimental effects on human health, and their mutagenic and carcinogenic potential has been the subject of intensive investigation. It is well known that bioactivation plays a key role in the toxicity of HAs. Hence, we will provide an overview of HAs bioactivation, followed by their probable toxic effects in humans.

5.4.1   BIOACTIVATION OF HAS

Most HAs are not mutagenic/carcinogenic in their native forms but acquire the capability of forming DNA adducts after metabolic activation. HA activation involves phase I N-hydroxylation followed by phase II esterification, both of which take place at the exocyclic amino group. The treatment *in vitro* of isolated rat mammary epithelial cells with N-hydroxy-PhIP produced PhIP–DNA adducts, whereas no adduct was detected for incubation using PhIP, even after incubation at a much higher dose for longer periods of time (Ghoshal et al., 1995). For some HAs, phase II activation may be critical for efficient formation of DNA adducts, as studies have reported increase in DNA-adduct formation with phase I and II enzymes compared with that achieved using phase I enzyme(s) alone (Liu and Levy, 1998).

The cytochrome P450 system, in particular cytochrome P450 1A2, is responsible for the phase I conversion of the aromatic amines to their corresponding hydroxylamines (Edwards et al., 1994). Four mammalian phase II enzymes, such as

N-acetyltransferase (NAT), sulfotransferase, prolyl tRNA synthetase, and phosphor-ylase, catalyze the conversion of hydroxylamines to form *N*-acetoxy, *N*-sulfonyloxy, *N*-prolyloxy, and *N*-phosphatyl ester derivatives, respectively. Esterification reactions catalyzed by these four enzymes generate momentary metabolites whose ester moieties are better leaving groups, thus greatly favor the formation of aryl-nitrenium ions (carcinogenic form of HAs) (Kadlubar and Beland, 1985; Schut and Snyderwine, 1999). Among these, NAT (especially NAT2) appears to play a dominant role, at least in phase II bioactivation of IQ, MeIQx, and PhIP (Ghoshal et al., 1995; Minchin et al., 1993). Individuals with both highly active NAT2 and CYP 1A2 may be predisposed with earlier onset of colon cancer than the general population (odd ratio = 2.86) as a result of enhanced HA bioactivation (Minchin et al., 1993).

The majority of HAs is metabolized in the liver. Bioactivated HA metabolites and their DNA adducts have also been reported in extrahepatic tissues such as lung, kidney, mammary gland, colon, and pancreas (Holme et al., 1987; Kaderlik et al., 1994).

5.4.2 Mutagenicity of HAs

DNA mutations play a key role in the initiation, promotion, and progression of cancer. HAs are potent mutagens and evidence is substantial to state that they form DNA adducts *in vitro* and *in vivo*. The formation of HA-DNA adducts is believed to be mediated via electrophilic attack of guanine bases at the N-2, N-7, or C-8 position by the HA nitrenium ions (Jamin et al., 2007), which induces different levels of DNA damage depending on the parent HA (Baranczewski et al., 2004). IQ can also form an adduct with adenine, which may be relevant to the observation about mutations involving A or T nucleobases *in vivo* (Jamin et al., 2007; Kakiuchi et al., 1995).

The Ames mutagenicity test, developed in the 1970s (Ames et al., 1975), was first used to assess the *in vitro* genotoxicity of HAs in prokaryotic cells. The human S9 liver homogenate is required for the metabolic bioactivation of HAs (Sugimura, 1997). Many HAs demonstrated potent mutagenicity in Ame's test, although there is significant variation between the most potent and the weakest HAs in these tests. Apart from *Salmonella* strains, certain genes (*lacZ*, *lacZa*, and *lacI*) of *Escherichia coli* have also been used to analyze the mutational specificity of some HAs (Kosakarn et al., 1993).

Mutagenicity of HAs has also been extensively evaluated in mammalian cells to derive information relating to their effect in eukaryotic cells. Different genes, such as hypoxanthine-guanine phosphoribosyl transferase, dihydrofolate reductase, and adenine phosphoribosyltransferase (*aprt*), have been selected to study the mutational characteristics of HAs in mammalian cells. For some HAs, contrasting results were obtained between Ame's test and mammalian cells-based assay. This is exemplified by PhIP, which, in bacterial cells, exhibited weak mutagenicity but is highly mutagenic in eukaryotic cells (Felton and Knize, 1990; Gooderham et al., 2001; Thompson et al., 1987). Such variations in mutagenicities of HAs in *Salmonella typhimurium* and in mammalian cells could arise from their different genotoxic behaviors in these cells (frameshift in the bacteria and base substitution in mammalian cells) (Schut and Snyderwine, 1999) besides other factors such as their DNA repairing mechanisms.

These are manifested in studies relating to PhIP and IQ or their metabolites. For example, using Chinese hamster ovary (CHO) cells, it was found that N-OH-PhIP, PhIP (in cells capable of expressing P450 1A2), and IQ primarily induced single-base transversion, mostly GC → TA, but also AT → TA and CG → AT (for PhIP-induced mutants) (Lee and Shih, 1995; Wu et al., 1995). Moreover, the observation that 75% of the PhIP-induced mutations occurred at the 3′ end of exon 2 and the beginning of exon 3 of the *aprt* gene suggested higher susceptibility to mutation in genes which harbor these sequences (Wu et al., 1995). Interestingly, HA-induced mutation in mammalian cell lines appears to be biased toward the non-transcribed strand of the gene (Morgenthaler and Holzhauser, 1995).

While HAs are highly mutagenic *in vitro*, data regarding their mutagenicity *in vivo* have been less consistent. The mutagenic potential of HAs observed *in vivo* can be affected by the type of animals, sex, and tissues chosen for evaluation. In particular, studies that use the parent HAs may be affected by the ability of the animal model in activating the administrated HA(s) (Gooderham et al., 2001). As an example, the metabolism of PhIP by CYP1A2 differs substantially between humans and mice, with more N_2-hydroxylation (activation) and less 4′-hydroxylation (detoxication) in humans (Cheung et al., 2005). The duration of exposure and time allowed for DNA-adduct formation is also critical to the mutagenic activities quantified. Quantification of DNA-adduct formation would require a relatively short-time period (24 h); while at least a few weeks are suggested for the adduct(s) to induce a permanent base sequence change (Davis et al., 1996). Therefore, reducing the proportion and the frequency of consumption of well-done meats in the diets is an important protective measure against HA-associated mutagenesis.

5.4.3 CARCINOGENICITY OF HAs

Carcinogenicity of HAs has been well documented in many organs/tissues using animal studies (Eisenbrand and Tang, 1993; Ohgaki et al., 1991), which led to the classification of eight HAs (MeIQ, MeIQx, PhIP, AαC, MeAαC, Trp-P-1, Trp-P-2, and Glu-P-2) by IARC in 1993 as possible (Group 2B) carcinogen, and IQ as a probable human carcinogen (Group 2A). In rodents (mice and rats), liver is the target organ that frequently develops cancer following the administration of many HAs, such as Trp-P-1, Trp-P-2, 2-amino-6-methyl-dipyrido[1,2-a:3′,2′-d]imidazole (Glu-P-1), Glu-P-2, AαC, MeAαC, IQ, MeIQ, and MeIQx (Sugimura et al., 2004). A key factor contributing to an increased incidence of liver carcinogenesis is the high levels of cytochrome P450 system in this metabolic organ, leading to elevated levels of bioactivated HAs. IQ has even been shown to induce hepatocarcinoma in non-human primates (Adamson et al., 1994). Other organs that develop cancer in rat and mouse models include the alimentary canals, blood vessels, mammary gland, prostate gland, Zymbal gland, clitoral gland, lung, lymphoid tissue, skin, urinary bladder, and pancreas (Yoshimoto et al., 1999). A large number of carcinogenic HAs and the multiple target sites may, therefore, imply a high cancer risk for human populations.

Species, sex, and age variation in susceptibility to HA-induced carcinogenesis have also been documented. For example, AαC-induced tumors in the liver and blood vessels in mice, but not in F344 rats (Sugimura et al., 2004). MeIQx (0.06% in diet for 30 weeks), however, was found to cause hepatocellular carcinoma in male mice, but not in females (Ryu et al., 1999). In addition, neonatal animals are considerably more sensitive to HA-induced carcinogenesis compared to adult mice (Jägerstad and Skog, 2005; Shan et al., 2004). PhIP exposure also increased the risk of mammary carcinogenesis in the second generation, expectedly via its transplacental transport to the fetus and its excretion into the milk (Brittebo et al., 1994). Epidemiological studies have provided sound evidence between the genotoxicity of HA and carcinogenesis in humans. Population-based data have shown a positive correlation between HA exposure and the risk of certain cancers, especially colon cancer (Lang et al., 1994; Schiffman and Felton, 1990).

5.5   ANALYTICAL METHODS OF HETEROCYCLIC AMINES

HAs are typically present in food in minute amounts (ng/g). Efficient and robust analytic techniques are essential for their quantitative determination, especially for complex matrices. Prior to analysis, liquid–liquid extraction or solid-phase extraction (SPE) is necessary for cleaning and for concentration of the target analytes. Liquid chromatography (LC) and gas chromatography (GC) have been used for the quantitative determination of HAs, with the former being the most popular technique. Coupling of LC and GC to MS enables satisfactory separation and detection with minute starting materials. The focus of this section will be on the most widely used methods for the analysis of HAs.

5.5.1   Solid-Phase Extraction

SPE applied for analysis of HAs is a gross cleaning process that serves to selectively concentrate target analytes from a complex mixture (Henry, 2000). It offers enhanced sensitivity that facilitates subsequent detection (Simpson and Wynne, 2000). The protocol adopted depends largely on the complexity of the sample matrix. For simple model systems, an aliquot of the reacted mixture is dissolved in 5 M HCl before spiking with authentic standards (Shin et al., 2002). More complex systems such as real meat or meat juice require homogenization in sodium hydroxide solution (Fay et al., 1997; Pais et al., 1999). After extraction/dissolution, diatomaceous earth is used for the preliminary cleaning of the crude extract.

Most SPE of HAs use a combination of a strong cationic exchange cartridge (PRS) and a reverse phase (C-18) cartridge. Toribio and coworkers' (Toribio et al., 2000) comprehensive study in 2000 provides valuable information about the suitability of different commercial brands and the structures of sorbents (PRS and C-18 cartridges) in terms of operation and recoveries of analytes from a lyophilized beef extract. An alternative SPE method employs the LiChrolut EN material (Vollenbroker and Eichner, 2000), which enables very quick determination of trace HAs in meat extracts. This may be most suitable for routine screening of food samples. However, the recoveries for nonpolar HAs, harman and norharman, were poor (20–30%) (Pesek and Matyska, 2000).

The recoveries of target analytes should always be checked to ensure the effectiveness and reliability of the available extraction and cleaning method. It is generally done by spiking the sample with specific amounts of known HA standards at the beginning and allows the spiked and unspiked samples to go through the same extraction process.

5.5.2   HIGH-PERFORMANCE LIQUID CHROMATOGRAPHY-UV, HPLC-FLUORESCENCE, AND HPLC-ELECTROCHEMICAL DETECTOR

High-Performance Liquid Chromatography-UV has undoubtedly been the most common coupled technique for the analysis of HAs formed in foods. For known HAs, the identification is achieved by comparison of UV spectra and retention time with those of an authentic standard. For fluorescent HAs (nonpolar HAs and PhIP), detection can be achieved by fluorescence detection, which has been found to be 100–400 times more sensitive than UV detection (Loprieno et al., 1991; Schwarzenbach and Gubler, 1992). Electrochemical detectors (ECD), with their relatively high selectivity and sensitivity, are suitable for the analyses of HAs in complex sample matrices (Billedeau et al., 1991). However, the EC detection is limited by the lack of online peak confirmation (Skog, 2002). In addition, simultaneous analysis of all HAs in one run is difficult using ECD as gradient elution cannot be applied in the high sensitivity range (Kataoka, 1997).

5.5.3   HIGH-PERFORMANCE LIQUID CHROMATOGRAPHY/MASS SPECTROMETRY

Liquid chromatography/mass spectrometry (LC/MS), by virtue of its elegant combination of LC's gentle separation and high sensitivity and selectivity of MS, is suggested as the most suitable technique for the analysis of multiple HAs, especially in complex food matrices (Kataoka, 1997; Santos et al., 2004). As HAs are stable during the ionization process, the protonated molecular ion peak may thus serve as the marker in mass selective detection of HAs (Fay et al., 1997). The selectivity of MS detection can be further enhanced by tandem MS (MS/MS), which has also been shown to be the most satisfactory for analyzing samples containing trace amounts of HAs (Santos et al., 2004).

The quantitative LC/MS analyses typically incorporate an internal standard, which provides a control over variability over LC injection and ionization. It is important to ensure that the internal standard chosen is not present in the target sample. TriMeIQx has been an internal standard of choice, and it is added to the sample directly before being injected into the LC/MS system (Messner and Murkovic, 2004; Santos et al., 2004). Some studies also used isotopically labeled internal standards, typically with $^2H_3$, the so-called isotope dilution technique (Fay et al., 1997). Although the latter can improve quantitative analysis, only few isotope-labeled analogues are commercially available (Santos et al., 2004).

Different types of soft ionization interfaces have been developed for coupling LC to MS in the analysis of HAs: thermospray ionization (TSI), atmospheric pressure chemical ionization (APCI), and electrospray ionization (ESI). TSI was used in earlier studies, and is capable of detecting a single HA from small amounts of samples from fish and beef (0.3 ng/g) and tryptophan pyrolysates (0.1 ng/g) (Turesky et al., 1988). In recent years, soft ionization techniques such as ESI and APCI have largely replaced TSI in the analysis of HAs, with the former being the most popular. ESI is suitable for analyzing more polar Has while APCI is suitable for less polar ones, so they may be considered complementary for HAs analysis (Christian, 2004; Pais et al., 1997). Due to its "softness," LC/ESI/MS is particularly useful for the detection of protonated molecular ions [M + H]$^+$ (Pais et al., 1997). Using LC/ESI/MS, 15 HAs were simultaneously measured and their identity is confirmed (Fay et al., 1997). Hence, LC/ESI/MS is a powerful technique for HAs analysis.

Quadrupole and ion trap have been the most popular mass analyzers for LC/MS of HAs. For quantitative purpose, peak area is suggested to be a better parameter (Santos et al., 2004). Recently, quadruple time of flight (Q-TOF), which enables data acquisition with a much higher mass accuracy (<2 mDa), has also been applied in both V-Optics™ and W-Optics™ configurations (Barceló-Barrachina et al., 2004). The former can achieve better sensitivity and lower detection limits, while the latter provides much higher mass resolution (~2 times the former), thus valuable for quantitation and studying fragmentation of HAs, respectively (Barceló-Barrachina et al., 2004).

5.5.4 Gas Chromatography/Mass Spectrometry

GC/MS is a powerful, yet inexpensive tool for the analysis of volatile compounds. GC/MS has been used for the analysis of HAs in different matrices, such as food products, cooking fumes, or process flavors. However, as most HAs are nonvolatile, their determination typically requires a derivatization step prior to GC/MS analysis. Although derivatization implies extra sample preparation, this step actually offers multiple advantages, improved selectivity, sensitivity, and separation, in addition to reduced polarity and improved volatility (Kataoka, 1996, 1997). Alternatively, HAs can be directly determined, without prior derivatization, by incorporating isotopically labeled internal standards in the samples. However, appropriate standards may not be always available (Pais and Knize, 2000).

Derivatization reactions for HAs can be divided into four types: alkylation, acylation, condensation to the Schiff base, and halogenation (Janoszka et al., 2003; Kataoka, 1997). Schiff base-type reaction has been most popular for the analysis of HAs from real food samples. Its application for the derivatization of HAs was first developed by Kataoka and Kijima (1997), who used *N,N*-dimethylformamide dimethyl acetal (DMFA-DMA) as the derivatization agent, giving rise to *N,N*-dimethylaminomethylene derivatives of HAs. Apart from DMFA-DMA, other dimethyl formamide dialkyl acetals (DMFA-diethylacetal, DMFA-diisopropylacetal, and DMFA-di-*tert*-butylacetal) have also been tested and it was found that derivatization with DMFA-di-*tert*-butylacetal provided the best yield and thus highest sensitivity in mass spectrometric determination (Barceló-Barrachina et al., 2005). Schiff-base type condensation is fast, simple, and economical, and forms stable derivatives (for 6 months at 4°C) (Janoszka et al., 2003), and is thus an ideal method for the analysis of HAs.

Positive ion electron ionization (EI) and negative ion chemical ionization (CI) are popular ionization interfaces. As HAs are all aromatic compounds, ionization with EI may be able to retain their molecular ion peak in their mass spectra in addition to providing information about their characteristic fragmentation patterns (Christian, 2004). Using GC/EI/MS, five AIAs were qualitatively confirmed based on their retention time and mass spectra (as PFPA derivatives) (Warzecha et al., 2004). However, GC/MS detection of PhIP was found to be problematic due to the low yield of derivatization and the presence of coeluted substances. CI is a softer ionization technique and the molecular ion peak is usually the dominant one in the mass spectra (Christian, 2004). Negative ion CI (NICI) is highly sensitive and selective, and in the selected ion monitoring (SIM) mode it detects HAs in high-temperature cooked meats with the detection limits of 0.05–0.2 ng/g (Murray et al., 1993; Pais and Knize, 2000). Similar detection limits (0.03–0.2 ng/g) were also obtained for grilled chicken (Tikkanen et al., 1996). Nevertheless, NICI has poor reproducibility, which prevents its wide utilization in the analysis of HAs in meat samples.

5.6 POSSIBLE APPROACHES TO CONTROL LEVELS OF HETEROCYCLIC AMINES IN FOODS

The findings that HAs are potent mutagens and probable human carcinogens have led to extensive research efforts to develop strategies to reduce risks of certain cancers in relation to this group of toxicants. There are mainly two approaches to reduce their adverse health effect. First is to abrogate their mutagenic/carcinogenic activity after entry of the purported mutagens into biological systems. This approach ranges from reducing the bioavailability of HAs, modulating the enzyme systems involved in HA metabolic activation, to regulation at genetic levels (Dashwood, 2002). The second one is related to the inhibition of HA formation in food. Approaches to control levels of HAs in foods are more desirable since they directly reduce our dietary exposure to mutagenic HAs. As cooked (especially high-temperature heated) meat and fish are considered principal sources of HAs (Murkovic, 2004), reducing the intake of these foods (Dashwood, 2002) is suggested as one of the feasible protective strategies in populations that have alternative primary food choices. Given that HA formation is affected by heating temperature, cooking time, and precursor concentrations, these parameters have been targets for developing inhibitory strategies. However, alternations of these parameters may be impractical due to their effect on organoleptic properties.

Recent studies, however, have explored the use of food additives, especially phytochemicals or plant extracts or tissues to reduce the formation of HAs in foods. Indeed, the identification of potent inhibitors has become one of the dominant directions in relation to protection from food-derived HAs. Inhibitors should fulfill the following criteria: (a) capable of reducing the total HA content at low doses; (b) not leading to the formation of new HAs; (c) not inducing the formation of other genotoxic substances; and (d) not negatively affecting the organoleptic properties. As antioxidant activity has been the most popular mechanism of inhibition proposed, polyphenols occupy the largest proportion of putative phytochemical inhibitors investigated (Dashwood, 2002; Oguri et al., 1998; Vitaglione et al., 2002). In this part of the chapter, we will provide an overview of potential inhibitors that can minimize HA occurrence in foods.

5.6.1 Synthetic Antioxidants

The impact of synthetic antioxidants such as butylated hydroxyanisole (BHA), butylated hydroxytoluene (BHT), propyl gallate (PG), and *tert*-butylhydroquinone (TBHQ) on the formation of HAs has been extensively investigated. Interestingly, these synthetic antioxidants were found to inhibit or enhance HA formation, depending on the experimental system employed. In real food systems, BHA, PG, and TBHQ at 100 ppm were found to inhibit HA formation (Chen, 1988). BHT also inhibits HA formation in another study using a real food system (Lan et al., 2004). In contrast, in simple model systems constituting pure HA precursors, BHT and TBHQ exerted a promoting effect on HA formation. In particular, TBHQ increased MeIQx formation by more than 200% at 100 ppm (Johansson and Jagerstad, 1996; Pearson et al., 1992; Vitaglione and Fogliano, 2004). Investigation of BHA and PG in simple model systems showed that they suppressed the formation of IQx-type HAs (Kato et al., 1996). Overall, the inconsistent effects of synthetic antioxidants do not warrant their use in foods to reduce HA formation.

5.6.2 Natural Antioxidants

A wide range of natural compounds have been evaluated for their effects on the formation of HAs in model systems and in real foods. These phytochemicals included antioxidant vitamins, phenolics, and carotenoids. Vitamin C and α-tocopherol did not demonstrate consistent effects when added to different real food systems (Kikugawa et al., 2000; Tai et al., 2001). Polyphenols, however, are among the most promising class of compounds in reducing the HA content in foods. Cheng et al. (2007) studied the effect of 12 food-derived phenolic antioxidants on the formation of HAs in simple chemical model systems and in fried beef patties. Remarkably, theaflavin 3, 3′-digallate, epicatechin gallate, rosmarinic acid, and naringenin were capable of simultaneously reducing the levels of PhIP, MeIQx, and 4,8-DiMeIQx. Among these polyphenols, naringenin was the most promising inhibitor. Lee and coworkers (Lee et al., 1992) also found that flavones inhibited IQ-type mutagen formation in simple model systems. Another example is curcumin, the active principle of turmeric, exhibited a dose-dependent inhibition of mutagenic HA formation in several model systems (Kolpe et al., 2002; Persson et al., 2003). Interestingly, no significant correlation was found between the antioxidant activity of polyphenols (assessed by Trolox equivalent antioxidant capacity assay) and their inhibitory effect on PhIP formation in chemical model systems (Cheng et al., 2007). This suggests that alternative mechanisms underlie the inhibitory effect of polyphenols, and also partly explains the relatively weak effect of synthetic antioxidants.

5.6.3 Plant Extracts

Given that many plants are rich sources of polyphenols, it is not surprising that certain plant extracts have been reported to effectively reduce the formation of HAs. Cheng et al. (2007) showed that apple and grape seed extracts (0.1%) were very effective in reducing both the individual (MeIQx, 4,8-DiMeIQx, and PhIP) and total HA content (>70%) in fried beef patties. Activity-guided fractionation using a chemical model system identified proanthocyanidins and phloridzin to be the dominant inhibitors in grape seed and apple extract, respectively. Murkovic and coworkers (1998) reported that application of dried rosemary, thyme, sage, and garlic to the surface of meat prior to heating resulted in significant reduction in HA contents, but to various extents with respect to different HAs. A recent study by Ahn and Gruen (2005) also demonstrated the effectiveness of rosemary and pycnogenol extracts in suppressing the formation of both polar and nonpolar HAs. Tea extracts have been extensively studied, but mainly for their effects on HA-induced mutagenicity or carcinogenicity (Bu-Abbas et al., 1994; Stavric et al., 1996) and bioavailability (Krul et al., 2001). However, some of these extracts were also shown to exert an enhancing effect on PhIP formation in chemical model systems (Zochling et al., 2002). The apparent discrepancy in the effects of antioxidant phytochemicals on HA formation might further emphasize the importance of clarifying the mechanism(s) involved in their inhibitory activity. A possible interpretation could be that in addition to antioxidant activity, they possess other concurring modulating activities which may account for their activity, but have not been taken into consideration during analysis of the experimental results.

5.7 MECHANISM OF INHIBITION OF HETEROCYCLIC AMINES FORMATION

Polyphenols are the most promising class of natural products that strongly inhibit the formation of HAs. Most mechanistic studies have ascribed inhibitory effects of polyphenols on HA formation to their free radical-scavenging activities (Pearson et al., 1992; Tsen et al., 2006). However, these studies neither examined a wide range of polyphenols with differential antioxidant activity, nor attempted to understand the relationship between these two parameters. As mentioned previously, there is a lack of correlation between the HA-inhibitory potency and antioxidant activity of 12 structurally diverse polyphenols, indicating that alternative, antioxidant-independent mechanisms may exist (Cheng et al., 2007). For example, all tea polyphenols exhibit high radical-scavenging capacities but not all of them are capable of inhibiting PhIP formation. Conversely, naringenin, a weak antioxidant, is a more potent inhibitor of PhIP formation than all the tea polyphenols evaluated. These data suggest that radical scavenging may not be the principal mechanism of intervention executed by these polyphenols.

Apart from free radicals, the Maillard reaction intermediates, such as reactive carbonyl species (RCS), generated from thermal and/or Strecker degradation reactions are also key intermediates in the formation of HAs (Cheng et al., 2006; Jagerstad et al., 1998; Kikugawa, 1999; Pearson et al., 1992). Cheng et al. (2008) hypothesized that the scavenging of RCS by polyphenols to form stable products may divert them from pathways that lead to the formation of HAs. To evaluate this hypothesis, key intermediates formed in PhIP-chemical model systems in the presence of naringenin were extensively characterized by HPLC-MS and NMR. This led to the identification of two novel adducts, 8-C-(E-phenylethenyl)naringenin and 6-C-(E-phenylethenyl) naringenin, corresponding to stable electrophilic substitution products between naringenin and phenylacetaldehyde, a key intermediate in the formation of PhIP. Stable isotope-labeling revealed phenylalanine as the source of reactive fragments that form adducts with naringenin, which prevents their further conversion into PhIP (Figure 5.3). Notably, these adducts were identified in fried beef patties treated with narin-genin, suggesting that the phenylacetaldehyde-scavenging mechanism is also opera-tive in real food systems and contributes to inhibited PhIP formation. Subsequently, Cheng et al. (2009) demonstrated that EGCG, a well-known antioxidant in tea, was also capable of directly trapping phenylacetaldehyde to form stable adducts in experiment models of PhIP formation, suggesting that EGCG also inhibited PhIP formation via a free radical-scavenging-independent mechanism. Crucially, it was found that EGCG peracetate, an EGCG derivative devoid of free radical-scavenging activity, also retained the capability to scavenge phenylacetaldehyde and was equally effective in inhibiting PhIP formation as EGCG. These data convincingly demon-strate that hydroxyl substituents, essential for the free radical-scavenging function-ality, do not significantly contribute to inhibition of PhIP formation by the studied polyphenols. Instead, the dominant mechanism(s) may be mediated via scavenging/trapping of PhIP Maillard intermediates.

These recent mechanistic breakthroughs showed that certain polyphenols inhibit HA formation by directly participating in the Maillard reaction. It would be of inter-est to thus research whether similar mechanism(s) of action also account for the potent negative influence of the HA-formation inhibitors (e.g., naringenin, EGCG,

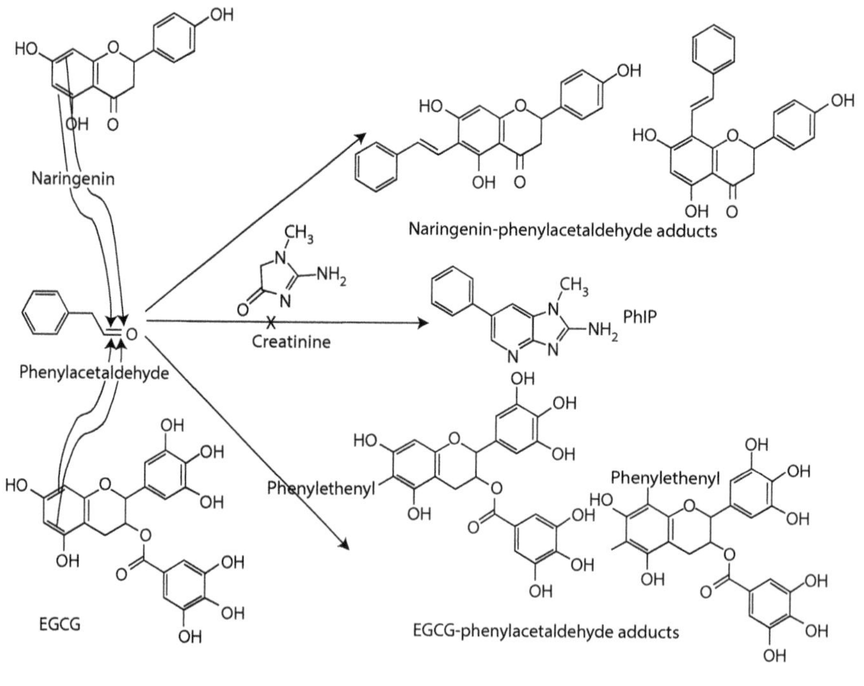

FIGURE 5.3 Postulated pathways for the inhibition of naringenin and EGCG on PhIP formation.

phloridzin, etc.) on the formation of other mutagenic HAs, such as imidazole quinolines, and imidazole quinoxalines. Also important is to investigate whether these novel polyphenol-RCS derivatives are safe and nongenotoxic.

REFERENCES

Abdulkarim, B. G., and Smith, J. S. 1998. Heterocyclic amines in fresh and processed meat products. *J. Agric. Food Chem., 46*, 4680–4687.

Adamson, R. H., Takayama, S., Sugimura, T., and Thorgeirsson, U. P. 1994. Induction of hepatocellular carcinoma in nonhuman primates by the food mutagen 2-amino-3-methylimidazo-[4,5-f]quinoline. *Environ. Health Perspect., 102*, 190–193.

Ahn, J., and Grun, I. U. 2005. Heterocyclic amines: 2. Inhibitory effects of natural extracts on the formation of polar and nonpolar heterocyclic amines in cooked beef. *J. Food Sci., 70*(4), C263–C268.

Ames, B. N., McCann, J., and Yamasaki, E. 1975. Methods for detecting carcinogens and mutagens with the Salmonella/mammalian-microsome mutagenicity test. *Mutat. Res., 31*, 347–364.

Baranczewski, P., Ustafsson, J.-A., and Moller, L. 2004. DNA adduct formation of 14 heterocyclic aromatic amines in mouse tissue after oral administration and characterization of the DNA adduct formed by 2-amino-9H-pyrido[2,3-b]indole (AαC), analysed by 32PHPLC. *Biomarkers, 9*, 243–257.

Barceló-Barrachina, E., Moyano, E., and Galceran, M. T. 2004. Determination of heterocyclic amines by liquid chromatography–quadrupole time-of-flight mass spectrometry. *J. Chromatogr. A, 1054*, 409–418.

Barceló-Barrachina, E., Santos, F. J., Puignou, L., and Galceran, M. T. 2005. Comparison of dimethyl-formamide dialkylacetal derivatization reagents for the analysis of heterocyclic amines in meat extracts by gas chromatography–mass spectrometry. *Anal. Chim. Acta., 545*, 209–217.

Billedeau, S. M., Bryant, M. S., and Holder, C. L. 1991. Analysis of heterocyclic amines using reversed-phase high performance liquid chromatography with electrochemical detection. *LC-GC, 9*, 116–120.

Brittebo, E. B., Karlsson, A. A., Skog, K. I., and Jagerstad, I. M. 1994. Transfer of the food mutagen PhIP to fetuses and newborn mice following maternal exposure. *Food Chem. Toxicol., 32*, 717–726.

Bu-Abbas, A., Clifford, M. N., Walker, R., and Ioannides, C. 1994. Marked antimutagenic potential of aqueous green tea extracts—Mechanism of action. *Mutagenesis, 9*, 325–331.

Chen, C. 1988. Factors influencing mutagen formation during frying of ground beef. Ph.D. Thesis, Michigan State University, East Lansing, Michigan.

Cheng, K. W., Chen, F., and Wang, M. 2006. Heterocyclic amines: Chemistry and health. *Mol. Nutr. Food Res., 50*, 1150–1170.

Cheng, K. W., Chen, F., and Wang, M. 2007. Inhibitory activities of dietary phenolic compounds on heterocyclic amine formation in both chemical model system and beef patties. *Mol. Nutr. Food Res., 51*(8), 969–976.

Cheng, K. W., Wong, C. C., Chao, J., Lo, C., Chen, F., Chu, I. K., Che, C. M., Ho, C. T., and Wang, M. 2009. Inhibition of mutagenic PhIP formation by epigallocatechin gallate via scavenging of phenylacetaldehyde. *Mol. Nutr. Food Res., 53*(6), 716–725.

Cheng, K. W., Wong, C. C., Cho, C. K., Chu, I. K., Sze, K. H., Lo, C., Chen, F., and Wang, M. 2008. Trapping of phenylacetaldehyde as a key mechanism responsible for naringenin's inhibitory activity in mutagenic 2-amino-1-methyl-6-phenylimidazo[4,5-b]pyridine formation. *Chem. Res. Toxicol., 21*(10), 2026–2034.

Cheng, K. W., Wu, Q., Zheng, Z. P., Peng, X., Simon, J. E., Chen, F., and Wang, M. 2007. Inhibitory effect of fruit extracts on the formation of heterocyclic amines. *J. Agric. Food Chem., 55*(25), 10359–10365.

Cheung, C., Ma, X., Krausz, K. W., Kimura, S., Feigenbaum, L., Dalton, T. P., Nebert, D. W., Idle, J. R., and Gonzalez, F. J. 2005. Differential metabolism of 2-amino-1-methyl-6-phenylimidazo[4,5-b]pyridine (PhIP) in mice humanized for CYP1A1 and CYP1A2. *Chem. Res. Toxicol., 18*(9), 1471–1478.

Christian, G. D. 2004. *Analytical Chemistry*, pp. 604–642. Hoboken, NJ: Wiley.

Dashwood, R. H. 2002. Modulation of Heterocyclic amine-induced mutagenicity and carcinogenecity: An "A-to-Z" guide to chemopreventive agents, promoters, and transgenic models. *Mutat. Res., 511*, 89–112.

Davis, C. D., Dacquel, E. J., Schut, H. A. J., Thorgeirsson, S. S., and Snyderwine, E. G. 1996. In vivo mutagenicity and DNA adduct levels of heterocyclic amines in Muta mice and c myc/lacZ double transgenic mice. *Mutat. Res., 356*, 287–296.

Edwards, R. J., Murray, B. P., Murray, S., Schulz, T., Neubert, D., Gant, T. W., Thorgeirsson, S. S., Boobis, A. R., and Davies, D. S. 1994. Contribution of CYP1A1 and CYP1A2 to the activation of heterocyclic amines in monkeys and human. *Carcinogenesis, 15*, 829–836.

Eisenbrand, G., and Tang, W. 1993. Food-borne Heterocyclic amines—Chemistry, formation, occurrence and biological activities—A literature review. *Toxicology, 84*, 1–82.

Fay, L. B., Ali, S., and Gross, G. A. 1997. Determination of Heterocyclic aromatic amines in food products: Automation of the sample preparation method prior to HPLC and HPLC-MS quantification. *Mutat. Res., 376*, 29–35.

Felton, J. S., and Knize, M. G. 1990. Heterocyclic amine mutagens/carcinogens in foods. In C. S. Cooper and P. L. Grover (Eds.), *Chemical-Carcinogenesis and mutagenesis I*, pp. 471–502. New York: Springer.

Ghoshal, A., Davis, C. D., Schut, H. A., and Snyderwine, E. G. 1995. Possible mechanisms for PhIP-DNA adduct formation in the mammary gland of female Sprague–Dawley rats. *Carcinogenesis, 16*, 2725–2731.

Gooderham, N. J., Murray, S., Lynch, A. M., Yadollahi-Farsani, M., Zhao, K., Boobis, A. R., and Davies, D. S. 2001. Food-derived heterocyclic amine mutagens: Variable metabolism and significance to humans. *Drug Metab. Disp., 29*, 529–534.

Henry, M. 2000. N. J. K. Simpson (Ed.), *Solid-Phase Extraction: Principles, Techniques, and Applications*, pp. 125–182. New York: Marcel Dekker Inc.

Holme, J. A., Alexander, J., and Dybing, E. 1987. Mutagenic activation of 2-amino-3-methylimidazo[4,5-f]-quinoline (IQ) and 2-amino-3,4-dimethylimidazo[4,5-f]-quinoline (MeIQ) by subcellular fractions and cells isolated from small intestine, kidney and liver of the rat. *Cell Biol. Toxicol., 3*(1), 51–61.

Jägerstad, M., and Skog, K. 2005. Review—Genotoxicity of heat-processed foods. *Mutat. Res., 574*, 156–172.

Jagerstad, M., Skog, K., Arvidsson, P., and Solyakov, A. 1998. Chemistry, formation and occurrence of genotoxic heterocyclic amines identified in model systems and cooked foods. *Zeitschrift Fur Lebensmittel-Untersuchung Und-Forschung a—Food Res. Technol., 207*, 419–427.

Jamin, E. L., Arquier, D., Canlet, C., Rathahao, E., Tulliez, J., and Debrauwer, L. 2007. New insights in the formation of deoxynucleoside adducts with the heterocyclic aromatic amines PhIP and IQ by means of ion trap MSn and accurate mass measurement of fragment ions. *J. Am. Soc. Mass Spectrom., 18*(12), 2107–2118.

Janoszka, B., Blaszczyk, U., Warzecha, L., Luks-Betlej, K., and Strozyk, M. 2003. The analysis of aminoazaarenes as their derivatives with GC-MS method in the heat-processed meat samples. *Chem. Anal., 48*, 707–721.

Johansson, M. 1995. Influence of lipids, and pro- and antioxidants on the yield of carcinogenic heterocyclic amines in cooked foods and model systems. Lund University, Sweden.

Johansson, M. A. E., and Jagerstad, M. 1996. Influence of pro- and antioxidants on the formation of mutagenic-carcinogenic heterocyclic amines in a model system. *Food Chem.,* *56,* 69–75.

Kaderlik, K. R., Minchin, R. F., Mulder, G. J., Ilett, K. F., Daugaard-Jenson, M., Teitel, C. H., and Kadlubar, F. F. 1994. Metabolic activation pathway for the formation of DNA adducts of the carcinogen 2-amino-1-methyl-6-phenylimidazo[4,5-b]pyridine (PhIP) in rat extrahepatic tissues. *Carcinogenesis, 15*(8), 1703–1709.

Kadlubar, F. F., and Beland, F. A. 1985. Chemical properties of ultimate carcinogenic metabolites of arylamines and arylamides. In R. G. Harvey (Ed.), *Polycyclic Hydrocarbons and Carcinogenesis,* vol. 283, pp. 341–370. Washington, DC: American Chemical Society.

Kakiuchi, H., Watanabe, M., Ushijima, T., Toyota, M., Imai, K., Weisburger, J. H., Sugimura, T., and Nagao, M. 1995. Specific 5′-GGGA-3′ → 5′-GGA-3′ mutation of the Apc gene in rat colon tumors induced by 2-amino-1-methyl-6-phenylimidazo[4,5-b]pyridine. *Proc. Natl. Acad. Sci. USA, 92*(3), 910–914.

Kataoka, H. 1996. Derivatization reactions for the determination of amines by gas chromatography and their applications in environmental analysis. *J. Chromatogr. A, 733,* 19–34.

Kataoka, H. 1997. Methods for the determination of mutagenic heterocyclic amines and their applications in environmental analysis. *J. Chromatogr. A, 774,* 121–142.

Kataoka, H., and Kijima, K. 1997. Analysis of heterocyclic amines as their N-dimethylaminothylene derivatives by gas chromatography with nitrogen-phosphorus selective detection. *J. Chromatogr. A, 767,* 187–194.

Kato, T., Harashima, T., Moriya, N., Kikugawa, K., and Hiramoto, K. 1996. Formation of the mutagenic/carcinogenic imidazo quinoxaline-type heterocyclic amines through the unstable free radical Maillard intermediates and its inhibition by phenolic antioxidants. *Carcinogenesis, 17*(11), 2469–2476.

Kikugawa, K. 1999. Involvement of free radicals in the formation of heterocyclic amines and prevention by antioxidants. *Cancer Lett., 143,* 123–126.

Kikugawa, K., Hiramoto, K., and Kato, T. 2000. Prevention of the formation of Mutagenic and/or carcinogenic heterocyclic amines by food factors. *Biofactors, 12,* 123–127.

Kolpe, U., Ramaswamy, V., Satish Rao, B. S., and Nagabhushan, M. 2002. Turmeric and curcumin prevents the formation of mutagenic Maillard reaction products. In *Int. Congr. Ser.,* vol. 1245, pp. 327–334.

Kosakarn, P., Halliday, J. A., Glickman, B. W., and Josephy, P. D. 1993. Mutational specificity of 2-nitro-3,4-dimethylimidazo[4,5-f]quinoline in the lacI gene of *Escherichia coli. Carcinogenesis, 14*(3), 511–517.

Krul, C., Luiten-Schuite, A., Tenfelde, A., van Ommen, B., Verhagen, H., and Havenaar, R. 2001. Antimutagenic activity of green tea and black tea extracts studied in a dynamic in vitro gastrointestinal model. *Mutat. Res., 474,* 71–85.

Lan, C. M., Kao, T. H., and Chen, B. H. 2004. Effects of heating time and antioxidants on the formation of heterocyclic amines in marinated foods. *J. Chromatogr. B—Anal. Technol. Biomed. Life Sci., 802,* 27–37.

Lang, N. P., Butler, M. A., Massengill, J., Lawson, M., Stotts, R. C., Hauer-Jensen, M., and Kadlubar, F. F. 1994. Rapid metabolic phenotypes for acetyltransferase and cytochrome P4501A2 and putative exposure to food-borne heterocyclic amines increase the risk for colorectal cancer or polyps. *Cancer Epidemiol. Biomarkers Prev., 3*(8), 675–682.

Lee, H., Chyr-Yir, J., and Tsai, S.-J. 1992. Flavone inhibits mutagen formation during heating in a glycine creatine glucose model system. *Food Chem. Toxicol., 45,* 235–238.

Lee, H., and Shih, M.-K. 1995. Mutational specificity of 2-amino-3-methylimidazo-[4,5-f] quinoline in the hprt locus of CHO-K1 cells. *Mol. Carcinogen., 13,* 122–127.

Liu, Y., and Levy, G. N. 1998. Activation of heterocyclic amines by combinations of prostaglandin H synthase-1 and -2 with N-acetyltransferase 1 and 2. *Cancer Lett., 133,* 115–123.

Loprieno, N., Boncristiani, G., and Loprieno, G. 1991. An experimental approach to identifying the genotoxic risk from cooked meat mutagens. *Food Chem. Toxicol., 29*(6), 377–386.

Messner, C., and Murkovic, M. 2004. Evaluation of a new model system for studying the formation of heterocyclic amines. *J. Chromatogr. B, 802,* 19–26.

Milic, B. L., Djilas, S. M., and Canadanovicbrunet, J. M. 1993. Synthesis of some heterocyclic aminoimidazoazarenes. *Food Chem., 46*(3), 273–276.

Minchin, R. F., Kadlubar, F. F., and Ilett, K. F. 1993. Role of acetylation in colorectal cancer. *Mutat. Res., 290*(1), 35–42.

Morgenthaler, P. M., and Holzhauser, D. 1995. Analysis of mutations induced by 2-amino-1-methyl-6-phenylimidazo[4,5-b]pyridine (PhIP) in human lymphoblastoid cells. *Carcinogenesis, 16*(4), 713–718.

Murkovic, M. 2004. Formation of heterocyclic aromatic amines in model systems. *J. Chromatogr. B, 802,* 3–10.

Murkovic, M., Steinberger, D., and Pfannhauser, W. 1998. Antioxidant spices reduce the formation of heterocyclic amines in fried meat. *Zeitschrift Fur Lebensmittel-Untersuchung Und-Forschung a—Food Res. Technol., 207,* 477–480.

Murray, S., Lynch, A. M., Knize, M. G., and Gooderham, N. J. 1993. Quantification of the carcinogens 2-amino-3,8-dimethyland 2-amino-3,4,8-trimethylimidazo [4,5-f]quinoxaline and 2-amino-1-methyl-6-phenylimidazo[4,5-b]pyridine in food using a combined assay based on gas chromatography-negative ion mass spectrometry. *J. Chromatogr. A, 616,* 211–219.

Oguri, A., Suda, M., Totsuka, Y., Sugimura, T., and Wakabayashi, K. 1998. Inhibitory effects of antioxidants on formation of heterocyclic amines. *Mutat. Res.—Fund. Mol. Mech. Mutagenesis, 402,* 237–245.

Ohgaki, H., Takayama, S., and Sugimura, T. 1991. Carcinogenicities of heterocyclic amines in cooked food. *Mutat. Res., 259*(3–4), 399–410.

Pais, P., and Knize, M. G. 2000. Chromatographic and related techniques for the determination of aromatic heterocyclic amines in foods. *J. Chromatogr. B, 747,* 139–169.

Pais, P., Moyano, E., Puignou, L., and Galceran, M. T. 1997. Liquid chromatography-atmospheric pressure chemical ionization mass spectrometry as a routine method for the analysis of mutagenic amines in beef extracts. *J. Chromatogr. A, 778,* 207–218.

Pais, P., Salmon, C. P., Knize, M. G., and Felton, J. S. 1999. Formation of mutagenic/carcinogenic heterocyclic amines in dry-heated model systems, meats, and meat drippings. *J. Agric. Food Chem., 47*(3), 1098–1108.

Pearson, A. M., Chen, C., Gray, J. I., and Aust, S. D. 1992. Mechanism(s) involved in meat mutagen formation and inhibition. *Free Radic. Biol. Med., 13,* 161–167.

Persson, E., Graziani, G., Ferracane, R., Fogliano, V., and Skog, K. 2003. Influence of antioxidants in virgin olive oil on the formation of heterocyclic amines in fried beefburgers. *Food Chem. Toxicol., 41,* 1587–1597.

Pesek, J. J., and Matyska, M. T. 2000. In N. J. K. Simpson (Ed.), *Solid-Phase Extraction: Principles, Techniques, and Applications,* pp. 19–38. New York: Marcel Dekker Inc.

Ryu, D. Y., Pratt, V. S., Davis, C. D., Schut, H. A., and Snyderwine, E. G. 1999. In vivo mutagenicity and hepatocarcinogenicity of 2-amino-3,8-dimethylimidazo[4,5-f]quinoxaline (MeIQx) in bitransgenic c-myc/lambda lacZ mice. *Cancer Res., 59,* 2587–2592.

Santos, F. S., Barceló-Barrachina, E., Toribio, F., Puignou, L., Galceran, M. T., Persson, E., Skog, K., Messner, C., Murkovic, M., Nabinger, U., and Ristic, A. 2004. Analysis of heterocyclic amines in food products: Interlaboratory studies. *J. Chromatogr. B, 802,* 69–78.

Schiffman, M. H., and Felton, J. S. 1990. Re: "Fried foods and the risk of colon cancer". *Am. J. Epidemiol., 131*(2), 376–378.

Schut, H. A., and Snyderwine, E. G. 1999. DNA adducts of heterocyclic amine food mutagens: Implications for mutagenesis and carcinogenesis. *Carcinogenesis, 20*(3), 353–368.

Schwarzenbach, R., and Gubler, D. 1992. Detection of heterocyclic aromatic amines in food flavors. *J. Chromatogr. A, 624*, 491–495.

Shan, L., Yu, M., Schut, H. A., and Snyderwine, E. G. 2004. Susceptibility of rats to mammary gland carcinogenesis by the food-derived carcinogen 2-amino-1-methyl-6-phenylimidazo[4,5-b]pyridine (PhIP) varies with age and is associated with the induction of differential gene expression. *Am. J. Pathol., 165*, 191–202.

Shin, H. S., Strasburg, G. M., and Gray, J. I. 2002. A model system study of the inhibition of heterocyclic aromatic amine formation by organosulfur compounds. *J. Agric. Food Chem., 50*(26), 7684–7690.

Simpson, N. J. K., and Wynne, P. M. 2000. The sample matrix and its influence on method development. In N. J. K. Simpson (Ed.), *Solid-Phase Extraction: Principles, Techniques, and Applications*, pp. 39–95. New York: Marcel Dekker Inc.

Skog, K. 1993. Cooking procedures and food mutagens: A literature review. *Food Chem. Toxicol., 31*, 655–675.

Skog, K. 2002. Problems associated with the determination of heterocyclic amines in cooked foods and human exposure. *Food Chem. Toxicol., 40*, 1197–1203.

Skog, K., Augustsson, K., Steineck, G., Stenberg, M., and Jagerstad, M. 1997. Polar and non-polar heterocyclic amines in cooked fish and meat products and their corresponding pan residues. *Food Chem. Toxicol., 35*, 555–565.

Skog, K., and Jagerstad, M. 1991. Effects of glucose on the formation of PhIP in a model system. *Carcinogenesis, 12*(12), 2297–2300.

Skog, K., Johansson, M. A., and Jägerstad, M. I. 1998. Carcinogenic heterocyclic amines in model systems and cooked food: A review on formation, occurrence and intake. *Food Chem. Toxicol., 36*, 879–896.

Skog, K., Solyakov, A., and Jagerstad, M. 2000. Effects of heating conditions and additives on the formation of Heterocyclic amines with reference to amino-carbolines in a meat juice model system. *Food Chem. Toxicol., 68*, 299–308.

Stavric, B., Matula, T. I., Klassen, R., and Downie, R. H. 1996. The effect of teas on the in vitro mutagenic potential of heterocyclic aromatic amines. *Food Chem. Toxicol., 34*, 515–523.

Sugimura, T. 1997. Overview of carcinogenic heterocyclic amines. *Mutat Res, 376*(1–2), 211–219.

Sugimura, T., Nagao, M., Kawachi, T., Honda, M., Yahagi, T., Seino, Y., Sato, S., Matsukara, N., Shirai, A., Sawamura, M., and Matsumoto, H. 1977. Mutagens-carcinogens in food, with special reference to highly mutagenic pyrolytic products in broiled foods. In H. H. Hiatt, J. D. Watson and W. J. A. (Eds.), *Origins of Human Cancer*, pp. 1561–1577. Cold Spring Harbour Laboratory.

Sugimura, T., Nagao, M., and Wakabayashi, K. 1982. Metabolic aspects of the comutagenic action of norharman. *Adv. Exp. Med. Biol. 136b*, 1011–1025.

Sugimura, T., Wakabayashi, K., Nakagama, H., and Nagao, M. 2004. Heterocyclic amines: Mutagens/carcinogens produced during coking of meat and fish. *Cancer Sci., 95*, 290–299.

Tai, C. Y., Lee, K. H., and Chen, B. H. 2001. Effects of various additives on the formation of heterocyclic amines in fried fish fibre. *Food Chem. Toxicol., 75*, 309–316.

Thompson, L. H., Tucker, J. D., Stewart, S. A., Christensen, M. L., Salazar, E. P., Carrano, A. V., and Felton, J. S. 1987. Genotoxicity of compounds from cooked beef in repair-deficient CHO cells versus Salmonella mutagenicity. *Mutagenesis, 2*(6), 483–487.

Tikkanen, L. M., Latva-Kala, K. J., and Heinio, R. L. 1996. Effect of commercial marinades on the mutagenic activity, sensory quality and amount of heterocyclic amines in chicken grilled under different conditions. *Food Chem. Toxicol., 34*, 725–730.

Toribio, F., Moyano, E., Puignou, L., and Galceran, M. L. 2000. Comparison of different commercial solid-phase extraction cartridges used to extract heterocyclic amines from a lyophilized meat extract. *J. Chromatogr. A, 880,* 101–112.

Tsen, S. Y., Ameri, F., and Smith, J. S. 2006. Effects of rosemary extracts on the reduction of heterocyclic amines in beef patties. *J. Food Sci., 71,* C469–C473.

Turesky, R. J., Bur, H., Huynh-Ba, T., Aeschbacher, H. U., and Milon, H. 1988. Analysis of mutagenic heterocyclic amines in cooked beef products by high-performance liquid chromatography in combination with mass spectrometry. *Food Chem. Toxicol., 26,* 501–509.

Vitaglione, P., and Fogliano, V. 2004. Use of antioxidants to minimize the human health risk associated to mutagenic/carcinogenic heterocyclic amines in food. *J. Chromatogr. B, 802,* 189–199.

Vitaglione, P., Monti, S. M., Ambrosino, P., Skog, K., and Fogliano, V. 2002. Carotenoids from tomatoes inhibit heterocyclic amine formation. *Eur. Food Res. Technol., 215*(2), 108–113.

Vollenbroker, M., and Eichner, K. 2000. A new quick solid-phase extraction method for the quantification of heterocyclic aromatic amines. *Eur. Food Res. Technol., 211,* 122–125.

Warzecha, L., Janoszka, B., Blaszczyk, U., Strozyk, M., Bodzek, D., and Dobosz, C. 2004. Determination of heterocyclic aromatic amines (HAs) content in samples of household-prepared meat dishes. *J. Chromatogr. B-Anal. Technol. Biomed. Life Sci., 802,* 95–106.

Weisburger, J. H. 1994. Specific Maillard reactions yield powerful mutagens and carcinogens. In T. P. Labuza, G. A. Reineccius, V. M. Monnier, J. O'Brien and J. W. Baynes (Eds.), *Maillard Reactions in Chemistry, Food, and Health,* pp. 335–340. Cambridge: The Royal Soc. Chem.

Wu, R. W., Wu, E. M., Thompson, L. H., and Felton, J. S. 1995. Identification of aprt gene mutations induced in repair-deficient and P450-expressing CHO cells by the food-related mutagen/carcinogen, PhIP. *Carcinogenesis, 16*(5), 1207–1213.

Yoshimoto, M., Tsutsumi, M., Iki, K., Sasaki, Y., Tsujiuchi, T., Sugimura, T., Wakabayashi, K., and Konishi, Y. 1999. Carcinogenicity of heterocyclic amines for the pancreatic duct epithelium in hamsters. *Cancer Lett., 143*(2), 235–239.

Zochling, S., Murkovic, M., and Pfannhauser, W. 2002. Effects of industrially produced flavours with pro- and antioxidative properties on the formation of the heterocyclic amine PhIP in a model system. *J. Biochem. Biophys. Meth., 53,* 37–44.

6 Chemistry and Safety of 3-MCPD

Zi Teng and Qin Wang

CONTENTS

6.1 INTRODUCTION

3-Monochloropropane-1,2-diol (3-MCPD) is an organic toxin that belongs to the chloropropanol family. It could be formed from a number of precursors, many of which are common food components and additives. Although it was known originally as a major contaminant in soy sauces, its occurrence was identified by later studies in a variety of other foods, such as bread, meat, and cheese. Although the content of 3-MCPD is generally low among these foods, its risk to human health should not be overlooked, because it demonstrates significant toxicity to the endocrine, urinal, lymphatic, nervous, and reproductive systems. This chapter describes the physicochemical properties, formation, occurrence, determination, toxicity, and mitigation of 3-MCPD in different food systems.

6.2 PHYSICOCHEMICAL PROPERTIES OF 3-MCPD

3-MCPD (3-monochloropropane-1,2-diol, 3-chloropropane-1,2-diol, or 3-monochloropropanediol) belongs to the chloropropanol or glycerol chlorohydrin category. It is so named because of the substitution of the 3-hydroxyl group in glycerol with a chlorine atom (Figure 6.1). 3-MCPD is a colorless, slightly oily liquid with a faint and pleasant odor, and it is readily soluble in water and ethanol. The boiling point and density of 3-MCPD are 213°C and 1.3204 g/cm^3 (20°C), respectively. 3-MCPD is a chiral molecule due to the chirality of the central carbon atom. It has been reported that the two enantiomers of 3-MCPD exhibit different biological activities: the

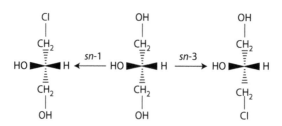

FIGURE 6.1 Chemical structure of (S)-(+)-3-chloro-propane-1,2-diol (left), glycerol (middle), and (R)-(−)-3-chloro-propane-1,2-diol (right).

(R)-isomer has a detrimental effect on the kidneys,[1] whereas the (S)-isomer exhibits antifertility activity.[2,3] The term "3-MCPD" that appears in regulatory documents usually refers to the racemic mixture.

3-MCPD is relatively unstable in aqueous media at basic pH, and it is decomposed to glycerol via the intermediate epoxide glycidol. This reaction is commonly utilized to reduce the level of 3-MCPD in commercial acid hydrolyzed vegetable proteins (acid-HVPs).

6.3 3-MCPD IN FOOD SYSTEMS

As shown in Figure 6.2, 3-MCPD can be formed by several food ingredients and additives under certain circumstances. While it is mostly found in abundance in acid-HVPs, its existence has been revealed in some conventional foods such as bread, meat, and beer. The formation of 3-MCPD requires one donor for its carbon backbone and another one for chlorine. Table 6.1 summarizes some common food systems that contain potential precursors for 3-MCPD. The following sections will discuss the major mechanism of 3-MCPD formation in food products, as well as its occurrence in different food systems.

6.3.1 FORMATION OF 3-MCPD

6.3.1.1 From Glycerol

Glycerol is a common humectant and sweetener in a wide range of food products, including baked foods, confectionary products, and beverages. It is also formed in processed foods through high-temperature hydrolysis of triglycerides. Glycerol can be converted into 3-MCPD by reacting with hydrochloric acid, sodium chloride,

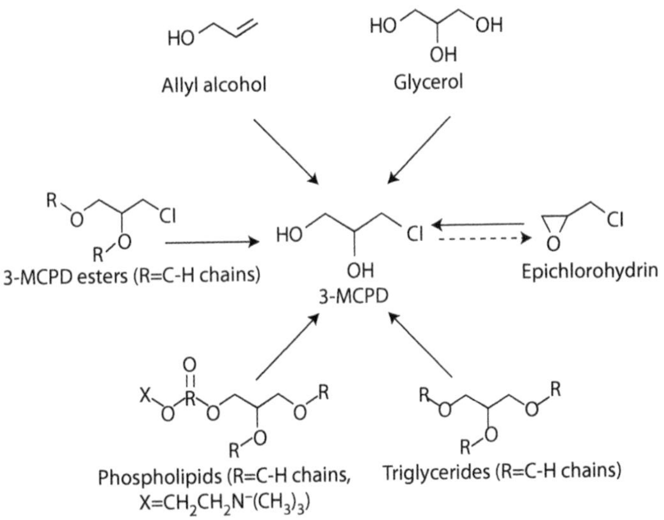

FIGURE 6.2 Possible precursors for 3-MCPD formation in food systems.

TABLE 6.1

Occurrence of 3-MCPD in Different Foods and Corresponding Precursors

Food Type	Carbon Source	Chlorine Source
Acid-HVPs, soy sauce, and oyster sauce[5]	Residual lipid	Hydrochloride
Bread, doughnuts, and hamburgers[12]	Glycerol and lipid	Added salt
Malt-derived foods[13]	Endogenous lipid	Endogenous chlorine
Coffee[13]	Endogenous lipid	Endogenous chlorine
Cheese[14]	Endogenous lipid	Endogenous chlorine
Cold-smoked products[4]	Cracked cellulose from wood	Added salt
Meat[14]	PAAE packaging	PAAE packaging
Salted fish[15]	Enzymatically degraded fats	Added salt
Drinking water[16]	PAAE flocculants	PAAE flocculants or hypochlorous acid

or dry hydrogen chloride gas (HCl).[4] Such reaction requires prolonged heating at temperature of about 100°C. Organic acids such as acetic acid may catalyze this reaction, but they are not necessary. The mechanism of such conversion in aqueous systems is most likely the nucleophilic substitution reaction (Sn2) of chloride anion on the backbone of glycerol.[5]

6.3.1.2 From Allyl Alcohol

3-MCPD may be formed by addition of hypochlorous acid (HOCl) across the double bond of allyl alcohol (prop-2-en-1-ol). Hypochlorous acid is found in chlorinated water, which is used in many food-processing procedures. Allyl alcohol is a thermal decomposition product of alliin [(S)-allyl-l-cysteine sulfoxide], a cysteine-derived amino acid existing in fresh garlic and onion.[6] The reaction proceeds rapidly at 50–60°C, producing 88% monochloropropanediols (preferably 3-MCPD according to Markovnikov's rules) and 9% dichloropropanols.[7] In the dehydration process to produce dry garlic or onion powders from the fresh produce (98% water), there is a considerable increase in the 3-MCPD level.

6.3.1.3 From Lipids

Acid-HVPs are a major source of 3-MCPD contamination, and the abundance of this compound is ascribed largely to the reaction between HCl and the residual lipids bound to vegetable proteins. The formation of 3-MCPD in acid-HVPs was studied by Collier et al.[5] by monitoring the reactions of glycerol, triacyl glycerols, and phospholipids with 5.5 M HCl. The yield of chloropropanediols, including 3-MCPD and 2-MCPD, was found to decrease in the following order: triacyl glycerols > phospholipids > glycerol. The latter two reactions might be initiated by partial hydrolysis of the ester group, followed by the substitution of the residual ester moiety by a chlorine anion. This mechanism is supported by the fact that the ratio of the isomers 3-MCPD/2-MCPD is in agreement with statistical substitution at two primary methylenes (CH_2) and one secondary methine (CH), that is, about 2:1.

6.3.1.4   From Carbohydrates and Their Derivatives

The existence of 3-MCPD in soy protein products was previously explained by the possible reaction between the carbohydrate in soya meal and the added HCl. However, Collier et al.[5] reported that the yields of 3-MCPD from two pure carbohydrates (soluble arabinoxylan and pectin with a degree of methylation at 33.7) were 98% lower than that obtained from crude soya meal. These substrates were incubated with HCl at 107°C for 16 h before testing. It was showed that the two sugars provided a comparable amount of carbohydrate with soya meal, but the lipid content of the pure carbohydrates was significantly lower. Therefore, it was proposed that carbohydrate is not a good precursor for 3-MCPD formation, and that the formation of 3-MCPD in carbohydrate-rich systems is probably due to the existence of residual lipids.

However, derivatives of sugar such as sucralose could be converted into 3-MCPD in the presence of glycerol.[8] Therefore, caution should be taken in the use of sucralose as a sweetening agent in baked foods, which contain glycerol or lipids, due to the potential formation of toxic chloropropanols.

6.3.1.5   Formation in Polyamidoamine-Epichlorohydrin Resins

Polyamidoamine-epichlorohydrin (PAAE) is a commercially produced chemical used in resins, textiles, and paper. In 2009, the annual global production of epichlorohydrin is estimated to be 2 billion tons. In food industry, PAAE serves mainly as a wet strength resin that improves the tensile strength of paper packages. A certain level of 2-chloromethyl-oxirane (epichlorohydrin, one of the monomers that constitute PAAE) is detected in PAAE-containing packages, and it exhibits an equilibrium with 1,3-dichloropropanol (1,3-DCP) in aqueous systems. A shift of the equilibrium toward 1,3-DCP can be induced in the presence of chlorine anion at acidic pH conditions. According to Collier et al.[5] 1,3-DCP may be formed from 3-MCPD with chloride anion in the presence of acetic acid as a catalyst. In addition, Boden et al. observed an interchange of chloropropanol species, including 1,3-DCP and 3-MCPD, which was affected with changing pH from 5 to 9.[9] These findings suggest a possible pathway of 3-MCPD formation from epichlorohydrin, with 1,3-DCP as an intermediate.

6.3.1.6   From Degradation of Esters In Vivo

Compared to the free chloropropanol that exists in limited categories of food products, esters of 3-MCPD are found in virtually all food products containing edible oils or fats, such as fried food items, margarine, and infant formulas. These esters may undergo degradation in vivo to produce free 3-MCPD under lipase-catalyzed hydrolysis. For instance, significant levels of 3-MCPD were released from bread on treatment with bakery-grade lipase in the baking process. In a model study that consisted of lipase, vegetable oil or fat, water, and NaCl, the generation of 3-MCPD was directly proportional to the lipase activity in the mixture.[10,11] Therefore, the residual lipase activities in certain foodstuffs containing salt and fat should be monitored, since they may be responsible for the formation of unwanted 3-MCPD during storage and processing. The fatty acid esters of 3-MCPD are discussed in a separate chapter of this book.

6.3.2   Occurrence of 3-MCPD in Commercial Food Systems

6.3.2.1   Acid-HVPs and Soy Sauce

By far, acid-HVPs have been the most abundant source of 3-MCPD among others. They are frequently added in condiments such as the soy sauce and oyster sauce as a flavor enhancer. Acid-HVPs are produced by boiling cereal (corn or wheat) or legume (such as soybean) proteins in HCl solution at a high temperature, followed by neutralization with sodium hydroxide. The resultant liquid contains free amino acids such as glutamic acid, which confers the food with an umami flavor. Chloropropanols (mainly 3-MCPD) are formed via the reaction of hydrochloric acid with the protein-bound lipids in the raw materials. Although few soy sauce manufacturers declare the addition of extraneous acid-HVPs, the boiling of raw cereals or beans with HCl is ubiquitously involved in the manufacturing process. Such procedure is similar to the production of acid-HVPs and, as a consequence, generates considerable amount of 3-MCPD.

Continuous surveys have been conducted to assess the level of 3-MCPD in commercial soy sauces and oyster sauces. In a Singaporean investigation carried out in 2005 (Table 6.2),[17] only three domestically manufactured sauces have declared the presence of HVP, two of which did not have quantifiable levels of 3-MCPD. Of the remaining 418 sauces that did not declare the usage of HVP, 61 sauces had 3-MCPD levels above 0.01 mg/kg, and 45 of them contained 3-MCPD above 0.02 mg/kg. The upper bound average 3-MCPD concentration for all the soy and oyster sauces was 0.45 mg/kg. By combining this value and the national average intake of soy sauce (at most 22 g/day), it was calculated that the daily intake of 3-MCPD from soy sauce was less than 8% of the tolerable daily intake (TDI). The highest 3-MCPD level of 110.8 mg/kg was found in an imported soy sauce.

A similar study was undertaken in the United States in 2003.[18] Of the 55 samples (soy sauce or oyster sauce) analyzed for 3-MCPD, 33 samples contained more than 0.025 mg/kg 3-MCPD, and 10 products (all of which were manufactured in Asia) contained 3-MCPD at above 1 mg/kg. None of the eight US-manufactured samples was detected with 1 mg/kg 3-MCPD. The highest level found was 876 mg/kg 3-MCPD, which was attributed to direct acid treatment of the soy sauce ingredients, high levels of residual lipids in the raw material before acid hydrolysis, and/or poor manufacturing control.

In the United Kingdom, three consecutive studies conducted in 1999, 2000, and 2002 revealed a steady reduction in the number of soy sauce samples containing more than 1 mg/kg 3-MCPD. In the 1999 survey, 48% and 23% of the samples contained more than 0.10 mg/kg and 1.0 mg/kg 3-MCPD, respectively.[19] These numbers were reduced to 31% and 17% in 2000[19] and to 8% and 2% in 2002. In the 2002 survey,[20] only eight of the 99 samples contained quantifiable 3-MCPD (>0.01 mg/kg).

6.3.2.2   Foods Not Related with Acid-HVPs

Besides acid-HVPs and soy sauces, several other food products have been detected with 3-MCPD levels greater than 0.02 mg/kg (Table 6.3). These include cereal and dairy products as well as coffee. The occurrence of 3-MCPD in these foods,

TABLE 6.2
Estimated Dietary Intake of 3-MCPD from Soy and Oyster Sauces in Singapore

Population Group	Sauce Intake	Average Intake[a]		Highest Intake[a]	
	g/day	ng/kg bw/day	%TDI	ng/kg bw/day	%TDI
Average respondents (18–69 years)	0.43	0.14	0.01	3.10	0.15
Average consumers (18–69 years)	8.90	2.97	0.15	66.45	3.32
High consumers (18–69 years)	19.80	6.59	0.33	147.52	7.38
Average consumers (18 to <30 years)	9.61	3.29	0.16	73.53	3.68
High consumers (18 to <30 years)	19.50	6.68	0.33	149.39	7.47
Average consumers (30 to <40 years)	10.64	3.53	0.18	79.05	3.95
High consumers (30 to <40 years)	22.0	7.23	0.36	161.82	8.09
Average consumers (40 to <50 years)	7.14	2.34	0.12	52.59	2.61
High consumers (40 to <50 years)	18.50	6.14	0.31	137.28	6.86
Average consumers (50–69 years)	8.07	2.7	0.13	60.32	3.02
High consumers (50–69 years)	18.80	6.28	0.31	140.46	7.02

Source: Adapted from Wong et al., *Food Control*, 2006, 17, 408–413.

[a] Average and highest intakes were calculated by multiplying the sauce intake of a population group by two levels of 3-MCPD, that is, 0.02 and 0.45 mg/kg, respectively. %TDI was based on the TDI value established by SFC/JECFA (2 µg/kg body weight/day).

although not as frequent as it is in acid-HVPs-related products, should not be overlooked, since the consumption of the former foodstuffs are greater than that of the latter ones.

6.3.2.2.1 Leavened Cereal-Derived Products

In a leavened dough, NaCl and glycerol are the major precursors for the formation of 3-MCPD. Although phospholipid showed higher reactivity than glycerol in the conversion to 3-MCPD as discussed before, its overall contribution to the final 3-MCPD content is not as high due to its low concentration. The highest level of 3-MCPD in a baked dough is approximately 0.4 mg/kg, which is typically found in the crust. This is likely because of the fact that the crust gains greatest exposure to high temperatures during the baking process, which promotes the formation of 3-MCPD.

TABLE 6.3

Contents of 3-MCPD in Different Foods in a 1999–2000 UK Survey

Food Group	Number of Samples	Number with 3-MCPD >0.01 mg/kg	Percentage with 3-MCPD >0.01 mg/kg	Range of 3-MCPD (mg/kg)
Cereals	106	49	46	<0.010–0.134
Dairy	35	4	11	0.016–0.031
Meats	63	26	41	<0.010–0.081
Soups	34	0	0	N/A
Others	62	10	16	<0.010–0.024

Source: Adapted from Crew et al., *Food Additives and Contaminants*, 2002, 19(1), 22–27.

Breitling-Utzmann et al. tested several bread ingredients for their influence on 3-MCPD formation.[21] Among all these ingredients, baking agent exhibited the greatest contribution to 3-MCPD generation. This was probably due to the existence of mono- and di-acylglycerols as emulsifiers in the baking agent. The effect of yeast on the content of 3-MCPD was more complicated.[22] The addition of yeast led to a continuous increase in the content of glycerol, which was favorable for 3-MCPD formation. However, high yeast content (8.1% of the flour mass) exhibited an obvious inhibitory effect on the generation of 3-MCPD, which arose possibly from the cytosol protein of the yeast. Other influencing factors included pH (organic acids facilitate the formation of 3-MCPD), temperature, and sugar content.

6.3.2.2.2 Malt-Derived Food Products

Malt-derived food products include food-grade malted grains, malt flours, and malt extract products, which are used for coloring and flavoring purposes. The formation of 3-MCPD is most likely attributed to the dry-kilning of malted and unmalted barley at temperatures higher than 170°C. It was also revealed that the endogenous components of the grain could suffice the formation of 3-MCPD; no additional fat, acid, or salt is needed.[23] In addition, compared to malt flours and pale brewing malt, dark brewing malt contains significantly higher 3-MCPD levels, sometimes exceeding 0.25 mg/kg.[21,24] However, the level of 3-MCPD in these foods is generally low, and it can be further lowered upon dilution when these ingredients are used for brewing.

6.3.2.2.3 Coffee

Coffee may contain 3-MCPD, but at low levels. The highest levels of 3-MCPD were seen in instant coffee and in products with prolonged roasting. The final color of the coffee beans is directly linked to 3-MCPD formation, with the darker beans having higher concentration. Chloride from salt and lipids naturally present in coffee beans are responsible for 3-MCPD formation during the roasting process.[25] However, 3-MCPD is not detected in coffee beverages probably because of the dilution with water.

6.3.2.2.4 Cheese

Although the level of 3-MCPD in uncooked cheese is found to be rather low, it could be elevated in melted or grilled cheese. Crews et al. found that grilling produced 3-MCPD in all studied cheese samples,[14] whereas only Parmesan cheese exhibited increased 3-MCPD level upon microwave processing. The reason for the occurrence of 3-MCPD in cheese is not yet known, but it can be speculated that the considerable amount of salt and lipid in the cheese may be contributing factors to the formation of this compound, and that the disparity in moisture contents among different cheese products might account for the difference in 3-MCPD production under microwave.

6.3.2.2.5 Smoked Food Products

A German study showed that 3-MCPD existed in smoked meat at a high level. The smoking time, type of wood used for smoking, and the salt content in the meat had significant influences on the 3-MCPD content in the product. In addition, 3-MCPD was not detected in the initial pellet for generating smoke, but it was found in abundance in the smoke itself and the samples scraped from the wall.[26] In a follow-up study, the addition of 20% calcium carbonate to the pellets prior to smoking significantly reduced the production of 3-MCPD in the smoke. Other experiments suggested that neither endogenous nor added lipids were involved in the synthesis of the contaminant. Based on these results, it was proposed that 3-MCPD was formed via 3-hydroxyacetone that was produced through the cracking of cellulose, and the added salt served as a major donor of chlorine.[27]

A similar study in the UK revealed the existence of 3-MCPD in cooked meat, prepared cheese, smoked food, and so on, although the content was not as high as that in the German study. This may be ascribed to different preparation/smoking procedures. It is worth mentioning that 3-MCPD was only found in food products prepared by "cold-smoking," that is, the temperature applied in the process was as low as 28°C, while it was not detected in foodstuffs smoked at high temperatures. Such findings suggested a different mechanism for 3-MCPD formation, which obviously needs further exploration.

6.3.2.2.6 Meat

3-MCPD is also detected in cooked meat products, such as salami, bacon, hamburgers, and so on. However, the influence of cooking condition on the formation of contaminant remained unknown. It is found that cooking sometimes encourages 3-MCPD formation and sometimes has no observable effect at all. Moreover, there is no confirmed link between 3-MCPD synthesis and precursors like glycerol. One possible explanation is that the contaminant may have originated from the residual epichlorhydrin in the PAAE coating of the meat. In a study by Crews et al., 3-MCPD was not found in boiled or stewed meat, which suggested that temperatures above 100°C might be necessary and that "wet" cooking may inhibit its formation.[14]

6.3.2.2.7 Salted Fish

3-MCPD was also detected in some salted fish samples, such as anchovy fillets in olive oil. Follow-up study[4] revealed that it was not formed during the processing procedure, that is, the maturation with salt, but was produced later on, during packaging and storing. The way in which 3-MCPD is formed in the anchovy fillets is still unknown, but it may be generated from enzymatic action on fats. These enzymes could trigger the release of glycerol-related precursors from fats, which was then able to react with chlorine ions. Another possible explanation is that fats may interact with chlorides to form chloroesters, which release 3-MCPD upon hydrolysis. This type of fat/enzyme-related reaction may be quite widespread, but the exact mechanism is yet to be discovered.

6.3.2.2.8 Drinking Water

Very low levels of 3-MCPD is detected in drinking water in some countries such as the United Kingdom. It is formed in two possible ways. On the one hand, epichlorohydrin-linked cationic polymer resins are sometimes used as flocculants in water purification, and 3-MCPD may be thus formed as a by-product.[28,29] According to UK's maximum allowed usage rate of flocculants in water (2.5 mg/L water), the maximum theoretical content of 3-MCPD would be 0.1 μg/L water. However, no data have been released by now on the actual 3-MCPD concentration in drinking water. On the other hand, several chlorohydrin species other than 3-MCPD, for example, 1,3-dichloro-2-propanol, have been identified in some water samples that have been treated with chlorine, chloramines, or a combination of ozone and chlorine. Therefore, it is possible that 3-MCPD also arises at a certain level as a disinfection by-product.[28]

6.3.3 REGULATIONS AND RECOMMENDATIONS ON 3-MCPD

According to the Scientific Committee on Food (SCF) and the Joint FAO/WHO Expert Committee on Food Additives (JECFA), the TDI of 3-MCPD is 2 μg/kg body weight per day.[30] However, in another study using different methodologies, the TDI for 3-MCPD was reported as 6.6–8.7 μg/kg body weight per day.[31] Both TDIs are above the highest intake levels of 3-MCPD that has been reported. According to the first TDI value, the maximum 3-MCPD concentration in soy sauce should not exceed 0.02 mg/kg, based on a 40% dry matter content. This value has been adopted by Singapore, EU, Australia, and New Zealand since 2002. In the United States, it is recommended by FDA that the content of 3-MCPD in HVPs and soy sauce should not exceed 1 mg/kg,[26] but there is no mandatory requirement.

6.4 DETERMINATION METHODS FOR 3-MCPD

Although the chemical structure of 3-MCPD is simple, the detection at sub mg/kg level is challenging. The absence of a suitable chromophore impedes the effective detection of 3-MCPD by high-performance liquid chromatography (HPLC) with ultraviolet or fluorescence detectors; the high boiling point and high polarity hurdles the analysis by gas chromatography (GC); and the low-molecular weight complicates the mass detection of 3-MCPD, since diagnostic ions cannot be reliably distinguished from background chemical noise. In spite of these challenges, sensitive detection methods for 3-MCPD have been developed in recent years. These methods are compared in Table 6.4 and will be discussed in the following section.

TABLE 6.4
Analytical Methods for 3-MCPD[a]

Extraction	Derivatization	Detection	LOD[b]	Comments
Multistep extraction with organic solvents[32]	HFBI/HFBA	GC/MS	5–50	AOAC standard. Accurate but time consuming.
Multistep extraction with organic solvents[33]	HFBI/HFBA	*GC/MS with PTV/ LVI and tandem MS*	0.044	Very low LOD. Expensive instruments.
Solid-phase microextraction[34]	HFBI/HFBA	GC/MS	3.87	Extraction time reduced from hours to 30 min.
Alumina column extraction[35]	HFBI/HFBA	GC/MS	1	Easy extraction. Simultaneous quantification of 3-MCPD and 1,2-DCP.
Multistep extraction with organic solvents[36]	*4-Heptanone*	GC/MS	0.48	Inexpensive and simple derivatization. Low LOD.
Multistep extraction with organic solvents[37]	*Phenylboronic acid*	GC/MS	1–2	Simple derivatization. Simultaneous quantification of 3-MCPD and esters.
Multistep extraction with organic solvents[38]	HFBI/HFBA	*Capillary GC/MS*	5	Simultaneous quantification of 3-MCPD and 1,3-DCP.
None[39]	*None*	*Molecularly imprinted polymer*	NG[c]	One-step measurement within 1 h. Interfered by other food ingredients.
None[40]	*None*	*Capillary electrophoresis*	130	One-step measurement. No extraction or derivatization needed.

[a] Texts in italics format indicate the differences from the AOAC standard method. The numbers in the superscript indicate for the references.
[b] LOD: Limit of detection, in ng/g.
[c] NG: Not given.

6.4.1 Gas Chromatography/Mass Spectrometry Analysis

6.4.1.1 AOAC Official Method (AOAC 2001.01)

Gas chromatography/mass spectrometry (GC/MS) analysis is adopted as an official method for quantifying 3-MCPD by AOAC, and it is accepted as a universal and normative approach. A typical GC/MS procedure includes extraction with salt solution, mixing the resultant solution with a diatomaceous matrix such as "Extrelut," removal of nonpolar components with organic solvents, derivatization with heptafluorobutyrylimidazole (HFBI) or heptafluorobutricanhydride (HFBA), addition of deuterated 3-MCPD as an internal standard, and measurement under GC/MS condition. An international collaborative study involving 12 laboratories[32] revealed satisfactory reproducibility of this method in all of the six tested food categories, that is, acid-HVPs, malt extract, soup powders, bread crumbs, salami, and cheese alternative. The detection limit of this method was reported to be 5–50 ng/g, depending on the type of samples.

Despite its desirable repeatability, the AOAC method suffers from several drawbacks. The extraction is tedious and laborious, typically consisting hours of salt extraction and multiple elutions with organic solvents. The derivatization shows limited selectivity because the derivatizing agents may react with nucleophilic molecules besides chloropropanols. Furthermore, the derivatization must be carried out in anhydrous environment, and the derivatized products are not stable for long-term use. Lastly, a simultaneous detection of 3-MCPD with other chloropropanols such as 1,3-dichloropropanol (1,3-DCP) is commonly needed in practice to avoid repetitive labors. A number of improved methods have been proposed to address these issues.

6.4.1.2 GC/MS with Programmable Temperature Vaporization

León et al.[33] developed a sensitive method using programmable temperature vaporization (PTV) with large volume injection (LVI) GC coupled with tandem mass spectrometry (MS/MS). By increasing the sample volume of the final extract from 5 to 70 μL, this method exhibited a low detection limit at 0.044 ng/g. It was also reported that 1,3-DCP can be detected by this method, although that was not the main focus of their study.

6.4.1.3 GC/MS with Solid-Phase Microextraction

Huang et al. substituted the conventional salt solution extraction with solid-phase microextraction (SPME).[34] In this method, the analytes from the headspace of samples were extracted onto a fiber matrix, such as polydimethylsiloxane and carbowax/divinylbenzene, and then desorbed in the injector of GC for separation and detection. The extraction procedure took only one step that lasted for 30 min, after which the fiber was transferred immediately to the GC injection port for thermal desorption. This method demonstrated a linear range of 0.0194–394 μg/g, with a detection limit of 3.87 ng/g.

6.4.1.4 GC/MS with Alumina Column Extraction

Abu-E-Haj[35] described a method for simultaneously determining 3-MCPD and 1,3-DCP using a small amount of food samples. Sample portions (1–2 g) spiked with deuterated 3-MCPD-d5 and 1,3-DCP-d5 as internal standards were mixed with 2 g of aluminum oxide and extracted in a disposable column. The compounds were eluted with 25 mL dichloromethane, concentrated, derivatized with HFBA, and then subjected to GC/MS measurement. The method was capable of quantifying 3-MCPD and 1,3-DCP simultaneously in samples such as soy sauce, cereal products, malt extracts, and soup powders. The limit of detection and limit of quantitation were 1 and 3 ng/g, respectively, and the recovery was around 80%.

6.4.1.5 GC/MS Using 4-Heptanone as a Derivatizing Agent

As an alternative to HFBA or HFBI, ketones are used as derivatizing agents because of their capability to react with diols to form cyclic acetals. Early procedures involving ketones showed advantage in inexpensive agents and simple operation, but they were not adopted until recently because of the errant results observed in these assays. Dayrit et al.[36] developed an optimized procedure for the analysis of 3-MCPD in soy sauce at a concentration range of 1–5000 ng/g, using 4-heptanone as the derivatizing ketone and 3-MCPD-d5 as the internal standard. Reliable quantification was achieved for soy sauce matrices, with a detection limit of 0.48 ng/g, and a lowest limit of quantitation of 1.2 ng/g.

6.4.1.6   GC/MS Using Phenylboronic Acid as a Derivatizing Agent

Küsters et al.[37] proposed an approach for simultaneous determination of free 3-MCPD and its esters. The samples were firstly extracted briefly with NaCl solution and methyl-tert-butyl ether. The lower aqueous layer containing free 3-MCPD was derivatized with phenylboronic acid (PBA) directly. The esters in the upper oil layer were cleaved with sodium methoxide and mixed with a buffer solution consisting of acetic acid and sodium chloride. An aqueous layer was separated from the above-mentioned mixture and then derivatized with PBA for 30 min. The two parts of PBA derivatives were subjected to GC/MS measurement. The method was validated for various foodstuffs such as bakery products, meat and fish products, and soups, as well as seasonings. The detection limits were 1–2 ng/g for 3-MCPD and 6 ng/g for 3-MCPD esters. The average recoveries were reported to be 95% and 85% for free 3-MCPD and its esters, respectively.

6.4.1.7   Capillary GC/MS Measurement

Chung et al. reported a highly selective and sensitive method for the simultaneous determination of 3-MCPD and 1,3-DCP at ng/g levels in soy sauce, using capillary GC with mass spectrometric detection.[38] In brief, samples were homogenized, mixed with sodium chloride solution, and then adsorbed on silica gel. Chloropropanols were then eluted with ethyl acetate, derivatized with HFBA, and analyzed with capillary GC/MS. The linear range of detection was established at concentrations from 10 to 1000 ng/g, and the detection limit was 5 ng/g. Satisfactory precision was achieved at about 5%, and recoveries of 1,3-DCP and 3-MCPD from soy sauce samples spiked at 25 mg/kg were 77% and 98%, respectively.

6.4.2   ANALYTICAL METHODS OTHER THAN GC/MS

A covalent interaction-based molecularly imprinted polymer (MIP) material for 3-MCPD detection was successfully fabricated by Leung et al.[39] This material employed 4-vinylphenylboronic acid as the functional monomer, which experienced an increase in Lewis acidity due to reaction of the arylboronic acid with 3-MCPD. A simple pH glass electrode is sufficient to monitor such change and thus indicate the presence of 3-MCPD. A linear potentiometric response was observed in unbuffered media containing 0–350 mg/kg 3-MCPD. In an attempt to monitor this contaminant in soy sauce samples, however, the chemosensing response of the MIP material is much reduced, probably due to the blocking or deactivation of receptor sites by interferents in the sample. Nonetheless, the work of Leung et al. demonstrated the potential of MIP-based chemosensors as a time-efficient, semiquantitative analytical tool for the detection of 3-MCPD in food products.

A capillary electrophoresis-based method was described by Xing et al.[40] to measure 3-MCPD in soy sauce samples. This procedure was highlighted for the simplicity of operation. The only pretreatment step was brief dilution with a borate buffer; no extraction or derivatization was required, thus saving hours to days of labor. The linear range for 3-MCPD was 6.6–200 µg/mL, and the recovery averaged at 93.8–106.8%. The detection limit of this novel method was 0.13 µg/mL.

6.5 PHARMACOKINETICS OF 3-MCPD

After oral consumption, 3-MCPD is rapidly distributed in different organs throughout the human body. It is then degraded into a few metabolites, each of which possesses considerable toxicity to certain organs or tissues. Understanding the pharmacokinetics of 3-MCPD is important for elucidating its toxicology.

6.5.1 BIO-DISTRIBUTION

Jones et al.[41] studied the tissue distribution, metabolism, and excretion of [^{36}Cl] 3-MCPD (100 mg/kg by i.p.) in male Sprague–Dawley rats. 3-MCPD exhibited the capacity of crossing the blood–brain barrier, blood–testis barrier, and placenta.[41,42] Radioactivity was rapidly distributed throughout the body. The levels of 3-MCPD in the cauda epididymis, caput epididymis, testes, brain, liver, pituitary, muscle, and red blood cells ranged from 0.3% to 0.5% of the total dose (in percentage of radioactivity) within 30 min of administration, and they were lowered to 0.2–0.3% of the total dose after 24 h. No specific accumulation was detected in any of these organs.

6.5.2 METABOLISM

In an in vitro study using Caco-2 cells, no significant degradation of 3-MCPD was observed.[43] When studied in vivo, 3-MCPD is metabolized via an epoxidation pathway involving the formation of glycidol, or through an oxidation pathway comprising the formation of β-chlorolactic acid and oxalic acid (Figure 6.3).[41] The former pathway is essential for the survival of various bacteria on halogenated alcohol substrates. The latter route is found to be common among mammalian species such as rats and mice, and it is supported by several toxicological studies involving radioactive labeling. 3-MCPD is excreted in the breath as CO_2 and in the urine in the original form and as the metabolites.[41]

6.6 TOXICITY OF 3-MCPD

6.6.1 ACUTE TOXICITY

6.6.1.1 Threshold Doses for Acute Toxicity

Qian et al.[44] investigated the acute toxicity of two enantiomers of 3-MCPD, namely, R- and S-3-MCPD, as well as their racemic mixture. Using sexually mature ICR mice, the authors reported that the LD_{50} (95% CI) of R, S, and (R,S)-3-MCPD were 290.54, 117.57, and 190.73 mg/kg, respectively. The kidney/weight and brain/weight

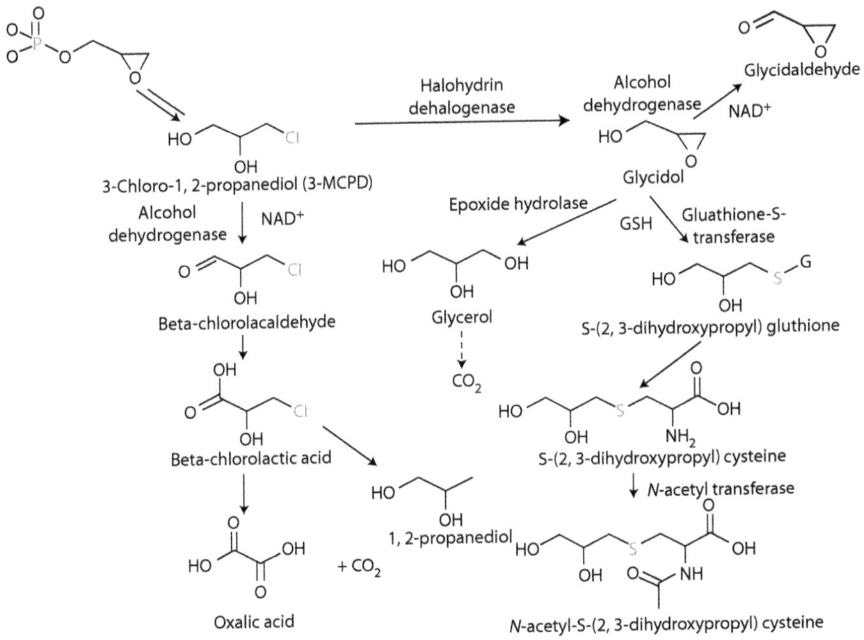

FIGURE 6.3 Metabolism pathways of 3-MCPD. (Adapted from Lynch et al., *International Journal of Toxicology*, January 1998, 17(1), 47–76.)

ratios in R-3-MCPD and (R,S)-3-MCPD treated groups were significantly higher ($P < 0.05$) than controls at the doses of 250–353.55 and >353.55 mg/kg, respectively, but this result was not observed in groups treated with S-3-MCPD. At the doses higher than 353.55 and 229.13 mg/kg, respectively, R-3-MCPD and (R,S)-3-MCPD caused obvious swell of liver cells and the swell and congestion of the liver sinus. Obvious morphological changes of kidney were not observed after administration of R, S, and (R,S)-3-MCPD. It is noteworthy that the LD_{50} values may vary significantly with the type of rats tested, but they generally fall in the range of 100–300 mg/kg body weight.

6.6.1.2 Immunotoxicity

Lee et al. found that 3-MCPD caused damage to the immune system of Balb/c mice after oral gavage at 100 mg/kg/day for 14 days.[45,46] The treated mice exhibited significantly reduced spleen cellularity, absolute and relative thymus weights, and thymus cellularity. Concomitantly, the antibody production against sheep red blood cells (a T cell-dependent antigen) was reduced in mice fed with high dose, and the natural killer cell activity was also reduced. The first change suggested that 3-MCPD might compromise the immune system's capacity to combat bacterial infection, while the latter effect implied an adverse effect on the immune system's ability to conduct tumor surveillance.

According to another relevant study by Byun (2005), less than 1 mM of 3-MCPD could reduce the functionality of lymphocytes and peritoneal macrophages in vitro.[47]

The treated Balb/c mice spleen cells showed a significant decrease in the lymphocyte blastogenesis, together with suppressed production of interferon (IFN)-c, interleukin (IL)-4, IL-10, nitric oxide (NO), and tumor necrosis factor. A follow-up study conducted by the same research group[48] revealed that the major metabolite of 3-MCPD, that is, β-chlorolactic acid, induced virtually the same toxic effect on immune response of lymphocytes and peritoneal macrophages in vitro. These results indicate that 3-MCPD might induce immunotoxic effect via its metabolites.

6.6.1.3 Antifertility

3-MCPD has demonstrated considerable reproductive toxicity to a wide variety of mammalian species that may include human.[49,50] When administrated continuously at low doses (5–10 mg/kg), 3-MCPD reduced spermatozoal motility, probably via energy deprivation. Low concentrations of 3-MCPD inhibited the activities of glyceraldehyde 3-phosphate dehydrogenase (GADPH), triose phosphate isomerase and adolase by more than 80%, thus blocking severely the conversion of glyceraldehyde 3-phosphate into 3-phosphoglycerate within 24 h of injection. In addition, the production of acetyl CoA from exogenous and endogenous fatty acids was impeded by 3-MCPD, although to a much lesser extent. These changes compromised the energy production by glycolysis in vivo, causing the conversion of more than 92% adenosine triphosphate (ATP) in spermatozoa into adenosine monophosphate (AMP).[51] In a recent study,[52,53] the spermatotoxic effect of 3-MCPD was also attributed to reduced H[+]-ATPase expression in the cauda epididymis. This led to an altered pH level in the cauda epididymis, which might cause a disruption of sperm maturation and the acquisition of motility. The enzyme activity and ATP level returned to normal 26 days after the withdrawal of 3-MCPD administration.

6.6.1.4 Neurotoxicity

In a study undertaken by Cavanagh et al., a single dose of S-(+)-3-MCPD (140 mg/kg body weight, I.P.) induced neurotoxic effects in male Fischer 433 rats.[54,55] The earlier neurotoxic changes were strictly confined to glial cells, particularly astrocytes; hemorrhages were not found. Within 6 h of astrocyte loss, microvessels in this area exhibited a loss of the normal paracellular localization of the transmembrane proteins occludin, claudin-5, and cytoplasmic zonula occludens-1. This phenomenon accompanied the focal vascular leak of dextran (10 kDa) and fibrinogen.[56] Minimal evidence of increased vascular leakage of horseradish peroxidase (HRP) was found in early stages, while later macrophage invasion and capillary proliferation was accompanied by rare focal leakage of HRP. Moreover, no loss of the endothelial lining was observed (Figure 6.4). These results suggested that the gross astrocytic damage did not necessarily impair the integrity of the blood–brain barrier. Although astrocytes were severely distended with fluid during early intoxication, and their organelles seriously disorganized, they did not die but regenerated rapidly. This was indicated by the recovery in the expression level of tight junction protein after 8 days.

In earlier studies, the neurotoxicity of 3-MCPD was ascribed to energy deprivation through GADPH inactivation, similar to the antifertility discussed above. However, Cavanagh et al.[54] pointed out that the patterns of damage to brain stem

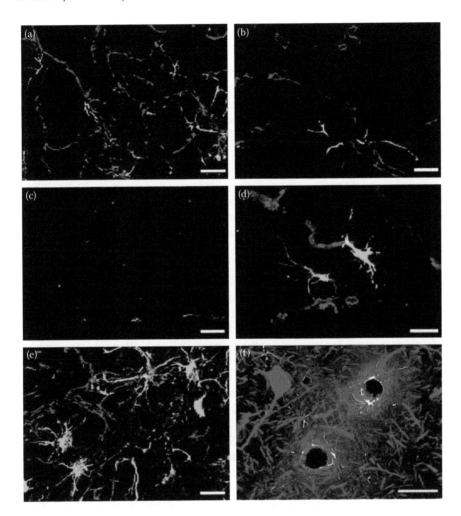

FIGURE 6.4 Confocal micrographs showing the damage and self-repair of inferior colliculus after injection of 3-MCPD. Green: glial fibrillary acidic protein (GFAP)-positive astrocytes; red: occludin immunoreactivity. (a,b,c): Damage observed 0, 2, and 6 days after 3-MCPD administration. (d,e): Recovery 8 and 14 days after 3-MCPD administration. (f): Leak of diffuse dextran (red) and loss of GFAP-positive astrocytes (yellow) 24 h after 3-MCPD administration. (Adapted from Willis, et al., *Glia*, 2004, 48, 1–13.)

centers caused by 3-MCPD were distinct from other acute energy deprivation syndromes. This finding suggested the existence of other unrecognized factors that determined whether a neuronal center is at risk or not. Skamarauskas et al.[57] sought alternative mechanisms to explain its toxicity to astrocytes. According to the authors, 3-MCPD might serve as a substrate for glutathione-S-transferase, and it generated an irreversible inhibitor of glutathione reductase after bioactivation by alcohol dehydrogenase. Both processes might lead to a decreased level of glutathione, followed by the disruption of redox state in the inferior colliculus.

6.6.2 Chronic and Subchronic Toxicity

6.6.2.1 Carcinogenicity

In a two-year drinking water study (3-MCPD content 0–500 mg/L) involving 50 male and 50 female Fischer 344 rats,[58] significant dose-dependent histopathological changes were found in the testis, mammary gland, kidneys, and pancreas in male rats, and kidneys in female rats. Among males, there was a significant dose-dependent decrease in Leydig cell (the interstitium of the testis, between the seminiferous tubules) hyperplasia. This phenomenon was accompanied by a dose-related increase in the incidence of Leydig cell adenoma to 94–100%, as well as a positive trend for carcinoma. Similar results were reported on Sprague–Dawley rats,[59] which exhibited higher mortality rate and lower body weight in response to 3-MCPD administration for 72 weeks. Severe testicular degeneration and atrophy, as well as occasional parathyroid adenomas, were observed on treated male rats. There were no neoplastic changes or toxicity observed in the treated females.

In a similar study[60] using 50 male and 50 female Sprague–Dawley rats, the incidences of combined renal tubular adenoma and carcinoma were significantly increased ($P < 0.01$) as a function of 3-MCPD concentration. There was also a highly significant increase in renal tubular hyperplasia in all male treatment groups as well as the females treated with highest dose. According to the author, renal tubular adenoma and carcinoma were first observed in males in the 78th and 74th week, respectively.

A significant ($P < 0.001$) dose-related increase of mammary gland hyperplasia was observed in male rats. A significant increase in mammary fibroadenoma was also observed at the highest dose ($P < 0.001$). One mammary adenoma and one mammary adenocarcinoma were observed in each of the mid- and high-dose groups.

6.6.2.2 Genotoxicity and Its Relevance to Carcinogenicity

While the carcinogenicity of 3-MCPD has been identified in a considerable number of literatures, the underlying mechanism remained controversial until recently. Quite a few studies in the past few decades demonstrated the mutagenic and genotoxic effects of 3-MCPD in vitro in bacteria, yeast, and mammalian cells.[61–63] These results gave rise to the postulation that the carcinogenicity of 3-MCPD arises from its genotoxicity. Although this hypothesis was challenged by the absence of genotoxicity in several in vivo studies, it was argued that those negative results might be originated from improper experimental methodologies.

However, with the aid of modern in vivo assays, it can now be confirmed that 3-MCPD does not exhibit any genotoxicity in vivo. In 2003, Robjohns et al. demonstrated the lack of genotoxic effect by 3-MCPD using a bone marrow micronucleus test and unscheduled DNA synthesis in the rat liver. Similarly,[64] El Ramy et al. evaluated DNA damages directly in selected target (kidneys and testes) and nontarget (blood leukocytes, liver, and bone marrow) male rat organs, using the in vivo alkaline single-cell gel electrophoresis (comet) assay. Genotoxic potential was not detected in vivo by 3-MCPD and β-chlorolactic acid, either in the target or nontarget organs. In addition, although glycidol is a known mutagen, β-chlorolactic acid did not exhibit any DNA-damaging effects in vitro in mammalian cells. Since 3-MCPD is mostly metabolized into β-chlorolactic acid by mammals, its carcinogenicity in vivo is probably related to a non-genotoxic mechanism.

Although the non-genotoxic property of 3-MCPD's carcinogenicity was confirmed not long ago, possible mechanisms other than genotoxicity have been explored extensively for more than 20 years. Lynch et al.[58] proposed that the metabolism of 3-MCPD to oxalate via β-chlorolactic acid may be responsible for the renal tumor formation in rats. Oxalate is known to possess severe cytotoxicity in the kidney.[41,65] In addition, crystals of oxalate were detected in the urine of rats treated with 3-MCPD (single dose of 100 mg/kg i.p.). These evidences support a role for sustained cytotoxicity resulted from oxalate as a possible mechanism for the induction of kidney tumors in rats. On the basis of these findings, it is plausible that 3-MCPD may now be considered carcinogenic to rodents via a non-genotoxic mechanism.

As for the incidence of sex-specific tumors in male rats (i.e., tumors in the testes, mammary gland, and preputial gland), it was postulated to originate from[58] the disturbance of sexual hormone level. 3-MCPD is known to induce a prolonged increase in circulating hormone levels, a single i.p. dose of 80 mg/kg body weight being able to increase serum levels of follicle-stimulating hormone, luteinizing hormone, and prolactin.[66,67] It is possible that the Leydig cell tumor is promoted by hormonal imbalance caused by 3-MCPD, and the increase in tumors at other hormone-responsive sites is secondary to further hormonal disturbances, which are possibly induced by proliferating Leydig cells.

6.6.2.3 Neurotoxicity

In addition to the acute neurotoxicity discussed in the previous section, 3-MCPD was also found to exhibit subchronic neurotoxic effects at the dose of 10 or 30 mg/kg body weight, in a 13-week study carried out by Kim.[68] The expression of two forms of nitric oxide synthase (NOS), namely, neuronal NOS (nNOS) and inducible NOS (iNOS), was determined using immunocytochemistry in rat cerebral cortex and striatum. The data suggested that subchronic 3-MCPD exposure might trigger compensatory mechanisms involving nNOS and iNOS expression to maintain the nitric oxide homeostasis in the rostral part of the neocortex and striatum, but such trend was not observed in the caudal brain. These results suggested that 3-MCPD induced neurotoxicity, at least in part, through disturbing the nitric oxide signaling pathway, and that it caused a rostrocaudal difference in the neocortex and striatum possibly through differential expression of nNOS and iNOS.

6.6.2.4 Antifertility

High levels of 3-MCPD may lead to prolonged or even permanent infertility in male rats. A single dose of 90–120 mg/kg results in an occlusion of the efferent ducts and/or the tubule of the caput epididymis. This pathological lesion can be caused by the development of either bilateral sperm retention cysts or spermatocoeles, and it may prevent immature testicular sperm from being passaged to the epididymal tract.[69] This produces a back-pressure of testicular fluid that causes edema, the inhibition of spermatogenesis and, ultimately, atrophies of the testes.

6.7 MITIGATION OF 3-MCPD

In dilute aqueous solution (pH 5–9), 3-MCPD degrades into glycerol at temperatures above 80°C.[70] This reaction is frequently used to lower the 3-MCPD level in acid-HVPs. In addition, 3-MCPD as well as its esters can be removed in aqueous and biphasic systems.[71] This is achieved by conversion of 3-MCPD to epoxide in the presence of halohydrin dehalogenase from *Arthrobacter* sp. AD2, followed by the generation of glycerol with the aid of an epoxide hydrolase from *Agrobacterium radiobacter* AD1. A nearly complete conversion can be achieved after sequential incubation of 10 mM 3-MCPD with these two enzymes in a binary solvent (water: oil = 1:20, v/v). In spite of the satisfactory conversion rate, the residual enzymes may limit the application of this method in food industry.

6.8 CONCLUSIONS

Great efforts have been made in the past few decades in research on the contamination of 3-MCPD in food systems. Mechanism studies have shown the formation of 3-MCPD from various precursors and explained its prevalence in the food industry. The rapid development of analytical techniques allowed the effective and accurate quantification of 3-MCPD in food products, providing reliable assessments on its occurrence in the food market. Modern biological methods enabled the in vitro and in vivo investigation on the toxicity and TDI of 3-MCPD. Relevant regulations and recommendations on allowable 3-MCPD contents were established in a number of countries, leading to a continuous decrease in the level of this contaminant. Future studies may include reliable detection of 3-MCPD in situ, feasible models for predicting and controlling the 3-MCPD generation during food process, and the assessment of long-term risks associated with continuous consumption. In addition, although there has been no reported outbreak of disease linked with this compound, prevention and therapy for 3-MCPD intoxication are necessary to minimize its threat on human health. Epidemiological studies are recommended to identify the population groups that are especially susceptible to 3-MCPD-related diseases. The synergetic effect of 3-MCPD, β-chlorolactic acid, and other toxic food components also needs further investigation.

REFERENCES

1. Porter, K. E., Jones, A. R. The effect of the isomers of alpha-chlorohydrin and racemic beta-chlorolactate on the rat-kidney. *Chemico-Biological Interactions*, 41(1), 1982: 95–104.
2. Jones, A. R., Ford, S. A. The action of (S)-α-chlorohydrin and 6-chloro-6-deoxyglucose on the metabolism of guinea pig spermatozoa. *Contraception*, 30(3), 1984: 261–269.
3. Ford, W. C. L., Waites, G. M. H. Activities of various 6-chloro-6-deoxysugars and (S)-α-chlorohydrin in producing spermatocoeles in rats and paralysis in mice and in inhibiting glucose metabolism in bull spermatozoa in vitro. *Journal of Reproduction and Fertility*, 65(1), 1982: 177–183.
4. Reece, P. The origin and formation of 3-MCPD in foods and food ingredients. Final project report. Food Standards Agency, London, 2005.

5. Collier, P. D., Cromie, D. D. O., Davies, A. P. Mechanism of formation of chloropropanols present in protein hydrolysates. *Journal of the American Oil Chemists Society*, 68(10), 1991: 785–790.
6. Kubec, R., Velisek, J., Dolezal, M., Kubelka, V. Sulfur-containing volatiles arising by thermal degradation of alliin and deoxyalliin. *Journal of Agricultural and Food Chemistry*, 45(9), 1997: 3580–3585.
7. Myszkowski, J., Zielinski, A. Z. Synthesis of glycerol monochlorohydrins from allyl alcohol. *Przemysl Chemiczny*, 44, 1965: 249–252.
8. Rahn, A., Yaylayan, V. A. Thermal degradation of sucralose and its potential in generating chloropropanols in the presence of glycerol. *Food Chemistry*, 118(1), 2010: 56–61.
9. Bodén, L., Lundgren, M., Stensio, K.-E., Gorzynski, M. Determination of 1, 3-dichloro-2-propanol and 3-chloro-1, 2-propanediol in papers treated with polyamidoamine-epichlorohydrin wet-strength resins by gas chromatography-mass spectrometry using selective ion monitoring. *Journal of Chromatography A*, 788(1), 1997: 195–203.
10. Seefelder, W., Varga, N., Studer, A., et al. Esters of 3-chloro-1,2-propanediol (3-MCPD) in vegetable oils: Significance in the formation of 3-MCPD. *Food Additives & Contaminants. Part A, Chemistry Analysis Control Exposure & Risk Assessment*, 25(4), 2008: 391–400.
11. Baer, I., de la Calle, B., Taylor, P. 3-MCPD in food other than soy sauce or hydrolysed vegetable protein (HVP). *Analytical and Bioanalytical Chemistry*, 396(1), 2010: 443–456.
12. Hamlet, C. G., Sadd, P. A., Gray, D. A. Generation of monochloropropanediols (MCPDs) in model dough systems. 1. Leavened doughs. *Journal of Agricultural and Food Chemistry*, 52(7), 2004: 2059–2066.
13. DiViNoVa, V., Dolezal, M., Velisek, J. Free and bound 3-chloropropane-1, 2-diol in coffee surrogates and malts. *Czech Journal of Food Sciences*, 25(1), 2007: 39.
14. Crews, C., Brereton, P., Davies, A. The effects of domestic cooking on the levels of 3-monochloropropanediol in foods. *Food Additives and Contaminants*, 18(4), 2001: 271–280.
15. Dolezal, M., Calta, P., Velisek, J. Formation and decomposition of 3-chloropropane-1, 2-diol in model systems. *Czech Journal of Food Sciences*, 22, 2004: 263–266.
16. World Health, O., Safety evaluation of certain food additives and contaminants. *Sixty-Seventh Meeting of the Joint FAO/WHO Expert Committee on Food Additives*. World Health Organization, 2007: 45–48.
17. Wong, K. O., Cheong, Y. H., Seah, H. L. 3-Monochloropropane-1, 2-diol (3-MCPD) in soy and oyster sauces: Occurrence and dietary intake assessment. *Food Control*, 17(5), 2006: 408–413.
18. Nyman, P. J., Diachenko, G. W., Perfetti, G. A. Survey of chloropropanols in soy sauces and related products. *Food Additives and Contaminants*, 20(10), 2003: 909–915.
19. Crews, C., Hough, P., Brereton, P., et al. Survey of 3-monochloropropane-1, 2-diol (3-MCPD) in selected food groups, 1999–2000. *Food Additives and Contaminants*, 19(1), 2002: 22–27.
20. Crews, C., Hasnip, S., Chapman, S., et al. Survey of chloropropanols in soy sauces and related products purchased in the UK in 2000 and 2002. *Food Additives and Contaminants*, 20(10), 2003: 916–922.
21. Breitling-Utzmann, C. M., Kobler, H., Herbolzheimer, D., Maier, A. 3-MCPD: occurrence in bread crust and various food groups as well as formation in toast. *Deutsche Lebensmittel-Rundschau*, 99(7), 2003: 280–285.
22. Hamlet, C. G., Sadd, P. A. Effects of yeast stress and pH on 3-monochloropropanediol (3-MCPD)-producing reactions in model dough systems. *Food Additives and Contaminants*, 22(7), 2005: 616–623.
23. Divinová, V., Dolezal, M., Velisek, J. Free and bound 3-chloropropane-1, 2-diol in coffee surrogates and malts. *Czech Journal of Food Sciences*, 25(1), 2007: 39.

24. Hamlet, C. G., Jayaratne, S. M., Matthews, W. 3-Monochloropropane-1, 2-diol (3-MCPD) in food ingredients from UK food producers and ingredient suppliers. *Food Additives and Contaminants* 19(1), 2002: 15–21.

25. Dolezal, M., Chaloupska, M., Divinova, V., Svejkovska, B., Velisek, J. Occurrence of 3-chloropropane-1, 2-diol and its esters in coffee. *European Food Research and Technology*, 221(3–4), 2005: 221–225.

26. FAO, WHO. Discussion paper on chloropropanols derived from the manufacture of acid-HVP and the heat processing of food. *Proceedings of the 1st Session of Codex Committee on Contaminants in Foods,* Beijing, China, 2007, pp. 16–20.

27. Kuntzer, J., Weisshaar, R. The smoking process: A potent source of 3-chloropropane-1, 2-diol (3-MCPD) in meat products. *Deutsche Lebensmittel-Rundschau*, 102(9), 2006: 397–400.

28. Nienow, A. M., Poyer, I. C., Hua, I., Jafvert, C. T. Hydrolysis and H_2O_2-assisted UV photolysis of 3-chloro-1,2-propanediol. *Chemosphere*, 75(8), 2009: 1015–1020.

29. Matthew, B. M., Anastasio, C. Determination of halogenated mono-alcohols and diols in water by gas chromatography with electron-capture detection. *Journal of Chromatography A*, 866(1), 2000: 65–77.

30. European, C. Commission Regulation (EC) No 1881/2006 of 19 December 2006 setting maximum levels for certain contaminants in foodstuffs. *Official Journal of the European Union L*, 364, 2006: 5–24.

31. Hwang, M., Yoon, E., Kim, J., Jang, D. D., Yoo, T. M. Toxicity value for 3-monochloropropane-1, 2-diol using a benchmark dose methodology. *Regulatory Toxicology and Pharmacology*, 53(2), 2009: 102–106.

32. Brereton, P., Kelly, J., Crews, C., et al. Determination of 3-chloro-1,2-propanediol in foods and food ingredients by gas chromatography with mass spectrometric detection: collaborative study. *J. AOAC Int.*, 84(2), 2001: 455–65.

33. León, N., Yusa, V., Pardo, O., Pastor, A. Determination of 3-MCPD by GC-MS/MS with PTV-LV injector used for a survey of Spanish foodstuffs. *Talanta*, 75(3), 2008: 824–831.

34. Huang, M., Jiang, G., He, B., et al. Determination of 3-chloropropane-1, 2-diol in liquid hydrolyzed vegetable proteins and soy sauce by solid-phase microextraction and as chromatography/mass spectrometry. *Analytical Sciences*, 21(11), 2005: 1343–1347.

35. Abu-El-Haj, S., Bogusz, M. J., Ibrahim, Z., Hassan, H., Al Tufail, M. Rapid and simple determination of chloropropanols (3-MCPD and 1, 3-DCP) in food products using isotope dilution GC/MS. *Food Control*, 18(1), 2007: 81–90.

36. Dayrit, F. M., Ninonuevo, M. R. Development of an analytical method for 3-monochloropropane-1, 2-diol in soy sauce using 4-heptanone as derivatizing agent. *Food Additives and Contaminants*, 21(3), 2004: 204–209.

37. Küsters, M., Bimber, U., Ossenbrüggen, A., et al. Rapid and simple micromethod for the simultaneous determination of 3-MCPD and 3-MCPD esters in different foodstuffs. *Journal of Agricultural and Food Chemistry*, 58(11), 2010: 6570–6577.

38. Chung, W.-C., Hui, K.-Y., Cheng, S.-C. Sensitive method for the determination of 1,3-dichloropropan-2-ol and 3-chloropropane-1,2-diol in soy sauce by capillary gas chromatography with mass spectrometric detection. *Journal of Chromatography A*, 952(1), 2002: 185–192.

39. Leung, M. K. P., Chiu, B. K. W., Lam, M. H. W. Molecular sensing of 3-chloro-1,2-propanediol by molecular imprinting. *Analytica Chimica Acta*, 491(1), 2003: 15–25.

40. Xing, X., Cao, Y. Determination of 3-chloro-1,2-propanediol in soy sauces by capillary electrophoresis with electrochemical detection. *Food Control*, 18(2), 2007: 167–172.

41. Jones, A. R., Milton, D. H., Murcott, C. The oxidative metabolism of alpha-chlorohydrin in the male rat and the formation of spermatocoeles. *Xenobiotica*, 8(9), 1978: 573–82.

42. El Ramy, R., Mostafa, O. E., Poul, M., et al. Lack of effect on rat testicular organogenesis after in utero exposure to 3-monochloropropane-1,2-diol (3-MCPD). *Reproductive Toxicology*, 22(3), 2006: 485–492.

43. Buhrke, T., Weißhaar, R., Lampen, A. Absorption and metabolism of the food contaminant 3-chloro-1,2-propanediol (3-MCPD) and its fatty acid esters by human intestinal Caco-2 cells. *Archives of Toxicology*, 85(10), 2011: 1201–1208.

44. Qian, G., Zhang, H., Zhang, G., Yin, L. Study on acute toxicity of R, S and (R, S)-3-monchloropropane-1, 2-diol. *Journal of Hygiene Research*, 36(2), 2007: 137.

45. Lee, J. K., Byun, J. A., Park, S. H., et al. Evaluation of the potential immunotoxicity of 3-monochloro-1,2-propanediol in Balb/c mice: I. Effect on antibody forming cell, mitogen-stimulated lymphocyte proliferation, splenic subset, and natural killer cell activity. *Toxicology*, 204(1), 2004: 1–11.

46. Lee, J. K., Byun, J. A., Park, S. H., et al. Evaluation of the potential immunotoxicity of 3-monochloro-1,2-propanediol in Balb/c mice: II. Effect on thymic subset, delayed-type hypersensitivity, mixed-lymphocyte reaction, and peritoneal macrophage activity. *Toxicology*, 211(3), 2005: 187–196.

47. Byun, J. A., Ryu, M. H., Lee, J. K. The immunomodulatory effects of 3-monochloro-1, 2-propanediol on murine splenocyte and peritoneal macrophage function in vitro. *Toxicology* in vitro, 20(3), 2006: 272–278.

48. Lee, J. K., Ryu, M. H., Byun, J. A. Immunotoxic effect of β-chlorolactic acid on murine splenocyte and peritoneal macrophage function in vitro. *Toxicology*, 210(2), 2005: 175–187.

49. Brown-Woodman, P. D., White, I. G. Effect of alpha-chlorohydrin and vasoligation on epididymal and testicular blood flow in the rat and on sperm parameters. *Acta Europaea Fertilitatis*, 9(4), 1978: 189.

50. Brown-Woodman, P. D. C., Mohri, H., Mohri, T., Suter, D. A. I., White, I. G. Mode of action of alpha-chlorohydrin as a male anti-fertility agent. Inhibition of the metabolism of ram spermatozoa by alpha-chlorohydrin and location of block in glycolysis. *Biochemical Journal*, 170, 1978: 23–37.

51. Rjones, A. Antifertility actions of a-chlorohydrin in the male. *Australian Journal of Biological Sciences*, 36(4), 1983: 333–350.

52. Kwack, S. J., Kim, S. S., Choi, Y. W., et al. Mechanism of antifertility in male rats treated with 3-monochloro-1,2-propanediol (3-MCPD). *Journal of Toxicology and Environmental Health, Part A* 67(23–24), 2004: 2001–2004.

53. Woods, J., Garside, D. A. An in vivo and in vitro investigation into the effects of α-chlorohydrin on sperm motility and correlation with fertility in the Han Wistar rat. *Reproductive Toxicology*, 10(3), 1996: 199–207.

54. Cavanagh, J. B., Nolan, C. C. The neurotoxicity of α-chlorohydrin in rats and mice: II. Lesion topography and factors in selective vulnerability in acute energy deprivation syndromes. *Neuropathology and Applied Neurobiology*, 19(6), 1993: 471–479.

55. Cavanagh, J. B., Nolan, C. C., Seville, M. P. The neurotoxicity of α-chlorohydrin in rats and mice: I. Evolution of the cellular changes. *Neuropathology and Applied Neurobiology*, 19(3), 1993: 240–252.

56. Willis, C. L., Leach, L., Clarke, G. J., Nolan, C. C., Ray, D. E. Reversible disruption of tight junction complexes in the rat blood-brain barrier, following transitory focal astrocyte loss. *Glia*, 48(1), 2004: 1–13.

57. Skamarauskas, J., Carter, W., Fowler, M., et al. The selective neurotoxicity produced by 3-chloropropanediol in the rat is not a result of energy deprivation. *Toxicology*, 232(3), 2007: 268–276.

58. Lynch, B. S., Bryant, D. W., Hook, G. J., Nestmann, E. R., Munro, I. C. Carcinogenicity of monochloro-1,2-propanediol (α-chlorohydrin, 3-MCPD). *International Journal of Toxicology*, 17(1), 1998: 47–76.

59. Weisburger, E. K., Ulland, B. M., Nam, J.-M., Gart, J. J., Weisburger, J. H. Carcinogenicity tests of certain environmental and industrial chemicals. *Journal of the National Cancer Institute*, 67(1), 1981: 75–88.

60. Cho, W.-S., Han, B. S., Lee, H., et al. Subchronic toxicity study of 3-monochloropropane-1,2-diol administered by drinking water to B6C3F1 mice. *Food and Chemical Toxicology*, 46(5), 2008: 1666–1673.

61. Stolzenberg, S. J., Hine, C. H. Mutagenicity of halogenated and oxygenated three carbon compounds. *Journal of Toxicology and Environmental Health, Part A Current Issues,* 5(6), 1979: 1149–1158.

62. McCann, J., Choi, E., Yamasaki, E., Ames, B. N. Detection of carcinogens as mutagens in the Salmonella/microsome test: Assay of 300 chemicals. *Proceedings of the National Academy of Sciences*, 72(12), 1975: 5135–5139.

63. Rossi, A. M., Migliore, L., Lascialfari, D., et al. Genotoxity, metabolism and blood kinetics of epichlorohydrin in mice. *Mutation Research/Genetic Toxicology*, 118(3), 1983: 213–226.

64. El Ramy, R., Ould Elhkim, M., Lezmi, S., Poul, J. M. Evaluation of the genotoxic potential of 3-monochloropropane-1,2-diol (3-MCPD) and its metabolites, glycidol and beta-chlorolactic acid, using the single cell gel/comet assay. *Food and Chemical Toxicology*, 45(1), 2007: 41–8.

65. Jones, A. R., Gadiel, P., Murcott, C. The renal toxicity of the rodenticide alpha-chlorohydrin in the rat. *Naturwissenschaften*, 66(8), 1979: 425.

66. Morris, I. D., Jackson, C. M. Gonadotrophin changes in male rats following a sterilizing dose of α-chlorohydrin. *International Journal of Andrology*, 1(1), 1978: 86–95.

67. Morris, I. D., Jackson, C. M. Gonadotrophin response after castration and selective destruction of the testicular interstitium in the normal and aspermatogenic rat. *Acta Endocrinologica*, 88(1), 1978: 38–47.

68. Kim, K. Differential expression of neuronal and inducible nitric oxide synthase in rat brain after subchronic administration of 3-monochloro-1, 2-propanediol. *Food and Chemical Toxicology*, 46(3), 2008: 955–960.

69. Cooper, E. R. A., Jackson, H. Chemically induced sperm retention cysts in the rat. *Journal of Reproduction and Fertility*, 34(3), 1973: 445–449.

70. Dolezal, M., Velisek, J. Kinetics of 2-chloro-1, 3-propanediol degradation in model systems and in protein hydrolysates. *Potravinarske Vedy-UZPI*, 13, 1994: 85–91.

71. Bornscheuer, U. T., Hesseler, M. Enzymatic removal of 3-monochloro-1,2-chloropanediol (3-MCPD) and its esters from oils. *European Journal of Lipid Science and Technology*, 112(5), 2010: 552–556.

7 3-MCPD Fatty Acid Esters
Chemistry, Safety, and Technological Approaches for Their Reductions

Ram Chandra Reddy Jala, Xiaowei Zhang,
Haiqiu Huang, Boyan Gao, Liangli (Lucy) Yu,
and Xuebing Xu

CONTENTS

7.1 INTRODUCTION

3-Monochloropropane-1,2-diol (3-MCPD) is a well-known contaminant formed from glycerides and chlorides under thermal treatment. The compound can be found in free form as well as in its fatty acid ester forms. Although 3-MCPD had been identified in hydrolyzed vegetable protein approximately 30 years ago, the presence of its esters in processed foods was discovered only in 2004 and has been reported at much higher concentrations than those found in its free form, especially in vegetable oils and fats as well as in fried foods. This discovery has been considered a matter of importance in relation to food safety, considering the toxicity of 3-MCPD and the possibility that its intake exceeds the tolerable level currently established, in the case of complete hydrolysis of its esters during digestion, releasing free 3-MCPD. Free 3-MCPD is a known carcinogen, and its fatty acid esters have been detected in food stuffs, including bread, doughnuts, coffee, infant formula, salty crackers, dark malt, French fries, refined vegetable oils, pickled olives, and herrings (Hamlet et al., 2004; Svejkovská et al., 2004; Velíšek et al., 1980; Zelinková et al., 2006, 2008). In 2008, esters of 3-MCPD were also detected in human breast milk, indicating that 3-MCPD esters could be absorbed, distributed, and incorporated into human organs and tissues (Zelinková et al., 2008). These findings have raised a food safety concern, with the assumption that 3-MCPD esters could be completely converted into free 3-MCPD (Weißhaar, 2011). Additionally, the esters of 3-MCPD may also be a toxicological concern considering their presence in milk. The structures of 3-MCPD, glycidol, and their esters are provided in Figure 7.1. This chapter reviews the exposure of 3-MCPD esters in humans, the chemical mechanisms for their formation, factors altering their levels in the refined edible oils and foods, their toxic effects and the biochemical mechanisms behind the analytical methods for their detection, and the possible approaches to reduce their levels in foods.

At the time of preparation and processing of food, undesired compounds such as acrylamide, nitrosamines, PAH, or 3-MCPD fatty acid esters can be formed in addition to desired components that improve the taste and palatability of the product. The formation of such undesirable contaminants mainly depends on the processing

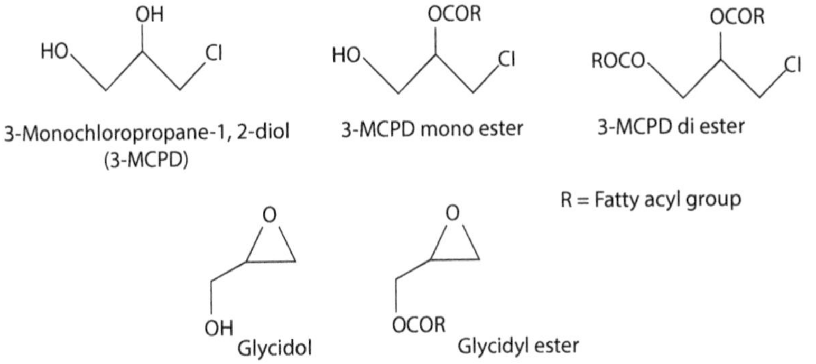

FIGURE 7.1 Structures of 3-MCPD, glycidol, and their esters.

conditions during the heating of the food. In most cases, it is very difficult to avoid the formation of such compounds without changing or impairing the quality of the product. In this connection, the occurrence of 3-MCPD esters or bound 3-MCPD in different refined edible fats and oils was discussed approximately 7–8 years ago. Despite the importance of lipids as precursors, 3-MCPD was not in the focus of lipid science and industry for a long time. However, when the method of analysis was improved and after the findings of higher concentrations of 3-MCPD esters in the fat component of instant infant milk, the activities of officials and industry gathered momentum. However, the problem of 3-MCPD esters is not as simple as the issue of acrylamide. The formation of acrylamide can be reduced by a set of relatively simple measures. However, a significant reduction of 3-MCPD esters is more complex, needing a big effort, and longer times on research and development. For this reason, recently, the issue of reduction of 3-MCPD and allied compounds in foods or edible vegetable oils is attracting a lot of attention of researchers.

7.2 HUMAN EXPOSURE/CONSUMPTION

Food ingredients such as refined oils and thermally processed cereals and retail foodstuffs, including cakes, bread, and infant formula, are the major dietary sources of 3-MCPD esters (Hamlet et al., 2011). The highest concentration of 3-MCPD esters found in refined oils ranges from 200 to 21,500 µg/kg, while in virgin and nonrefined oils, the concentration is below or slightly above the detection limit (<300 µg/kg) (Svejkovská et al., 2004; Zelinková et al., 2006). Moreover, evidence indicates that the physically refined oil may yield higher level of 3-MCPD esters than that of chemical methods (Hrncirik and van Duijn, 2011).

However, it seems that the level of 3-MCPD esters is also affected by the characteristics of the crude oils. Refined palm oil usually contains much higher amount of 3-MCPD esters than seed oils, such as rapeseed, soybean, sunflower seeds, and maize (Franke et al., 2009; Zelinková et al., 2006). Furthermore, reports indicate the occurrence of 3-MCPD esters in roasted cereals such as rye, wheat, malt, and barley (Divinová et al., 2007; Doležal et al., 2009; Stadler et al., 2007; Svejkovská et al., 2004), and the concentrations of 3-MCPD esters are relatively high ranging from 25 to 1896 µg/kg (Hamlet et al., 2011). Roasted cereals and malts are important constituents of cereal-based beverages and are widely used in the brewing and baking industries. In ready-to-eat foods, it is not surprising to find that the products using significant amounts of refined oils such as potato crisps, coffee creamer, biscuits, and doughnut contain greater level of 3-MCPD esters (Hamlet et al., 2011). Moreover, in 2008, 30–2195 µg per kg of fat or 6–76 µg/kg milk of 3-MCPD esters was found in human breast milk (Zelinková et al., 2008). As of today, there is little knowledge of 3-MCPD esters in nonthermal processed food products.

7.3 CHEMICAL MECHANISM FOR THE FORMATION OF 3-MCPD ESTERS

3-MCPD esters were first identified in 1980 by Velíšek et al. (1980). Thirty years later, their exact formation mechanism still remains unclear. It was reported that the 3-MCPD esters might be formed during the oil deodorization operation (Franke et al., 2009; Weißhaar, 2008). To date, there are five proposed mechanisms for the formation of 3-MCPD esters (Figure 7.2). Two of them involve direct substitution by the chlorine ion at the glycerol carbon atoms replacing either a hydroxyl group (pathway **a**) or an ester fatty acid group (pathway **b**) (Collier et al., 1991; Rahn and Yaylayan, 2011b). Another two mechanisms involve the formation of an epoxide intermediate (pathway **c**) (Sonnet, 1991) or an acyloxonium ion (pathway **d**) (Collier et al., 1991; Velíšek et al., 2002), followed by a nucleophilic attack by chlorine ion to form the final 3-MCPD esters. The last proposed mechanism postulated the formation of an intermediate cyclic acyloxonium-free radical, followed by a nucleophilic substitution with chlorine radicals (pathway **e**) (Zhang et al., 2013). The free radical mechanism was supported by solid evidences. Each mechanism will be discussed in more detail below.

7.3.1 3-MCPD Ester Formation through Direct Nucleophilic Substitution (the Pathways a and b)

Collier et al. (1991) proposed a mechanism in which chlorine ions could directly substitute either a hydroxyl group or an ester of fatty acid group in triolein through a S_N2 pathway to produce free 3-MCPD and 2-MCPD. Svejkovská et al. (2006) also reported similar observation that 3-MCPD esters could be produced by heating 1-monopalmitin, 1,3-dipalmitin, tripalmitin, and soybean oil. The amount of 3-MCPD esters produced from 1-monopalmitin was 10 times more than that from tripalmitin. These results lead to the proposed S_N2 substitution mechanism (Rahn and Yaylayan, 2011a), although they do not necessarily support the direct nucleophilic substitution (Figure 7.2, pathways a and b).

7.3.2 3-MCPD Ester Formation through Glycidol Esters (the Pathway c)

A number of reports have indicated that glycidol could be converted into 3-MCPD through a ring opening of the epoxide in the presence of a chloride reagent (Weißhaar and Perz, 2010). It is very reasonable that the corresponding esters of glycidol would undergo similar transformations to form 3-MCPD esters. Furthermore, Sonnet (1991)

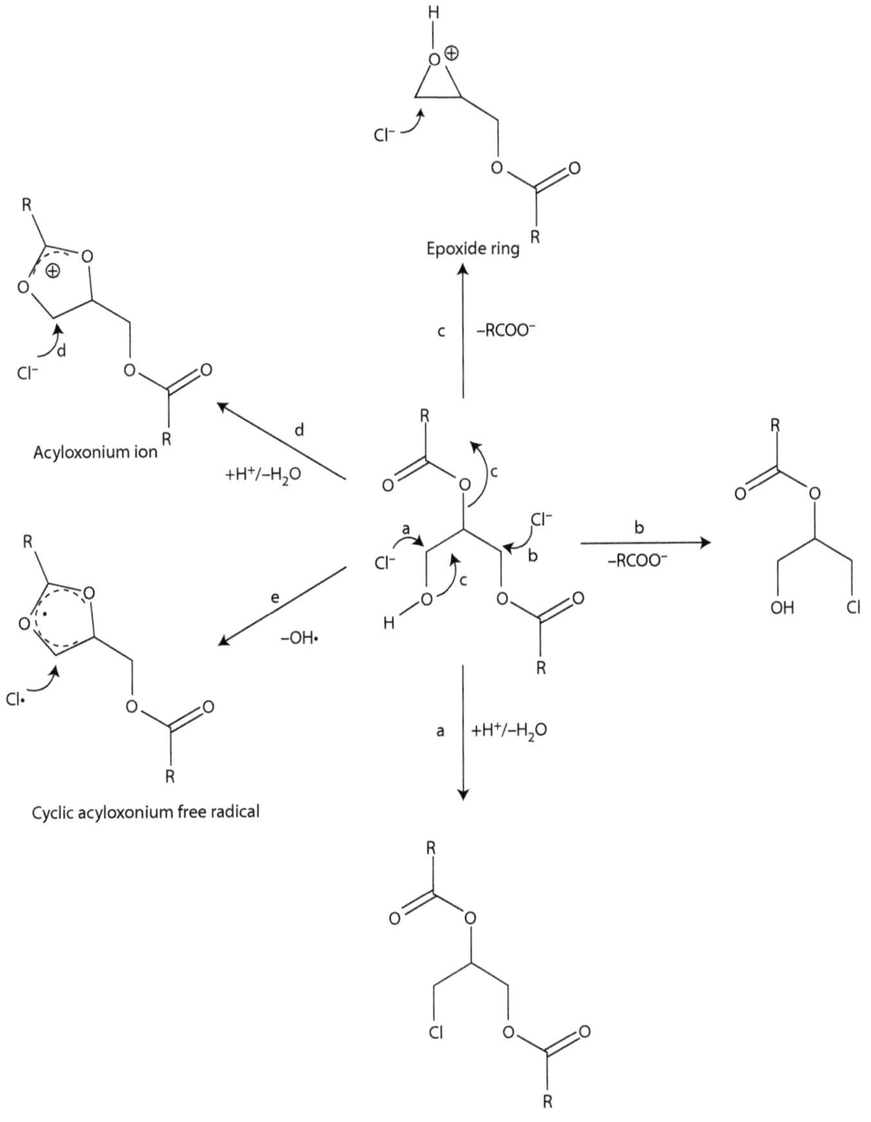

FIGURE 7.2 Summary of proposed pathways of 3-MCPD ester formation. Pathway (a): direct nucleophilic attack at glycerol carbon carrying a hydroxyl group; pathway (b): direct nucleophilic attack at glycerol carbon carrying an ester group; pathway (c): formation of an epoxide ring; pathway (d): formation of acyloxonium ion; and pathway (e): formation of cyclic acyloxonium free radical. R represents fatty acid side chain. (Adapted from Zhang, X. W. *et al.* 2013. *Journal of Agricultural and Food Chemistry, 61*(10), 2548–2555.)

reported that glycidol esters (GEs) could react with bromide ions to form 3-bromo-propanediol esters, and it is possible that chlorine ions would attack the least hindered carbon of GE to form 3-MCPD monoester (Figure 7.2, pathway c). However, the exact mechanism of GE formation was still unclear. Diacylglycerol (DAG) and monoacylglycerol (MAG) are considered to be the precursors of GE (Masukawa et al., 2010). They may first form acyloxonium ions and then generate GE (Figure 7.2). Based on the work of Svejkovská et al. (2006), Weißhaar et al. (2010) proposed that acyloxonium could be converted into glycidols and chlorine ions would react with the glycidol to form 3-MCPD esters through epoxide ring opening subsequently. This proposed mechanism seemed plausible; however, more evidence is needed to confirm it since no direct proof has been found to date.

7.3.3   3-MCPD ESTER FORMATION THROUGH ACYLOXONIUM CATION INTERMEDIATE (PATHWAY D)

Collier et al. (1991) proposed the acyloxonium cation intermediate in the formation of 3-MCPD esters. The ability of acetic acid to increase the ratio of 3-MCPD to 2-MCPD esters was attributed to the participation of C-2 carbon and due to the formation of a cyclic acyloxonium cation (Figure 7.2, pathway d). To further confirm this mechanism, triolein was chlorinated in an aqueous HCl system, and it was proposed that triolein would first be hydrolyzed into diolein, following which an acyloxonium ion could be formed. Velíšek et al. (2002) investigated the formation of optical chloropropanediol isomers and their decomposition into protein hydrolytes, and supported the acyloxonium ion mechanism. However, no direct evidence was available for this proposed mechanism until 2011, when Rahn and Yaylayan (2011a) first tried to monitor the formation of the acyloxonium cation using Fourier transform infrared (FT-IR). Tripalmitin or 1,2-dipalmitin samples were heated at 100°C in the presence of $ZnCl_2$, and a new band was observed at 1651 cm^{-1}. Under the low moisture thermal treatment conditions, 1,2-dipalmitoyl glycerol was much faster than either tripalmitin and 1-monopalmitin in generating 3-MCPD esters. Taking the information together, these results suggested that C-2 carbonyl group was involved in forming the acyloxonium cation intermediate for 3-MCPD diester synthesis with possible catalytic action of a free hydroxyl group as a proton transfer site. However, the FT-IR spectrum was not a direct evidence for the formation of the acyloxonium ions except for the involvement of C-2 carbon. More evidence is needed to prove the existence of acyloxonium cation.

7.3.4   3-MCPD ESTER FORMATION THROUGH CYCLIC ACYLOXONIUM FREE RADICAL INTERMEDIATE (PATHWAY E)

Zelinková et al. (2006) found that thermal treatment of rapeseed oils at 230°C time-dependently increased the level of 3-MCPD diesters. Destaillats et al. (2012) also observed that heating triglycerides with an organic chloride compound at 235°C significantly increased the formation of 3-MCPD diesters. These observations led to a hypothesis that free radicals might be generated and involved in the formation of

3-MCPD diesters. Zhang et al. (2013) reported that a free radical was formed and mediated the formation of 3-MCPD diesters based on the electron spin resonance (ESR) signals and quadrupole time of flight (Q-TOF) tandem mass spectrometry (MS/MS) detection of sn-1,2-stearoylglycerol (DSG). First, an ESR spectroscopy examination of the mixture of vegetable oils and 5,5-dimethyl-1-pyrroline N-oxide (DMPO), a radical trapping agent for ESR determination, showed the presence of free radicals. The ESR investigation also indicated the correlation between the pretreatment temperature and the free radical concentration. Then, a hypsochromic shift of carbonyl group was found through an FT-IR spectroscopy analysis of DSG at 25°C and 120°C. At 25°C, DSG exhibited two carbonyl absorbances, one band at 1733 cm^{-1} and the other at 1711 cm^{-1}, indicating the presence of two chemically different environments for the two ester carbonyl groups. Upon heating to 120°C, only a peak at 1744 cm^{-1} was observed, reflecting the possible involvement of an ester carbonyl group in forming 3-MCPD diesters. Based on these facts, a novel free radical mechanism was proposed that cyclic acyloxonium free radical (CAFR) mediated the formation of 3-MCPD diesters (Figure 7.2). Furthermore, the DMPO and CAFR adducts were detected through a Q-TOF MS/MS investigation. The observation of [CAFR]$^+$ (m/z 607.5629, cacl. 607.5665), [CAFR + DMPO + O + Na]$^+$ (m/z 760.6448, cacl. 760.6431), and [CAFR + DMPO + OH + H]$^+$ (m/z 739.6674, cacl. 739.6690) suggested the presence of CAFR. Finally, the free radical mechanism was validated by the formation of 3-MCPD diesters through reacting DSG with a number of organic and inorganic chlorine agents, including chlorine gas, at 120°C and 240°C under a low moisture condition. Additional research is needed to investigate the formation mechanism of 3-MCPD mono-ester or 3-MCPD diester from TAG.

In summary, five proposed mechanisms for the formation of 3-MCPD esters have been proposed to date. Certain direct evidences have been generated that supported the radical mechanism for 3-MCPD ester formation from DAG, while the formation mechanism from TAG precursor or MAG is still unclear. It is very important to know the exact mechanism of 3-MCPD ester formation to reduce their levels in refined edible oils and other related food products.

7.4   FACTORS ALTERING THEIR LEVELS IN REFINED EDIBLE OILS

Zelinková et al. (2006) showed for the first time that remarkably high amounts of 3-MCPD esters could be found in edible vegetable oils after refining. Later it was found that among the most commonly utilized edible oils, palm oil had a high potential to form these esters. Therefore, the present discussion uses palm oil as a model. In most of these cases, the same concepts/conclusions can be applied to other vegetable oils.

7.4.1   EFFECT OF WATER UTILIZED IN PROCESSING

Water used for the generation of the strip steam during deodorization was the first suspect to have an effect as a source of chloride; however, Pudel et al. (2011) proved that the type of water has no influence. Matthaus et al. (2011) for the first time assumed that the chlorine donor has to be present in the oil in an oil-soluble form to enable the reaction with other precursors. They found amounts between 1 and 6 mg water-soluble chloride/kg in different palm oils; however, it was unclear whether these water-soluble compounds participated in the 3-MCPD ester formation during heat treatment or whether the reactive chlorine donors have to be present in another form.

7.4.2 Effect of Chlorine Substances in Palm Fruit Extracts

Nagy et al. (2011) used holistic mass-defect filtering of isotopic signatures and found that various extracts of palm fruit, and partially and fully refined palm oil naturally contain a huge number of different unexpected and unknown chlorine containing substances. In addition, in many organic chlorine compounds they found several inorganic chlorides such as calcium chloride, magnesium chloride, iron [II] chloride, and iron [III] chloride in palm oil. According to them, during processing, the relatively polar chlorinated constituents in the palm oil could be converted into more and more lipophilic forms (such as MCPD esters) along the oil processing chain. Looking at the processing chain of palm oil, during the growing of the oil palm, the first input of inorganic chloride containing compounds takes place as fertilizers in the form of potassium chloride or ammonium chloride to improve oil palm growth and bunches yield. During the growth period, these chloride compounds are absorbed by the plant and accumulated in the fruits. However, a lot of drinking water is essential for the further processing where iron [III] chloride is being used as a coagulant in water treatment (Abdollah, 2010). The organochlorines seem to be endogenously produced by the oil palm during maturation (Craft et al., 2012; Nagy et al., 2011); however, they also showed that organochlorines, initially present in oil palm fruits, are transformed into more lipophilic forms during the further palm oil processing. Further, they stated that organochlorines might begin to decompose at temperature greater than 120°C and at higher temperature (>150°C) the formation of 3-MCPD ester starts.

According to Destaillats et al. (2012), 3-MCPD esters are formed during deodorization due to a thermally catalyzed decomposition of organochlorine compounds naturally occurring in palm oil into reactive chlorinated compounds such as hydrogen chloride. These compounds can react with acyl glycerols to yield MCPD diesters and also release the free fatty acid from the intact triglyceride molecule.

7.4.3 Effect of Raw Material (Palm Oil) Quality

Poor quality crude palm oil which contains high FFA, high phosphorous content, and high monoglycerides may result in a greater level of 3-MCPD ester formation, especially in its unrefined form. Zulkurnain et al. (2012) indicated that the intrinsic composition of the crude palm oil could contribute to 3-MCPD ester formation.

7.4.4 Effect of Deodorization Temperature

Zulkurnain et al. (2012) also studied the effect of increasing deodorization temperature on 3-MCPD ester formation in high-quality palm oil. The formation of 3-MCPD esters was significantly increased as the deodorization temperature was increased. The level of chloroesters reached 9 mg/kg when the sample was deodorized at 250°C. However, further increase in temperature led to the degradation of 3-MCPD esters (Matthaus et al., 2011; Pudel et al., 2011).

7.5 ABSORPTION AND DISTRIBUTION OF 3-MCPD FATTY ACID ESTERS

An in vivo study using male Wistar rats (strain Crl:WI) accessed the bioavailability of 3-MCPD esters. Studies on distribution of 3-MCPD esters upon oral administration of 53.2 mg/kg body weight (BW) 3-MCPD dipalmitate showed the presence of 3-MCPD esters in small and large intestine within 24 h, while only free 3-MCPD was detected in blood, liver, kidney, and fat tissues (Abraham et al., 2012).

It was determined that free 3-MCPD was able to migrate via a paracellular diffusion through a Caco-2 monolayer without being absorbed or metabolized by the cells (Buhrke et al., 2011). However, 3-MCPD monoesters were hydrolyzed by cellular lipases in Caco-2 cells, and released free 3-MCPD, which could then be absorbed through paracellular diffusion; while the diester, 3-MCPD dipalmitate, was metabolized by the cells over time (Buhrke et al., 2011).

7.6 TOXIC EFFECTS OF 3-MCPD FATTY ACID ESTERS

As a newly identified class of food process contaminants, there are few toxicological data available on fatty acid esters of 3-MCPD, and the primary toxicological concern is based on the potential release of 3-MCPD from the parent esters in the gastrointestinal system (Buhrke et al., 2011). A recent report indicated that the toxicological effect of 3-MCPD esters might affect kidney and testes in experimental animals (Liu et al., 2012). This part will review the current knowledge on toxicological effect of 3-MCPD esters in various organs and tissues.

7.6.1 ACUTE ORAL TOXICITY

A study in 2011 by Tee and colleagues using 5 male and 5 female Sprague–Dawley rats determined that the LD_{50} values of 3-MCPD esters were greater than 400 ppm (Tee et al., 2011). In the following year, acute oral toxicity of 3-MCPD 1-monopalmitate and 3-MCPD dipalmitate was studied in Swiss mice (male and female, ~4 weeks, 18–22 g). The LD_{50} value of 3-MCPD 1-monopalmitate was determined to be 2676.81 mg/kg, while the LD_{50} value of 3-MCPD dipalmitate was presumed to be greater than 5000 mg/kg (Liu et al., 2012).

7.6.2 LIVER

The liver of experimental animals (mouse and rat) was slightly affected by 3-MCPD ester treatments. Swelling and fatty liver were observed in Swiss mice (male and female, ~4 weeks, 18–22 g) (Liu et al., 2012) and centrolobular hepatocyte degeneration was observed in Wistar rats receiving 156.75 mg/kg BW per day of 3-MCPD dipalmitate (male and female, 200–220 g) (Barocelli et al., 2011).

7.6.3 KIDNEY

Wistar rats (male and female, 200–220 g) exposed to 156.75 mg/kg BW per day of 3-MCPD dipalmitate showed pale, yellowish, and enlarged kidney (Barocelli et al., 2011). Degenerative changes were shown to be induced by 2040 mg/kg BW and greater dose of 3-MCPD 1-monopalmitate and 5000 mg/kg BW dipalmitate in different tubular segments of the kidney in Swiss mice (male and female, ~4 weeks, 18–22 g) (Figure 7.3). The histopathology included focal degeneration (hydropic

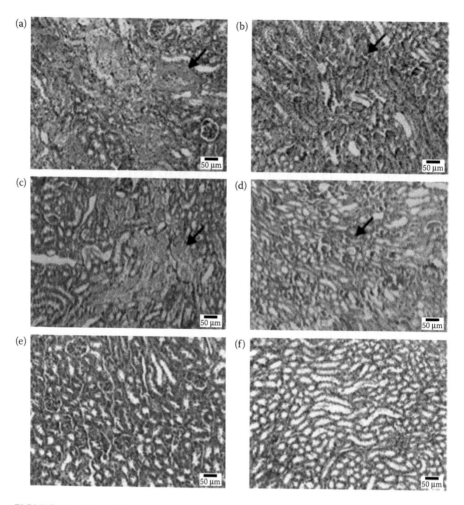

FIGURE 7.3 Mice kidney sections from control, 3-MCPD 1-monopalmitate and dipalmitate groups. (a) and (b) Mice fed 4162 mg/kg 3-MCPD 1-monopalmitate; (c) and (d) Mice fed 5000 mg/kg 3-MCPD dipalmitate; and (e) and (f) Control mice. Part figures (a) and (c) show renal tubular necrosis as compared with (e); and (b) and (d) show protein casts scatted among the renal tubules as compared with (f). Hematoxylin–eosin staining (×200). (Adapted from Liu, M. *et al.* 2012. *Food and Chemical Toxicology*, 50(10), 3785–3791.)

degeneration) and necrosis of tubular cells, along with protein casts scattered among tubules in the overlying cortex and subjacent medulla. Similar lesion was also observed in Wistar rat treated with 3-MCPD dipalmitate, which showed moderate to severe degeneration, hypertrophy, and swelling of epithelial cells of proximal and distal tubules, associated with hydropic-vacuolar degenerations of tubular epithelial cells and multiple foci of interstitial nephritis (Barocelli et al., 2011). The cytotoxicity of 3-MCPD esters was also tested in NRK-52E rat kidney cells (Figure 7.4). 3-MCPD 1-monopalmitate at initial concentrations of 10 and 100 µg/mL significantly suppressed NRK-52E rat kidney cell proliferation after 48 h of treatment (Liu et al., 2012).

7.6.4   TESTES

Male Swiss mice (~4 weeks, 18–22 g) exposed to 3-MCPD 4162 mg/kg BW 1-monopalmitate and 5000 mg/kg BW dipalmitate decreased spermatids in a single seminiferous tubule and Leydig cells, disperse array, and desquamation of the constructive cells (Liu et al., 2012). Mild-to-moderate decrease of spermatids, atrophy of spermatogenic and supporting cells in the seminiferous tubules were observed in Wistar rat (200–220 g) exposed to 156.75 mg/kg BW 3-MCPD dipalmitate (Barocelli et al., 2011).

7.6.5   OTHER TOXIC EFFECTS

Tee and colleagues (2011) evaluated the cytotoxicity of 1-palmitoyl, 1-steroyl, 2-oleoyl, and 1-palmitoy-2-oleoyl-3-chloropanediol. 2-Oleoyl-3-chloropanediol showed a toxic effect at 781 ppm with an IC_{50} value of 520 ppm. 1-Palmitoyl and 1-steroyl-3-chloropanediol showed cellular toxicity at 200,000 ppm level, with IC_{50} of 70,000 and 31,000 ppm, respectively. 1-Palmitoyl-2-oleoyl-3-chloropanediol showed no cytotoxic effect at the concentration of 24–781 ppm under the same experimental condition (Tee et al., 2011).

3-MCPD fatty acid esters may affect the body weight in experimental animals. Swiss mice fed with 3-MCPD 1-monopalmitate (>2040 mg/kg body weight) showed significant lower body weight over a 14-day feeding period (Liu et al., 2012). Wistar rats, male and female, both showed significant reduction of body weight gain upon the 3-MCPD dipalmitate treatment (Barocelli et al., 2011). Upon exposure to 3-MCPD esters, passive hyperemia and edema were observed in the lungs and the heart was frequently dilated and degenerated (Barocelli et al., 2011). A slight decrease of lymphocytes was seen in the thymus and spleen (Liu et al., 2012).

7.7   ANALYTICAL METHODS FOR THEIR DETECTION

In 2007, Weißhaar (2008) initiated the measuring of higher concentrations of 3-MCPD esters in refined fats and oils, and developed many methods for their determination. Initially, Weißhaar determined free 3-MCPD after transesterification of the 3-MCPD fatty acid esters with sodium methoxide. This and other methods were tested using multi-laboratories by BfR in 2008 (Pudel et al., 2011).

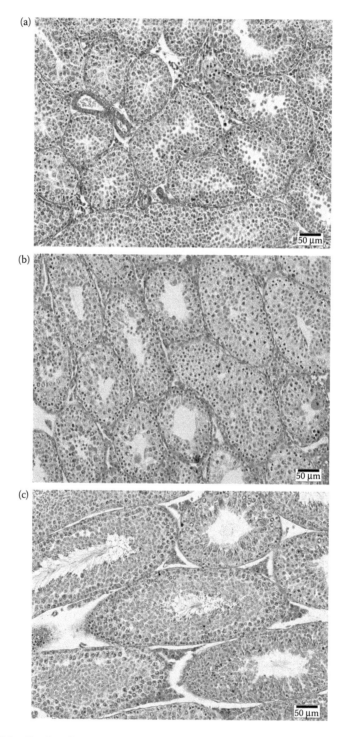

FIGURE 7.4 Continued

In March 2009, the validated method was published by the German Society for Fat Science (DGF) as German Standard Method C-III 18 (09). 3-MCPD can be formed or lost during the process of extraction from food products if strict protocols are not followed. For example, it is already proved that 3-MCPD is formed during the analysis if sodium chloride is present. From this it was concluded that together with 3-MCPD, other 3-MCPD forming substances such as glycidol were determined. For this reason, Weißhaar modified his method to determine both parameters (Weißhaar and Perz 2010). The amended DGF Standard Method C-III 18 (09) (German standard methods 2009) describes a procedure for the separate determination of 3-MCPD esters in fats and oils by means of GC/MS after cleavage of esters with methanol and sodium methoxide. Assuming that no other compounds are present, with the exception of glycidol, which forms 3-MCPD under the conditions of the analysis, the method also enables the indirect determination of glycidol or its esters, as glycidol is identified as one of these 3-MCPD forming substances. Feibig (2011) further developed a couple of robust methods for the determination of the total true 3-MCPD ester and induced 3-MCPD from ester-bound glycidol. Those methods were successfully validated by a collaborative trial conducted according to internationally agreed procedures. On the assumption that there were no other 3-MCPD forming substances except glycidol, both methods can be used to estimate the 3-MCPD levels in fats and oils. Considering the aim to reduce the content of 3-MCPD esters in fats and oils, the tested standards should be able to help in achieving this. Recently, a direct method using HPLC-MS/TOF is under test using multi-laboratories by AOCS. The method was initially proposed by ADM. The advantage of the method is not involving too much pre-treatment, which is believed to have potential impact on the real content of the compounds concerned. However, there is a concern of the availability of the facility in majority of the analytical systems. Therefore, further research is needed to find out the perfect and simple protocol for 3-MCPD determination.

7.8 POSSIBLE APPROACHES TO REDUCE THEIR LEVELS IN FOODS

Thorough knowledge on the chlorine donor and the decomposition of chlorine-containing compounds during deodorization are considered to be important points for the development of mitigation strategies to reduce the quantities of 3-MCPD esters in refined vegetable oils. The following strategies can be adapted to reduce their levels in foods; in particular, refined edible oils as these approaches were already examined in case of palm oil.

FIGURE 7.4 Mice testes sections from control, 3-MCPD 1-monopalmitate and dipalmitate groups. (a) Male mice in 4162 mg/kg 3-MCPD 1-monopalmitate group; (b) male mice in 5000 mg/kg 3-MCPD dipalmitate group; (c) control mice. Part figures (a) and (b) showed decrease of spermatids in a single seminiferous tubule, disperse array, and desquamation of the constructive cells as compared with (c). Hematoxylin–eosin staining (×200). (Adapted from Liu, M. et al. 2012. *Food and Chemical Toxicology, 50*(10), 3785–3791.)

7.8.1 To Remove Reactants from the Raw Materials

One of the better possible approaches to gain control on the formation of 3-MCPD esters and allied compounds seems to be the removal of reactants from the raw material before refining. The reason for the formation of DAG and MAG before processing of refining is the enzymatic degradation of TAGs after maturity. Activities such as deactivation of lipase by short harvest cycles and/or immediate pretreatment in the oil mill or breeding of palm fruits with a lower lipase activity may prevent the triglyceride hydrolysis. However, these are time- and labor-consuming possibilities. An enzymatic esterification of DAG and MAGs back to TAGs before further processing would also lower the amount of DAG and MAGs in the raw material; but it is not an easy task to bring this balance reaction into the direction of TAGs in a way that would make this possibility successful. Washing of the crude oils before processing in order to lower the amounts of precursors such as chlorides could be a possible better idea.

7.8.2 To Remove Phospholipids and FFA from the Raw Material

The pretreatment of the oil before deodorization can contribute to the reduction of the amount of contaminants such as 3-MCPD esters formed afterwards. There is no direct correlation between phospholipid removal during degumming and 3-MCPD and glycidyl ester formation during deodorization; similarly, no direct relation was observed between removal of FFA during neutralization and formation of 3-MCPD esters during deodorization. However, interestingly, controlling both steps has shown positive effects and led to a reduction of the potential to form MCPD esters and glycidyl esters during deodorization (Pudel et al., 2011).

7.8.3 To Remove Chlorine Substances from Palm Oil Using Water or Water/Alcohol Mixtures

The removal of chlorine substances from the crude oil or even from the pulp during the early stage of processing can definitely reduce the formation of the esters drastically. Washing crude palm oil with polar solvents such as water or water/alcohol mixtures was suggested (Matthaus et al., 2011). During the washing process chlorine-containing polar compounds were removed from the oil, resulting in significant lower formation of 3-MCPD esters. Craft et al. (2012) suggested that it would be better to remove the reactive chlorine species from the extracted palm fruit oil rather than from the crude palm oil as more polar chlorinate compounds convert into more apolar compounds during the processing. This reveals that the earlier the removal of the chlorinated species takes place, the better for the mitigation of 3-MCPD esters.

7.8.4 To Remove Chlorine Substances during Water Degumming and Bleaching

The formation of 3-MCPD esters during deodorization can be influenced by the degumming and bleaching conditions. Water degumming significantly reduced the formation of 3-MCPD esters from 9.79 to 1.55 mg/kg. This could be due to the removal of reactants such as chloride, which contributes to the formation of 3-MCPD esters during water degumming (Matthaus et al., 2011). The phosphoric acid dosage can have a great effect also on the development of chloroesters than the different types of bleaching adsorbents. Bleaching using magnesium silicate shows the greatest reduction of 3-MCPD esters compared to the natural clay, activated clay, or carbon, especially when water degumming was performed. Magnesium silicate possesses a relatively large surface area and the most active basic sites compared to alumina, silica, and activated carbon (Zhu et al., 1994). These properties may be responsible for the adsorption or degradation of particular precursors of 3-MCPD esters from bleached palm oil prior to deodorization. The combined adsorbents such as magnesium silicate and activated clay could be effective in reducing 3-MCPD ester formation and improving the refined oil quality (Zulkurnain et al., 2012). Bleaching using partly activated bleaching earth, namely, Tonsil 4191 FF at temperatures of 60°C and 90°C, can also reduce the formation of contaminants during the following deodorization (Pudel et al., 2011).

7.8.5 To Remove Chlorine Substances from Palm Oil Using Adsorption Materials

Adsorption materials that are already applied for the removal of polar components from frying oils can be examined for this purpose. Generally, these compounds result from degradation reactions in the oils due to the high temperatures during frying. Usually, 3-MCPD esters and allied compounds differ in their polarity compared to those of pure triglycerides as their structures are closer to those of mono- and diglycerides. Mono- and diglycerides form a major part of the polar compounds in edible oils (Arroyo et al., 1992). Several inorganic powders such as amorphous magnesium silicate, zeolite, silicon oxide, sodium aluminum, synthetic calcium silicate, and synthetic magnesium silicate were examined with respect to the ability of lowering polar compounds and related substances (Strijowski et al., 2011). Even though a few powders were able to reduce the concentration of polar compounds, in particular, a couple of powders were effective in reducing the contents of 3-MCPD esters and related substances. However, the potential adsorption materials should be characterized with respect to their surface properties to get more information about the correlation between reductions of polar/chlorine compounds and their properties.

7.8.6 To Remove 3-MCPD Esters and Related Compounds from Processed Oils Using Enzymatic Approach

The inclusion of a further purification step after deodorization in order to remove 3-MCPD esters and related compounds may reduce their contents in the edible oils. An enzymatic approach to remove 3-MCPD and its esters from aqueous and biphasic systems by converting it into glycerol was suggested by Bornscheuer and Hesseler (2010). 3-MCPD-1-monooleate was converted in a biphasic system in the presence of an edible oil by *Candida antarctica* lipase A to yield free 3-MCPD and corresponding fatty acids. In a further step, the conversion of 3-MCPD was performed to yield harmless glycerol by an enzyme cascade consisting of a halohydrin dehalogenase from *Arthobacter* sp. AD2 and an epoxide hydrolase from *Agrobacterium radiobacter* AD1.

7.8.7 To Manipulate the Deodorization Conditions

Deodorization at temperatures below 240°C seems to be advantageous to achieve low concentration of contaminants, especially glycidyl esters. Controlling temperature may be an effective approach to reduce the 3-MCPD ester level in the refined oils.

7.8.8 To Replace Higher Contaminated Fats by Lower Contaminated Fats

For sensitive food stuffs such as infant milk and children's food, reduction of 3-MCPD esters can be achieved much faster by changing fat composition of the products, that is, replacing higher contaminated fats by lower ones Weißhaar (2008).

7.9 CONCLUSION AND PROSPECTIVE

Even though a number of toxicological studies have been performed on 3-MCPD, few studies on the potential safety concern of 3-MCPD fatty acid esters have been reported to date. The primary safety concern is the possible hydrolysis of 3-MCPD esters in the gastrointestinal tract to free 3-MCPD. Considering the fact that 3-MCPD ester was found in human milk in significant amount, the possibility exists that 3-MCPD esters undergo the process of absorption, digestion, distribution, and re-esterification in the circulation system or in specific organs. Recent studies indicated that the toxicological properties of 3-MCPD esters might be affected by the type of fatty acids, position of esterification, and the number of substitutions. Thus, further study is needed to elucidate the toxicological effect of 3-MCPD esters.

However, food industry, particularly infant formula industry, has taken actions to implement controls of the content of 3-MCPD and its derivatives as well as the glycidyl esters, regardless of the regulatory requirements from different organizations. This commercial push gives significant impact on the oil and fat industry to take actions to meet the needs from those business sides.

Technology has been advanced in the last 3–5 years. Around three dozens of patent submissions in the last 3 years indicate that technical/industrial activities rose from the issues. The oil and fat industry has made ready to supply the required low 3-MCPD ester and low glycidol products even though there is still large space to fill for the cost-effective approaches. With the need to supply good oils and fats not only for infants but also for adults, this may bring further consideration for the sourcing of oils and fats in future.

REFERENCES

Abdollah, N. A. B. 2010. Some problem in chemical and physical treatment of water supply. Thesis, Johor Bahru (Johor Malaysia), Universiti Teknologi Malaysia.

Abraham, K., Appel, K., Berger-Preiss, E., Apel, E., Gerling, S., Mielke, H., Creutzenberg, O., and Lampen, A. 2013. Relative oral bioavailability of 3-MCPD from 3-MCPD fatty acid esters in rats. *Archives of Toxicology, 87,* 649–659.

Arroyo, R., Cuesta, C., Garrido-Polonlio, C., Lopez-Varela, S., and Sanchez-Muniz, F. J. 1992. High performance size exclusion chromatographic studies on polar components formed in sunflower oil used for frying. *Journal of the American Oil Chemists Society, 69,* 557–563.

Barocelli, E., Corradi, A., Mutti, A., and Petronimo, P. G. 2011. Comparison between 3-MCPD and its palmitic esters in a 90-day toxicological study. In: CFP/EFSA/CONTAM/ 2009/01 [Accepted for publication on 22 August 2011], University of Parma, Parma, Italy.

Bornscheuer, U.T., Hesseler, M. 2010. Enzymatic removal of 3-monochloro-1, 2-propanediol (3-MCPD) and its esters from oils. *European Journal of Lipid Science and Technology, 112,* 552–556.

Buhrke, T., Weißhaar, R., and Lampen, A. 2011. Absorption and metabolism of the food contaminant 3-chloro-1,2-propanediol (3-MCPD) and its fatty acid esters by human intestinal Caco-2 cells. *Archives of Toxicology, 85*(10), 1201–1208.

Collier, P. D., Cromie, D. D. O., and Davies, A. P. 1991. Mechanism of formation of chloropropanols present in protein hydrolysates. *Journal of the American Oil Chemists Society, 68*(10), 785–790.

Craft, B. D., Nagy, K., Sandoz, L., and Destaillat, F. 2012. Factors impacting the formation of monochloropropanediol (MCPD) fatty acid diesters during palm (*Elaeis guineensis*) oil production. *Food Additives and Contaminants, 29,* 354–361.

Destaillats, F., Craft, B. D., Sandoz, L., and Nagy, K. 2012. Formation mechanisms of monochloropropanediol (MCPD) fatty acid diesters in refined palm (*Elaeis guineensis*) oil and related fractions. *Food Additives and Contaminants Part a—Chemistry Analysis Control Exposure* and *Risk Assessment, 29*(1), 29–37.

Divinová, V., Doležal, M., and Velíšek, J. 2007. Free and bound 3-chloropropane-1,2-diol in coffee surrogates and malts. *Czech Journal of Food Sciences, 25*(1), 39–47.

Doležal, M., KertISoVá, J., ZelInKoVá, Z., and Velíšek, J. 2009. Analysis of bread lipids for 3-MCPD esters. *Czech Journal of Food Sciences, 27,* 417–420.

Feibig, H. J. 2011. Determination of ester-bound 3-chloro-1, 2-propanediol and glycidol in fats and oils—A collaborative study. *European Journal of Lipid Science and Technology, 113,* 393–399.

Franke, K., Strijowski, U., Fleck, G., and Pudel, F. 2009. Influence of chemical refining process and oil type on bound 3-chloro-1,2-propanediol contents in palm oil and rapeseed oil. *Lwt-Food Science and Technology, 42*(10), 1751–1754.

German Standard Methods for the Analysis of Fats and other Lipids. 2009. C-III 18 (09), Ester-bound 3-chloropropane-1, 2-diol (3-MCPD-esters) and glycidol (glycidyl esters)—Determination in fats and oils by GC-MS. *German Standard Methods for the Analysis of Fats and other Lipids.* WVG, Stuttgart.

Hamlet, C. G., Sadd, P. A., and Gray, D. A. 2004. Chloropropanols and their esters in baked cereal products. *Abstracts of Papers of the American Chemical Society, 227,* U38–U38.

Hamlet, C., Asuncion, L., Velíšek, J., Doležal, M., Zelinková, Z., and Crews, C. 2011. Formation and occurrence of esters of 3-chloropropane-1,2-diol (3-CPD) in foods: What we know and what we assume. *European Journal of Lipid Science and Technology, 113*(3), 279–303.

Hrncirik, K., and van Duijn, G. 2011. An initial study on the formation of 3-MCPD esters during oil refining. *European Journal of Lipid Science and Technology, 113*(3), 374–379.

Liu, M., Gao, B.-Y., Qin, F., Wu, P.-P., Shi, H.-M., Luo, W., Ma, A.-N., Jiang, Y.-R., Xu, X.-B., and Yu, L.-L. 2012. Acute oral toxicity of 3-MCPD mono- and di-palmitic esters in Swiss mice and their cytotoxicity in NRK-52E rat kidney cells. *Food and Chemical Toxicology, 50*(10), 3785–3791.

Masukawa, Y., Shiro, H., Nakamura, S., Kondo, N., Jin, N., Suzuki, N., Ooi, N., and Kudo, N. 2010. A new analytical method for the quantification of glycidol fatty acid esters in edible oils. *Journal of Oleo Science, 59*(2), 81–88.

Matthaus, B., Pudel, F., Fehling, P., Vosmann, K., and Freudenstein, A. 2011. Strategies for the reduction of 3-MCPD esters and related compounds in vegetable oils. *European Journal of Lipid Science and Technology, 113*, 380–386.

Nagy, K., Sandoz, L., Craft, B. D., and Destaillats, F. 2011. Mass–defect filtering of isotope signatures to reveal the source of chlorinated palm oil contaminants. *Food Additives and Contaminants, 28*, 1492–1500.

Pudel, F., Benecke, P., Fehling, P., Freudenstein, A., Matthaus, B., and Schwaf, A. 2011. On the necessity of edible oil refining and possible sources of 3-MCPD and glycidyl ethers. *European Journal of Lipid Science and Technology, 113*, 368–373.

Rahn, A. K. K., and Yaylayan, V. A. 2011a. Monitoring cyclic acyloxonium ion formation in palmitin systems using infrared spectroscopy and isotope labelling technique. *European Journal of Lipid Science and Technology, 113*(3), 330–334.

Rahn, A. K. K., and Yaylayan, V. A. 2011b. What do we know about the molecular mechanism of 3-MCPD ester formation? *European Journal of Lipid Science and Technology, 113*(3), 323–329.

Sonnet, P. E. 1991. A short highly regioselective and stereoselective synthesis of triacylglycerols. *Chemistry and Physics of Lipids, 58*(1–2), 35–39.

Stadler, R. H., Theurillat, V., Studer, A., Scanlan, F., and Seefelder, W. 2007. The formation of 3-monochloropropane-1,2-diol (3-MCPD) in food and potential measures of control. In *Thermal Processing of Food* (pp. 141–154): Wiley-VCH Verlag GmbH and Co. KGaA, Weinheim, Germany.

Strijowski, U., Heinz, V., and Franke, K. 2011. Removal of 3-MCPD esters and related substances after refining by adsorbent material. *European Journal of Lipid Science and Technology, 113*, 387–392.

Svejkovská, B., Doležal, M., and Velíšek, J. 2006. Formation and decomposition of 3-chloropropane-1,2-diol esters in models simulating processed foods. *Czech Journal of Food Sciences, 24*(4), 172–179.

Svejkovská, B., Novotný, O., Divinová, V., Réblová, Z., Doležal, M., and Velíšek, J. 2004. Esters of 3-chloropropane-1,2-diol in foodstuffs. *Czech Journal of Food Sciences, 22*(5), 190–192.

Tee, V. P., Shahrim, Z., and Nesaretnam, K. 2011. Cytotoxicity assays and acute oral toxicity of 3-MCPD esters. In *Abstracts of Papers, 9th Euro Fed Lipid Congress*. Rotterdam.

Velíšek, J., Davidek, J., Kubelka, V., Janíček, G., Svobodová, Z., and Simicová, Z. 1980. New chlorine-containing organic-compounds in protein hydrolysates. *Journal of Agricultural and Food Chemistry, 28*(6), 1142–1144.

Velíšek, J., Doležal, M., Crews, C., and Dvoøák, T. 2002. Optical isomers of chloropropanediols: Mechanisms of their formation and decomposition in protein hydrolysates. *Czech Journal of Food Sciences, 20*(5), 161–170.

Weißhaar, R. 2008. 3-MCPD-esters in edible fats and oils—A new and worldwide problem. *European Journal of Lipid Science and Technology, 110*(8), 671–672.

Weißhaar, R. 2011. Fatty acid esters of 3-MCPD: Overview of occurrence and exposure estimates. *European Journal of Lipid Science and Technology, 113*(3), 304–308.

Weißhaar, R., and Perz, R. 2010. Fatty acid esters of glycidol in refined fats and oils. *European Journal of Lipid Science and Technology, 112*(2), 158–165.

Zelinková, Z., Novotný, O., Schůrek, J., Velíšek, J., Hajšlová, J., and Doležal, M. 2008. Occurrence of 3-MCPD fatty acid esters in human breast milk. *Food Additives and Contaminants Part a—Chemistry Analysis Control Exposure* and *Risk Assessment, 25*(6), 669–676.

Zelinková, Z., Svejkovská, B., Velíšek, J., and Doležal, M. 2006. Fatty acid esters of 3-chlo-ropropane-1,2-diol in edible oils. *Food Additives and Contaminants, 23*, 1290–1298.

Zhang, X. W., Gao, B. Y., Qin, F., Shi, H. M., Jiang, Y. R., Xu, X. B., and Yu, L. L. 2013. Free radical mediated formation of 3-monochloropropanediol (3-MCPD) fatty acid diesters. *Journal of Agricultural and Food Chemistry, 61*(10), 2548–2555.

Zhu, Z. Y., Yates, R. A., and Caldwell, J. D. 1994. The determination of active filter aid adsorption sites by temperature-programmed desorption. *Journal of the American Oil Chemists Society, 71*, 189–194.

Zulkurnain, M., Lai, O M., Latip, R A., Nehdi, I A., Ling, T C., and Tan, C P. 2012. The effects of physical refining on the formation of 3-monochloro-propane-1, 2-diol esters in relation to palm oil minor components. *Food Chemistry, 135*, 799–805.

8 *Trans* Fatty Acids

Hongyan Li, Casimir C. Akoh, Jing Li, Huan Rao, and Zeyuan Deng

CONTENTS

8.1 INTRODUCTION

Saturated fatty acids (SFA) are solid at room temperature. Unsaturated fatty acids in nature are less tightly packed because of the *cis* configuration of the double bonds and they, generally, are liquids or oils at room temperature (Risérus, 2006). *Trans* fatty acids (TFA) consist of at least one isolated, nonconjugated, double bond in the *trans* geometric configuration. However, it excludes conjugated fatty acids such as conjugated linoleic acids (CLA) (Kim et al., 2008). The *trans* and *cis* configurations of TFA are different because of the positions of hydrogen atoms on the carbon atoms (Figure 8.1). The melting point of TFA is between that of saturated and *cis* unsaturated fatty acids since the *trans* double bond has a more rigid configuration and requires much less space than the *cis* double bond (see Figure 8.2).

TFA might be found in dairy fat because of ruminal activity, and in partially hydrogenated and heated vegetable oils. Almost 80–90% of dietary TFA are generated during oil refining and partial hydrogenation processes, whereas another 2–8% is from dairy products (Larqué et al., 2001). Hydrogenated vegetable oils have been used largely in the food industry because of their low cost and intermediate melting point. They provide favorable characteristics to the food and extend the shelf life of the products compared to unsaturated fatty acids. Foods such as margarine, bakery products, deep fried and frozen foods, packaged snacks, and cooking oils are major sources of dietary TFA.

TFA have been characterized by gas chromatography (GC), silver ion chromatography (SIC), Fourier transform infrared spectroscopy (FTIR), and capillary electrophoresis (CE). Although the reported data on the amount of TFA in food have great variation, epidemiological studies have revealed that TFA have a wide range of hazardous effects. These may include increased risk of cardiovascular diseases,

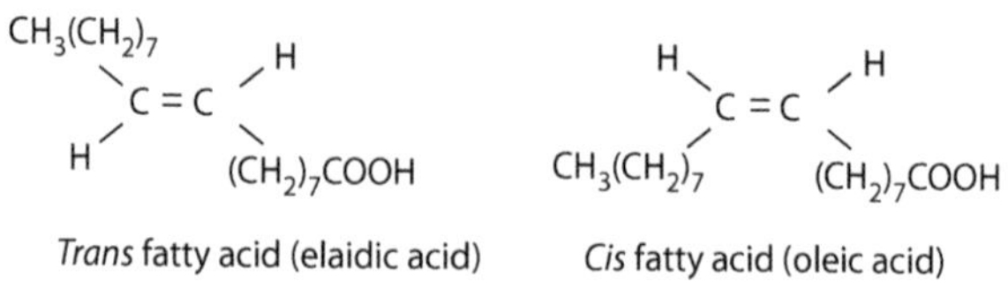

Trans fatty acid (elaidic acid)   Cis fatty acid (oleic acid)

FIGURE 8.1   Structure of *trans* and *cis* fatty acids.

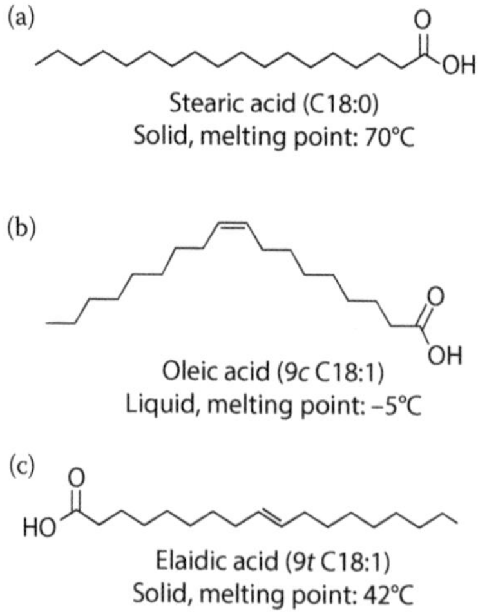

(a)

Stearic acid (C18:0)
Solid, melting point: 70°C

(b)

Oleic acid (9c C18:1)
Liquid, melting point: −5°C

(c)

Elaidic acid (9t C18:1)
Solid, melting point: 42°C

FIGURE 8.2   Structure and melting point of (a) saturated, (b) *cis* (18:1), and (c) *trans* (18:1) fatty acids.

adiposity, insulin resistance, compromised fetal growth and child development, and cancer. The US Food and Drug Administration (FDA) and Canadian Food Inspection Agency (CFIA) have required for the labeling of TFA content on food labels since 2006 (CFIA, 2008; FDA, 2010). Increased efforts also have been made to reduce TFA contents in processed foods.

This chapter will review the available information on the TFA occurrence, analytical methods, health implications, and regulations.

8.2 FORMATION AND OCCURRENCE

TFA may form during industrial partial hydrogenation of vegetable oils and the natural digestion process in ruminant animals. The structures of TFA from these two sources are different. For example, linoleic acid (18:2) forms primarily vaccenic acid (11*t* C18:1) by biohydrogenation and elaidic acid (9*t* C18:1) by partial hydrogenation (Craig-Schmidt, 2006). Partially hydrogenated vegetable oils may contain approximately 30% of TFA, whereas typical animal fats contain <5% of TFA (Daniel et al., 2005). It is apparent that the intake of naturally occurring TFA is less than the industrial TFA. Thus, the intake of natural TFA contributes <0.5% to total energy intake and the major source of dietary TFA is mostly the industrial TFA, contributing 2–4% or more to total energy intake (Micha and Mozaffarian, 2008).

8.2.1 HYDROGENATION OF OILS

Hydrogenation is the process of converting liquid oils into solid and/or improving their oxidative stability. Industrial hydrogenation uses a nickel catalyst to hydrogenate/saturate double bonds in vegetable oils. Chemically, complete reduction of all the double bonds would yield 100% SFA. However, when the hydrogenation is not complete, the partially hydrogenated oils may result in the addition of hydrogen atoms at some of the empty positions and form both *cis* and TFA (Figure 8.3). As shown in Figure 8.3, oleic acid could be partially hydrogenated to 8, 9, 10 *cis* or *trans* isomers. In other words, the scarcity of hydrogen on catalyst surface may cause the formation of TFA. During hydrogenation, a hydrogen atom can enter at any end of the double bond of unsaturated fatty acid and form a free-radical site, which is rather unstable. If this free-radical site is bound to the catalyst, and the surface of the catalyst is incompletely covered with hydrogen, then the possibility of positional and geometrical isomerization is high. An atom of hydrogen in the neighboring carbon can be removed from either side of the partially saturated bond, and therefore regenerating the double bond or leading to the formation of a positional isomer (Martin et al., 2007). As the formation of the free-radical site allows free rotation, the double bond is rearranged to produce different isomers (*cis* or *trans* configuration). Hydrogenation of oils became a common practice in the oil industries since the early twentieth century in both developed and developing countries, and resulted in the presence of most TFA in margarines, vegetable shortening, salad, and cooking oils.

Thermal processes such as vegetable oil deodorizing, refining, and frying may also generate TFA. Deodorization (180–270°C) and refining (60–100°C) need high temperatures (Martin et al., 2007). Similarly, during deep-fat frying (180°C or

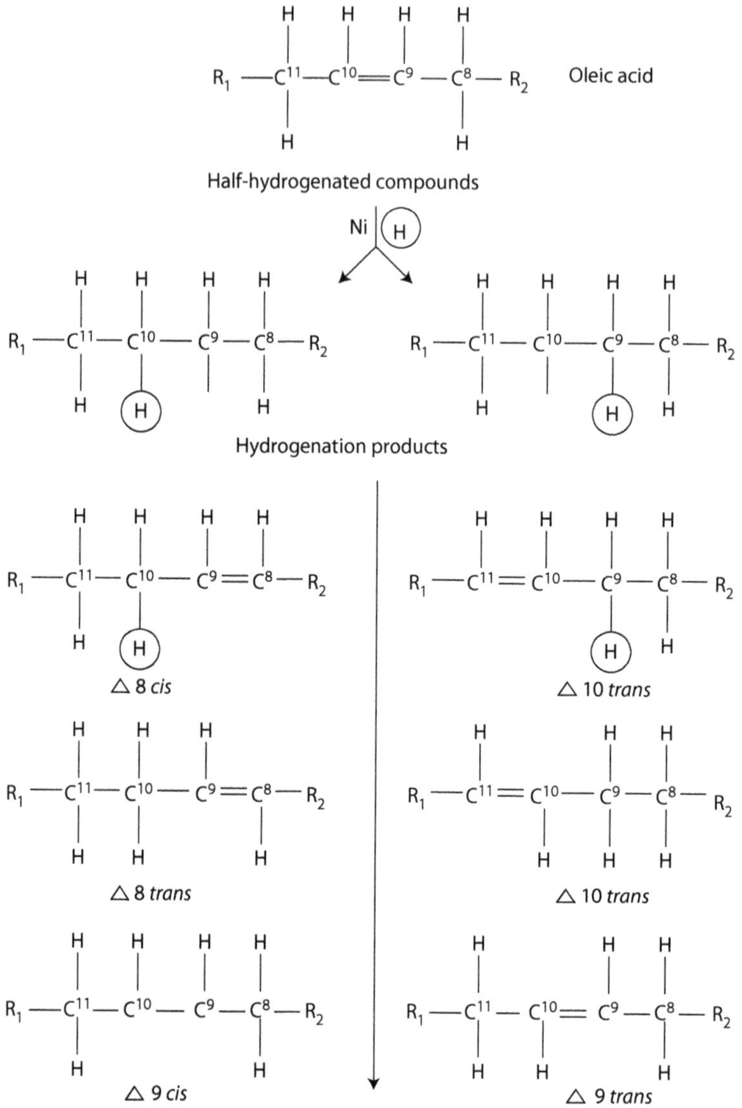

FIGURE 8.3 The mechanism of generation of *trans* isomers during hydrogenation. R_1 = methyl and R_2 is the COOH end of the fatty acid. (Adapted from Żbikowska, A. 2010. *Pol. J. Food Nutr. Sci.* 60, 107–114.)

above), complex reactions such as oxidation, hydrolysis, isomerization, polymerization, and cyclization take place in the edible oils (Bouchon, 2009). The formation of industrial TFA during thermal processes is due to the isomerization of *cis* fatty acids at high temperatures following a similar free-radical mechanism. It is easier for the unsaturated fatty acid to lose one hydrogen and transform double bond into *trans* configuration than accept another hydrogen and change into the saturated form because of thermodynamics (Żbikowska, 2010).

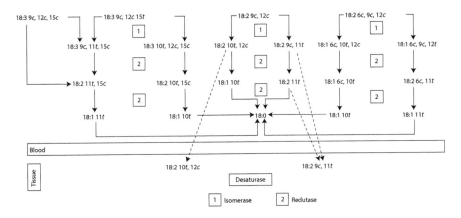

FIGURE 8.4 Formation of acids 11t 18:1 and 10t 18:1 in ruminants from linoleic, α-linolenic, and γ-linolenic acids. (Adapted from Martin, C.A., *et al.* 2007. *An. Acad. Bras. Cienc.* 79, 343–350.)

8.2.2 NATURAL DIGESTION OF RUMINANT ANIMALS

TFA are formed naturally in the rumens of ruminant animals such as cows, sheep, and goats through biohydrogenation by bacteria such as *Butyrivibrio fibrisolvens* and *Megasphaera esdenii* (Figure 8.4). The isomerase in the microflora of ruminants' alimentary tract leads to the isomerization of unsaturated fatty acids (from *cis* to *trans*). The polyunsaturated fatty acids could be isomerized to the mono TFA (the majority are 11*t* 18:1 and the minority are 10*t* 18:1), and then be absorbed or hydrogenated to 18:0 (Figure 8.4). In the tissue, 9*c*11*t* 18:2 can be formed through the action of enzyme Δ-9-desaturase. It is impossible to completely eliminate the natural TFA in the dairy and meat from these animals since the enzymatic isomerization is a good technology to reduce the industrial TFA. The natural TFA content usually ranges from about 2% to 7%, and varies between seasons and regions according to the different feeding practices of the animals (Larqué et al., 2001).

There are some results that have shown that industrial and natural TFA have similar severe effects on health; however, some of the literatures reported that the natural TFA (11*t* C18:1) could be transformed to the CLA (9*c*11*t* C18:2) by Δ-9-dehydrogenase, which may reduce atherosclerosis and adiposity (Jakobsen et al., 2006). Moreover, the adverse effects of natural TFA are limited because the content of TFA from natural source is relatively lower compared with higher amount of industrial TFA in the diet. Therefore, this chapter focuses on health effects of industrial TFA.

8.2.3 LEVELS OF *TRANS* FATTY ACIDS IN FOODS

The TFA levels have been evaluated over time by the methods mentioned earlier, and up-to-date contents of TFA in the main commercially available foods are summarized in Table 8.1. In general, TFA contents show a great variability among different food

TABLE 8.1

Contents of TFA (%) in the Main Commercially Available Foods

Foods	Country	TFA Content (%)[a]	Reference
Cookies and crackers	Argentina	2.85–28.95	Tavella et al. (2000)
Cookies and crackers	United States	2.87–8.18	Robinson et al. (2008)
Margarines	Argentina	18.15–31.84	Tavella et al. (2000)
Margarines	Swiss	0.3–2.0	Richter et al. (2009)
Margarines	Pakistan	2.2–34.8	Kandhro et al. (2008)
Margarines	UK	0.1	Roe et al. (2013)
Margarines	New Zealand	2.7–6.9	Saunders et al. (2008)
Snacks	Argentina	0–10.58	Tavella et al. (2000)
Snacks	United States	5.15–12.56	Robinson et al. (2008)
Snacks	Swiss	0.64–12.26	Richter et al. (2009)
Snacks	UK	0.10–0.29	Roe et al. (2013)
Snacks	New Zealand	0–0.8	Saunders et al. (2008)
Soybean oils	China	1.15	Hou et al. (2012)
Rapeseed oils	China	1.37	Hou et al. (2012)
Sunflower oils	China	1.41	Hou et al. (2012)
Corn oils	China	2.01	Hou et al. (2012)
Oils	Swiss	0.03–10.5	Richter et al. (2009)
Breakfast cereals	Swiss	0.4	Richter et al. (2009)
Breakfast cereals	United States	3.12	Robinson et al. (2008)
Breakfast cereals	Swiss	0.19–0.54	Richter et al. (2009)
Breakfast cereals	UK	0.04–0.06	Roe et al. (2013)
Ice cream	UK	0.25–2.27	Roe et al. (2013)
Sliced bread	Argentina	2.35–27.7	Tavella et al. (2000)
Butter	Argentina	4.63	Tavella et al. (2000)
Chocolate	New Zealand	0–34	Saunders et al. (2008)
Pies and pastry	New Zealand	2.1–7.1	Saunders et al. (2008)
Biscuits and cakes	New Zealand	0–3.5	Saunders et al. (2008)
Fried and fast food	Swiss	0.2–21.97	Richter et al. (2009)
Fine bakery products	Swiss	0.32–16.97	Richter et al. (2009)
Pizza	UK	1.22	Roe et al. (2013)

[a] The results were expressed in grams of fatty acid per 100 g total fatty acids (%).

groups and countries. In Argentina, the content of TFA, elaidic acid, was 2.35–27.7% in sliced bread, 2.85–28.95% in cookies and crackers, 18.15–31.84% in margarines, 4.63% in butter, and 0–10.58% in snacks, respectively (Tavella et al., 2000).

In China, 93 samples of soybean oil ($n = 29$), rapeseed oil ($n = 23$), sunflower oil ($n = 22$), and corn oil ($n = 19$) were analyzed by GC, and TFA (>2%) were detected in 17 (18%) samples, ranging from 0.14% to 4.76%. The overall TFA content was 1.15% for soybean oils, 1.37% for rapeseed oils, 1.41% for sunflower oils, and 2.01%

for corn oils (Hou et al., 2012). In addition, *trans* C18:2 and C18:3 fatty acids were predominant in the investigated edible oils (Hou et al., 2012).

In Switzerland, 119 food items were purchased and analyzed for their TFA content (Richter et al., 2009). Among the samples, nearly 40% had more than 2% TFA and 8 were found to contain more than 20% TFA. The highest mean value was observed in the fine bakery products (6% TFA) and the lowest in the breakfast cereals (<0.4% TFA) (Richter et al., 2009). Margarines had TFA contents between 0.3% and 2.0%. Among the TFA, C18:1 was the predominant TFA in all samples except for the plant oils, in which C18:2 and C18:3 contents were the highest.

Sixty-two composite samples including pizza, garlic bread, breakfast cereals, quiche, fat spreads, a range of fish and meat products, chips, savory snacks, confectionery, and ice cream were collected in the United Kingdom in 2010 and analyzed for TFA contents. The results showed that the concentration of *trans* elaidic acid (9*t* C18:1) in these samples were <0.2% food (Roe et al., 2013).

TFA levels in selected samples of foods (biscuits, cakes, margarines, chocolate, snacks, pies, pastry, and partially cooked chips) made in New Zealand were, on the average, below 10% but have declined significantly over the past decades (Saunders et al., 2008).

Ten margarine brands from Pakistan were analyzed for TFA contents by GC. Among the samples, only one contained a low level of TFA (2.2%), while the remaining contained very high amounts of TFA (11.5–34.8%) (Kandhro et al., 2008).

8.3 ANALYTICAL METHODS

Determination of TFA is an analytical challenge. Similar to common fatty acids, several techniques, such as GC, SIC, FTIR, and CE, have been used to measure TFA (Adlof, 1994; Molkentin and Precht, 1995; Mossoba et al., 1996). Among these separation and detection techniques for TFA in food, GC is the most popular and reliable system (Delmonte and Rader, 2007).

8.3.1 SAMPLE PREPARATION AND EXTRACTION

Sample preparation is a critical step of a successful analytical method. There are important common precautions that must be taken to prepare samples for subsequent analyses. Lipid extraction is traditionally carried out in different ways using organic solvents such as hexane, petroleum ether, a mixture of chloroform and methanol (2:1, v/v), isopropanol and cyclohexane (4:5, v/v), or a mixture of chloroform/methanol/ water (8:4:3, v/v/v) (Alves et al., 2008; Robinson et al., 2008; Seppänen-Laakso et al., 2002; Smedes, 1999). Soxhlet extraction has also been used. However, it needs a long extraction time at high temperature with a high solvent usage (Glew et al., 2006). New methods such as accelerated solvent extraction (Jiménez et al., 2009), dynamic ultrasound-assisted extraction (Ruiz-Jiménez et al., 2004), and microwave-assisted extraction (Priego-Capote et al., 2004) have been applied. These new methods are potential alternatives to conventional solvent extraction because they could significantly reduce both extraction time and solvent consumption.

8.3.2 Gas Chromatography

GC with flame ionization detector (GC-FID) is an excellent analytical tool for fatty acid methyl esters (FAMEs) that volatilize upon heating. GC could provide a complete fatty acid profile and is particularly an effective separation method and highly sensitive instrument for analyzing TFA. Long (100 m or above), flexible, fused-silica capillary columns coated with highly polar cyanopolysiloxane stationary phases containing various polar substituents are the best to analyze TFA (Kramer et al., 2002). However, GC does not allow direct individual separation, and need a mandatory derivatization step because TFA have relatively high polarity (Huang et al., 2006). Thus, TFA are generally derivatized to the corresponding FAMEs which are nonpolar derivatives (Delmonte et al., 2011). During preparation of FAMEs, most methods are carried out with saponification (methanolic sodium hydroxide, methanolic sodium methoxide) and transesterification (methanolic boron trifluoride) (Gunawan et al., 2010). However, GC may increase the risk of isomerization of double bonds because of the high temperature (Liu et al., 2007a).

Typically, in our laboratory, FAME profiles are analyzed with a capillary GC equipped with a flame ionization detector and a fused-silica capillary column (100 m, 0.25 mm I.D., 0.2 μm film thickness) coated with 100% cyanopropyl polysiloxane (Model 6890N, Agilent Technologies, Shanghai, China). The initial temperature of the program is 45°C held for 4 min, and then increased at a rate of 13°C/min to 175°C, and held for 27 min. The oven temperature is then further increased to 215°C at a rate of 4°C/min and held for 35 min. Hydrogen was chosen as the carrier gas here due to its high velocity, diffusivity, low resistance to mass transfer, as well as for avoiding the isomerization caused by high temperature. Quantification and identification of the individual FAME can be done using the external standards (#463) and internal standard (C21:0) (Li et al., 2013).

8.3.3 Silver Ion Chromatography

While other chromatographic techniques are used in the analysis of TFA, most researchers use SIC (Juanéda, 2002). The silver ion-impregnated stationary phases for thin-layer chromatography (Ag⁺-TLC), high-performance liquid chromatography (Ag⁺-HPLC), and solid-phase extraction (Ag⁺-SPE) are now available for the separation and determination of TFA (Momchilova et al., 1998). The complexation reaction between Ag^+ and TFA is reversible, which makes SIC highly suitable for use in chromatographic separation processes. However, contamination by trace silver salts and low reproducibility are the major difficulties (Mjøs, 2005).

Ag⁺-TLC is usually used to pre-fractionate *cis* and *trans* isomers as their methyl ester derivatives, and then the collected fractions are analyzed by GC (Precht and Molkentin, 1996). FAMEs were fractionated by TLC on silica gel plates impregnated with silver nitrate (Wolff et al., 2000). The plate was dipped into a 5% silver nitrate solution in acetonitrile for 15 min. Subsequently, the plate was activated at 100°C for 1 h. The sample was applied to the plate in a thin band of 10 cm and the plate was developed with a mixture of hexane and diethyl ether (90:10, v/v) in a saturated TLC chamber (Destaillats et al., 2007). The plate was air-dried and sprayed with a solution of 2′,7′-dichlorofluorescein and viewed under UV light (Dionisi et al., 2002).

Ag^+-HPLC, utilizing columns containing silver ions bonded to a silica or similar substrate, uses derivatization reactions with phenacyl and naphthacyl esters so that UV detection method at 242 nm could be used (Christie and Breckenridge, 1989). Two or three lipid columns were usually connected in series to improve sample capacity and peak-to-peak resolutions (Adlof and Lamm, 1998). Unlike GC, the separation and identification were at lower temperatures (25°C) (Adlof, 2007).

Ag^+-SPE cartridge (750 mg/6 mL) was used to obtain pure saturated, *trans* and *cis* monounsaturated and diunsaturated fatty acids (Kramer et al., 2008).

8.3.4 FOURIER TRANSFORM INFRARED SPECTROSCOPY

FTIR is an excellent analytical tool for the rapid determination of TFA because it does not require organic solvent extraction, derivatization, and costly internal standards (Sherazi et al., 2009a). FTIR is based on the unique CH out-of-plane deformation band produced at 966 cm^{-1} for analysis of *trans* double bonds (Bansal et al., 2009). Attenuated total reflectance (ATR) is a widely used sample-handling technique in mid-FTIR spectroscopy, which is based on the phenomenon of total internal reflection of light (Fritsche et al., 1998). However, FTIR could not quantify the contents of individual fatty acid and the results are inaccuracies particularly at low TFA levels (below 5%) because of potential sources of interferences and inaccuracies (Mossoba et al., 2007).

According to the method of Sherazi et al. (2009b), infrared spectra were obtained by a ThermoNicolet Avatar 330 FTIR spectrometer equipped with a removable ZnSe crystal, KBr optics, and deuterated triglycine sulfate (DTGS) detector. A single bounce-ATR accessory with heated demountable cell and time-proportional temperature controller (Spectra-Tech, Shelton, CT) was used. ATR quantitation was based on the measurement of integrated area in the range of 4000–650 cm^{-1}, centered at 966 cm^{-1}. The spectrum of each standard or sample was ratioed against a fresh background spectrum recorded from the bare ATR crystal.

8.3.5 CAPILLARY ELECTROPHORESIS

CE with indirect UV detection has been developed due to its high separation efficiency, short analysis time, and absence of derivatization reaction in sample preparation (Chi et al., 2005). In general, CE uses alkaline buffers with chromophoric anionic species or chromophoric additives for the indirect UV detection of TFA (de Oliveira et al., 2003). However, the sensitivity of CE is not good enough because of the high background noise, fluctuating baseline, and the short sample path length (de Castro Barra et al., 2013).

According to the method optimized by De Castro et al. (2010), the electrolyte consisted of 8.0 mmol/L Brij35, 0.5 mol/L of sodium hydroxide, 4.0 mmol/L SDBS, 15.0 mmol/L KH_2PO_4/Na_2HPO_4 buffer (pH 7.0), 45% v/v ACN, 8% methanol, and 1.5% v/v *n*-octanol. Samples were hydrodynamically injected (12.5 mbar for 4 s) and the electrophoretic system was operated under normal polarity and constant voltage

(+19 kV) (de Castro Barra et al., 2012). A fused-silica capillary tube (Polymicro Technologies, Phoenix, AZ), 48.5 cm long (40 cm effective length × 75 μm I.D. × 375 μm O.D.) 2 cm, was used. The total TFA contents were expressed as elaidic acid equivalents and were performed in a capillary electrophoresis system under indirect UV detection at 224 nm in 7.5 min.

8.4　EVIDENCE OF HEALTH IMPLICATIONS

TFA are considered to be nutritionally unnecessary, and evidence suggest that excessive TFA intake is a significant risk factor for chronic diseases such as cardiovascular events, diabetes, compromised fetal growth and child development, and cancers (Willett, 2006). Given the high prevalence of these diseases, the possible increase in risk caused by high TFA consumption should be carefully considered. It is nearly impossible here to cover all health implications of TFA and their mechanisms; therefore, for readers who are interested in detailed information, comprehensive reviews by others are recommended (Albuquerque et al., 2011; Mozaffarian et al., 2009; Remig et al., 2010; Stender and Dyerberg, 2004; Wang et al., 2012).

8.4.1　CARDIOVASCULAR HEALTH

The evidence is sufficient to suggest the potential harmful effects of TFA on cardiovascular disease (CVD) (Ascherio and Willett, 1997; Booker and Mann, 2008; Katan, 2006; Willett et al., 1993). CVD is the leading cause of death and it was reported that intake of 5 g TFA per day could cause an increase of 25% in the risk of ischemic heart disease (Stender et al., 2006a). The erythrocyte membrane TFA index was proven to have a strong and positive linear correlation with the 10-year coronary heart disease risk probability according to a cross-sectional study of 2713 individuals (581 with acute coronary syndrome, 631 with chronic coronary artery disease, 659 high-risk population, and 842 healthy volunteers) (Liu et al., 2013). Specifically, the diets of 667 men aged 64–84 years and free of coronary heart disease were investigated in Zutphen for 10 years (1985–1995). Their average TFA intakes were 1.9–4.3% of energy and there were 98 cases of fatal or nonfatal coronary heart disease (Oomen et al., 2001). The result showed that the relative risk for a difference of 2% of energy in TFA intake at baseline was 1.28 (95% confidence interval (CI = 1.01–1.61)), which means a high intake of TFA contributes to the risk of CVD.

　　TFA adversely affect CVD by changing the lipid profiles, systemic inflammation, and endothelial dysfunction. The mechanisms of these effects are not well established, but may involve inflammatory markers, circulating biomarkers of endothelial dysfunction, and Caspase activities (Mozaffarian et al., 2006).

8.4.1.1　Lipid Effects

TFA have adverse effects on blood lipids, including raising low-density lipoprotein (LDL) and triacylglycerol levels, and reducing high-density lipoprotein (HDL) levels. A high-TFA diet causes adverse changes in the plasma lipoprotein profile, with an increase in the plasma LDL/HDL cholesterol ratio, which is believed to be a risk factor for CVD (de Freitas et al., 2011).

Feeding TFA diet (5.1% energy) to 14 young healthy women for 4 weeks resulted in a higher total/HDL cholesterol ratio and an elevation in triacylglycerol and apolipoprotein B concentrations (Louheranta et al., 1999). Consumption of a single meal enriched with TFA (9.2% energy) or SFA for 4 weeks showed similar acute adverse effects on the decrease of HDL cholesterol (21% reduction) (de Roos et al., 2001). TFA consumption affected serum lipids and lipoproteins and significantly increased the total/HDL cholesterol ratio and apoprotein (Apo) B levels in 13 randomized controlled trials, suggesting that TFA consumption could affect both blood lipid concentrations and apolipoprotein levels (Mozaffarian and Clarke, 2009). Similar result was obtained in an experiment with 327 US women (Sun et al., 2007). The higher erythrocyte TFA levels (corresponding to a range of habitual TFA intake from 2.5 to 3.6 g/day) were associated with significantly higher LDL-C, lower HDL-C, and a higher LDL-C:HDL-C ratio ($p < 0.1$) (Sun et al., 2007).

8.4.1.2 Systemic Inflammation and the Suggested Mechanisms

Systemic inflammation and endothelial dysfunction may be involved in the pathogenesis of atherosclerosis, which is the most common cause of CVD. It was believed that atherosclerosis was a dynamic process with inflammatory changes in the endothelium of conduit arteries (Massaro et al., 2008). TFA intake could interfere with cell membrane functions by increasing inflammatory markers such as tumor necrosis factor (TNF), C-reactive protein (CRP), and interleukin-6 (IL-6) (Ridker et al., 2000).

In an analysis among 730 women aged 43–69 years (mean BMI = 26.3 kg/m^2), each 1% higher membrane TFA level was associated with 249% higher TNF-α levels (256 pg/mL), 41% higher TNF receptor 1 levels (537 pg/mL), and 247% higher TNF-α levels (39,242 pg/mL) (Lopez-Garcia et al., 2005). Moreover, the CRP levels of the highest quintile (2.1% energy) was 73% higher compared to individuals in the lowest quintile of TFA consumption (0.9% energy) (Lopez-Garcia et al., 2005). In a cross-sectional study of 823 generally healthy women, after adjustment for the factors that might influence inflammation (including age, smoking, physical activity, medication use, alcohol consumption, and other dietary habits), TFA intake was positively associated with sTNF-R1 (108 pg/mL; 95% CI: 50) and sTNF-R2 (258 pg/mL; 95% CI: 138) (Mozaffarian et al., 2004). In 50 healthy men, 5 weeks of TFA consumption (8% energy) increased plasma IL-6 (16%) and CRP levels (20%) compared to MUFA (Baer et al., 2004). The diet high in hydrogenated fat was proven to increase the production of inflammatory cytokines TNF-α and IL-6 in 19 hypercholesterolemic patients after 1 month of TFA intake (6.7% energy) (Han et al., 2002). The strong link between systemic inflammation (particularly as reflected by these inflammatory markers) and TFA intake indicated that TFA consumption is proinflammatory.

8.4.1.3 Endothelial Dysfunction and the Suggested Mechanisms

Endothelial dysfunction is one of the earliest hallmarks of vascular abnormality and intimately involved in the development of atherosclerosis (Grover-Páez and Zavalza-Gómez, 2009). The endothelium expresses an anticoagulant, antiadhesive, and vasodilatory phenotype, and it also plays an integral role in the regulation of vascular tone, leukocyte adhesion, platelet activity, and thrombosis (Vita and Keaney, 2002). TFA could induce apoptosis in human endothelial cells in vitro and affect the endothelial dysfunction by circulating biomarkers including soluble intercellular adhesion molecule1 (sICAM-1), soluble vascular-cell adhesion molecule1 (sVCAM-1), and E-selectin. In a controlled feeding trial among 50 men, 5 weeks of TFA intake (8% energy) raised levels of E-selectin by 14%, 6%, and 6% compared with equivalent calories from MUFA, stearic acid, and 12:0–16:0 SFA, respectively (Baer et al., 2004). TFA intake was assessed from 1986 to 1990 in 730 women aged 43–69 years and the result showed that greater TFA intake was associated with higher levels of several circulating biomarkers including sICAM-1, sVCAM-1, and E-selectin ($p < 0.01$) (Lopez-Garcia et al., 2005).

In addition, TFA could trigger caspase pathways and responses by incorporating into cell membrane phospholipids, altering the function of specific membrane receptors, and directly binding to and modulating nuclear receptors regulating gene transcription (Mozaffarian, 2006). The activities and mRNA expression of caspases 8, 9, and 3 were significantly increased after exposure to TFA (9t C18:1), which indicates that TFA could induce apoptosis of human umbilical vein endothelial cells through the activation of the initiators caspases 8 and 9 as well as the effector caspase 3 (Qiu et al., 2012). Furthermore, the death receptor pathway and the mitochondrial pathway were involved in the apoptosis course (Figure 8.5). Caspase 8 caused the release of cytochrome c from mitochondria and then activated caspase 9 and 3. The same effect on caspase pathway 8 and 3 was found in HepG2 cells after the addition of elaidic acid (Kondoh et al., 2007). It was proven that caspase 8-mediated death receptor pathway was the upstream signal pathway of TFA and caspase 9 was the downstream signal pathway, and could be activated by a mitochondrial pathway depending on Bid activation (Qiu et al., 2012).

Activation of inflammatory responses and endothelial dysfunction may represent important mediating steps between TFA consumption and the risk of CVD. All of the above experimental and observational evidence from a variety of perspectives provided a link between consumption of TFA and lipid profiles, systemic inflammation, and endothelial dysfunction.

8.4.2 Diabetes

There are some observational studies that suggest that high intakes of TFA could affect insulin sensitivity and consequently increase the risk of developing Type 2 diabetes. Insulin resistance (low insulin sensitivity) is a major contributor to Type 2 diabetes (Cryer et al., 2007). According to the reports, high intake level and a longer duration of exposure to TFA may affect insulin sensitivity (Bendsen et al., 2011). Moreover, overweight individuals were more susceptible to developing insulin resistance on TFA diets. Thirty-six overweight mildly hypercholesterolemic middle-aged and older men and women were fed TFA diet (3–7% energy) for 5 weeks. Their fasting insulin levels showed that high-TFA diet produced high insulin levels (Lichtenstein et al., 2003). For the overweight patients with Type 2 diabetes, high-TFA diet (20% energy) for 6 weeks caused a significantly elevated postprandial insulin response (Christiansen et al., 1997). In addition, the consumption of TFA (11.75% of total fat) by the mothers during the lactation period caused cardiac insulin resistance in the adult progeny in Wistar rats (Osso et al., 2008).

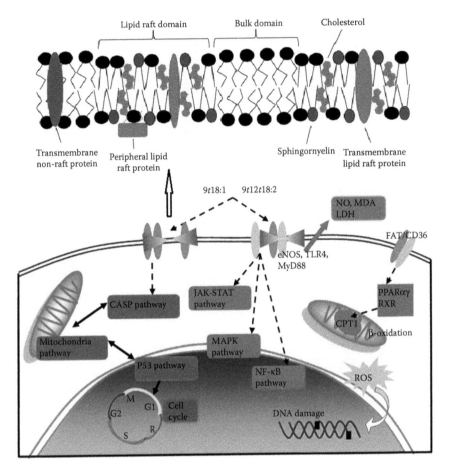

FIGURE 8.5 The mechanism of TFA inducing the apoptosis in human umbilical vein endothelial cells.

However, some studies showed no significant effect of TFA intake on insulin sensitivity, especially in lean healthy people. In a Finnish study, 14 young healthy women were given TFA diet (5.1% energy) for 4 weeks to investigate the effects of TFA on insulin sensitivity and the results showed that TFA intakes were not associated with insulin sensitivity (Louheranta et al., 1999). Similarly, a dose of TFA (elaidic, 9% energy), compared with SFA (palmitic) and MUFA (oleic), was given to 25 healthy American adults and there were no differences between TFA diet on insulin sensitivity or secretion in lean individuals (Risérus, 2006).

8.4.3 COMPROMISED FETAL DEVELOPMENT

Dietary TFA can be incorporated into both fetal and adult tissues and fluids including liver, adipose tissue, spleen, and plasma although the proportion is lower than they are in the diet (Larqué et al., 2001). TFA could also be transported across the placenta and are present in breast milk and the content is correlated to the mother's dietary TFA intake (Hayat et al., 1999). TFA have been inversely correlated to infantile birth weight and have adverse effects on the growth and development of children by interfering with essential fatty acid metabolism, membrane structures or metabolism, and the intake of the *cis* essential fatty acids (Innis, 2006). Reductions in TFA intake during pregnancy might reduce fetal loss (Morrison et al., 2008).

Using data from 700 infant–mother pairs in Maastricht, a significant and negative association was observed between TFA in cord erythrocytes and birth weight, which indicate that TFA may compromise fetal development (Dirix et al., 2009). In premature infants, TFA in blood plasma inversely correlated with birth weight, indicating that TFA may impair early human growth (Koletzko and Decsi, 1997). The birth weight, birth length, and head circumference were also negatively related to the TFA concentration in maternal plasma phospholipids for two longitudinal pregnancy and birth cohorts in the Netherlands (Hornstra et al., 2006). From a study of 80 adolescent healthy Brazilian mothers, results showed that there was no difference between TFA content in maternal plasma and in umbilical cord. Despite the low levels of TFA found in maternal blood (<2%), TFA have a negative correlation with gestational age at birth and may contribute to premature birth (Santos et al., 2012).

8.4.4 CANCERS

TFA have been hypothesized to be carcinogenic although it remains controversial (Nkondjock et al., 2003). The published data have several inconsistencies and are open to different interpretations. The methodological limitations and experimental considerations may cause the inconsistencies (Smith et al., 2009). In order to accurately assess the relationship between TFA and cancer risks, further research is warranted.

An ecological study across 11 European countries showed that significant and positive association was found between TFA and colorectal cancer risk (95% CI = 0.74–0.98) (Bakker et al., 1997). In a total of 1993 case studies conducted in Utah, Northern California, and Minnesota, a strong positive significant correlation was found between TFA and the incidence of colon cancer specifically in women (odds ratio [OR] = 1.5; 95% CI = 1.0–2.4) (Slattery et al., 2001). A case–control study ($n = 1012$) was performed and TFA intake showed a statistically significant positive association with prostate cancer, but only weakly increased risk for the isomers of *cis* fatty acids (Liu et al., 2007b). In addition, a positive association was found between serum 11*t* C18:1 concentration (95% CI = 1.03–2.77) and an increased risk for prostate cancer in a nested case–control study (272 cases, 426 controls) (King et al., 2005). An increased risk of breast cancer was associated with increased levels of *trans* palmitoleic acid and elaidic acid (95% CI = 1.08–2.83) in a 7-year study (363 cases with incidence of invasive breast cancer among 19,934 women). This indicated that high serum TFA level was probably one factor contributing to increased risk of invasive breast cancer in women (Chajès et al., 2008).

In a total of 516 cases aged 50–74 years and 551 hospital-based controls in the United States (OR = 0.90; 95% CI = 0.40–2.0), no significant association was found between TFA and colorectal adenomatous polyps (McKelvey et al., 1999). In another case–control study, among 14,916 apparently healthy men at the beginning with 476 of them diagnosed with prostate cancer during a 13-year follow-up, no correlation was found between blood TFA levels and risk of prostate cancer although the blood levels of *trans* isomers of oleic (95% CI = 1.12–4.17) and linoleic acids (95% CI = 1.03–3.75) were associated with an increased risk of nonaggressive prostate tumors (Chavarro et al., 2008). In a follow-up study of a total of 4052 postmenopausal women (71 cases were identified with invasive breast cancer after an average of 5.5 years), no correlation was found between erythrocyte membrane TFA and postmenopausal breast cancer risk (Pala et al., 2001).

8.5 MITIGATING THE INTAKE OF TFA

According to the estimates of the TFA content in the diet, there was less than 1 g/person/day TFA in Asian/Pacific countries and 10–20 g/person/day in some Western countries (Craig-Schmidt, 2006). The TRANSFAIR Study investigated the intake of TFA in western Europe based on market basket analysis of the diets (van Poppel, 1998). According to the data, mean daily intakes of TFA in 14 European countries were from 1.4 g/day (0.6% energy) to 5.4 g/day (2.0% energy). Large intakes of TFA should be avoided due to health implications. Although the average intake of TFA may have declined slightly in recent years due to a general reduction in fat content in foods, the average intake is still high (7–8 g/person/day) (Stender et al., 2006b). Therefore, the intake of TFA should be managed or reduced as part of a public health strategy. In general, the future trend of low-TFA foods could be achieved through appropriate guidance and education of consumers, voluntary or mandatory labeling/legislation, and adoption of alternative fats and oils. One of the best ways to reduce dietary TFA is to change individual eating habit.

8.5.1 REGULATION

In general, regulations on TFA vary from country to country. The Food and Agriculture Organization (FAO) and the World Health Organization (WHO) recommend that fats for human consumption should contain <4% of the total fat as *trans* (FDA, 2010). The US Food and Drug Agency (FDA) and Canadian Food Inspection Agency (CFIA) were the first to introduce the mandatory declaration of TFA (CFIA, 2008; FDA, 2010). The American Heart Association recommends limiting TFA to <1% energy (Leth et al., 2006). FDA decreed that by January 1, 2006, manufacturers should declare the amount of TFA present in foods on the nutrition label (Moss, 2006). Moreover, FDA labeling rules allow products containing <0.5 g *trans* fat per serving to be expressed as 0 g of *trans* fat (FDA, 2010). In Canada, the limit is 0.2 g of *trans* fat per serving and per reference amount (CFIA, 2008).

In order to reduce the health risk related to TFA, Denmark was the first country to prohibit the sale of foods containing >2% industrially produced TFA after January 1, 2004 (Astrup, 2006). Danish legislation was an effective intervention and the exposure to TFA by a high-*trans*-fat menu was reduced from 30 g (industrially produced TFA) in 2001 to <1 g in 2005 in Denmark (Leth et al., 2006).

The National Heart Foundation of New Zealand and the Australian National Heart Foundation approved that TFA should be <1% in margarines and table spreads since 2006 (Saunders et al., 2008). The United Kingdom has also taken active steps toward the possible banning of TFA. The European Commission has established that TFA content in infant formulas and follow-on formulas shall not exceed 3% of the total fat content (Aro, 2006). In China, the amount of TFA should be marked on the labels of prepackaged food, according to the country's first national standard for food nutrition, and the labeling took effect from January 1, 2013.

8.5.2 APPROACHES TO REDUCE TFA

Clearly, efforts have been made to search for alternatives to TFA. Some companies are now taking steps to develop TFA-reduced or TFA-free products. These include use of fully hydrogenated process instead of partial hydrogenation to produce fats with low TFA levels; use of crossbreeding and genetic engineering to produce modified plants which contain low-linoleic, mid-oleic, or high-oleic oils; use of high content of SFA oils (palm oil, palm kernel oil, coconut oil); and the most common modern-day techniques: interesterification of highly hydrogenated fats with liquid oils achieved chemically or enzymatically (Tarrago-Trani et al., 2006).

Interesterification is the hydrolysis of the ester bond between the fatty acid and glycerol and subsequent reformation of the ester bond among the mixed free-fatty acids and glycerol (Shin et al., 2010). It rearranges the fatty acids in the glycerol molecule but does not change the fatty acid profiles of the starting fats (Lee et al., 2008). Chemical interesterification is a random process and there is substantial oil loss (30%) due to the formation of soap and fatty acid ester, while enzymatic interesterification by the use of microbial lipases is more selective since lipases interact with specific triacylglycerol ester bonds (Da Silva et al., 2010). By this procedure, it will be possible to obtain margarine with virtually zero *trans* content (Adhikari et al., 2010; Kim et al., 2008). For example, *trans*-free structured margarine fat analogs were enzymatically prepared by interesterifying palm stearin and regular cottonseed oil or palm stearin and high-stearate soybean oil (Pande et al., 2013; Pande and Akoh, 2013).

8.6 CONCLUSION

There are sufficient data showing that TFA do not provide any known nutritional benefits. Rather, some adverse health effects including cardiovascular events, diabetes, compromised fetal growth and child development, and cancers have been acknowledged. Accordingly, greater transparency in labeling and active consumer education are needed to reduce the intake of TFA. However, there are alternatives that may considerably help decrease human exposure to TFA. New approaches such as developing new oil seeds with desirable fatty acid composition and oxidative stability for different food applications, improving hydrogenation technique, and producing novel cooking oils through enzymatic interesterification should continue to be studied and developed in the future.

REFERENCES

Adhikari, P., Zhu, X.-M., Gautam, A., Shin, J.-A., Hu, J.-N., Lee, J.-H., Akoh, C.C., Lee, K.-T. 2010. Scaled-up production of zero-*trans* margarine fat using pine nut oil and palm stearin. *Food Chem.* 119, 1332–1338.

Adlof, R., Lamm, T. 1998. Fractionation of *cis*- and *trans*-oleic, linoleic, and conjugated linoleic fatty acid methyl esters by silver ion high-performance liquid chromatography. *J. Chromatogr. A* 799, 329–332.

Adlof, R. 2007. Analysis of triacylglycerol and fatty acid isomers by low-temperature silver-ion high performance liquid chromatography with acetonitrile in hexane as solvent: Limitations of the methodology. *J. Chromatogr. A* 1148, 256–259.

Adlof, R.O. 1994. Separation of *cis* and *trans* unsaturated fatty acid methyl esters by silver ion high-performance liquid chromatography. *J. Chromatogr. A* 659, 95–99.

Albuquerque, T.G., Costa, H.S., Castilho, M.C., Sanches-Silva, A. 2011. Trends in the analytical methods for the determination of *trans* fatty acids content in foods. *Trends Food Sci. Technol.* 22, 543–560.

Alves, S.P., Cabrita, A.R.J., Fonseca, A.J.M., Bessa, R.J.B. 2008. Improved method for fatty acid analysis in herbage based on direct transesterification followed by solid-phase extraction. *J. Chromatogr. A* 1209, 212–219.

Aro, A. 2006. The scientific basis for *trans* fatty acid regulations—Is it sufficient?: A European perspective. *Atherosclerosis Suppl.* 7, 67–68.

Ascherio, A., Willett, W.C. 1997. Health effects of *trans* fatty acids. *Am. J. Clin. Nutr.* 66, 1006–1010.

Astrup, A. 2006. The *trans* fatty acid story in Denmark. *Atherosclerosis Suppl.* 7, 43–46.

Baer, D.J., Judd, J.T., Clevidence, B.A., Tracy, R.P. 2004. Dietary fatty acids affect plasma markers of inflammation in healthy men fed controlled diets: A randomized crossover study. *Am. J. Clin. Nutr.* 79, 969–973.

Bakker, N., Van't Veer, P., Zock, P.L. 1997. Adipose fatty acids and cancers of the breast, prostate and colon: An ecological study. *Int. J. Cancer* 72, 587–591.

Bansal, G., Zhou, W., Tan, T.-W., Neo, F.-L., Lo, H.-L. 2009. Analysis of trans fatty acids in deep frying oils by three different approaches. *Food Chem.* 116, 535–541.

Bendsen, N.T., Haugaard, S.B., Larsen, T.M., Chabanova, E., Stender, S., Astrup, A. 2011. Effect of *trans*-fatty acid intake on insulin sensitivity and intramuscular lipids—A randomized trial in overweight postmenopausal women. *Metabolism* 60, 906–913.

Booker, C.S., Mann, J.I. 2008. *Trans* fatty acids and cardiovascular health: Translation of the evidence base. *Nutr. Metab. Cardiovasc. Dis.* 18, 448–456.

Bouchon, P. (2009) Chapter 5: Understanding oil absorption during deep-fat frying. In *Advances in Food and Nutrition Research* (Ed.) Steve, L.T. (New York: Academic Press), pp. 209–234.

CFIA. Canadian Food Inspection Agency Information letter: Labeling of *trans* fatty acids (2008). [on-line]: http://www.inspection.gc.ca/english/fssa/labeti/inform/transe.shtml. Accessed 06.27.13.

Chajès, V., Thiébaut, A.C.M., Rotival, M., Gauthier, E., Maillard, V., Boutron-Ruault, M.-C., Joulin, V., Lenoir, G.M., Clavel-Chapelon, F. 2008. Association between serum *trans*-monounsaturated fatty acids and breast cancer risk in the E3N-EPIC study. *Am. J. Epidemiol.* 167, 1312–1320.

Chavarro, J.E., Stampfer, M.J., Campos, H., Kurth, T., Willett, W.C., Ma, J. 2008. A prospective study of *trans*-fatty acid levels in blood and risk of prostate cancer. *Cancer Epidemiol. Biomark. Prev.* 17, 95–101.

Chi, F.-H., Lin, P.H.-P., Leu, M.-H. 2005. Quick determination of malodor-causing fatty acids in manure by capillary electrophoresis. *Chemosphere* 60, 1262–1269.

National standard for *trans* fat labelling. [on-line]: http://www.china.org.cn/china/2011-11/04/content_23820814.htm. Accessed 06.27.13.

Christiansen, E., Schnider, S., Palmvig, B., Tauber-Lassen, E., Pedersen, O. 1997. Intake of a diet high in trans monounsaturated fatty acids or saturated fatty acids. Effects on postprandial insulinemia and glycemia in obese patients with NIDDM. *Diabetes Care.* 20, 881–887.

Christie, W.W., Breckenridge, G.H.M. 1989. Separation of *cis* and *trans* isomers of unsaturated fatty acids by high-performance liquid chromatography in the silver ion mode. *J. Chromatogr. A* 469, 261–269.

Craig-Schmidt, M.C. 2006. World-wide consumption of *trans* fatty acids. *Atherosclerosis Suppl.* 7, 1–4.

Cryer MD, P.E., Irene, E., and Karl, M.M. 2007. Insulin therapy and hypoglycemia in Type 2 diabetes mellitus. *Insulin.* 2, 127–133.

Da Silva, R.C., Soares, D.F., Lourenço, M.B., Soares, F.A.S.M., Da Silva, K.G., Gonçalves, M.I.A., Gioielli, L.A. 2010. Structured lipids obtained by chemical interesterification of olive oil and palm stearin. *LWT-Food Sci. Technol.* 43, 752–758.

Daniel, D.R., Thompson, L.D., Shriver, B.J., Wu, C.-K., Hoover, L.C. 2005. Nonhydrogenated cottonseed oil can be used as a deep fat frying medium to reduce *trans*-fatty acid content in French fries. *J. Am. Diet. Assoc.* 105, 1927–1932.

De Castro Barra, P.M., Barra, M.M., Azevedo, M.S., Fett, R., Micke, G.A., Costa, A.C.O., De Oliveira, M.A.L. 2012. A rapid method for monitoring total *trans* fatty acids (TTFA) during industrial manufacturing of Brazilian spreadable processed cheese by capillary zone electrophoresis. *Food Control.* 23, 456–461.

De Castro Barra, P.M., Castro, R.D.J.C., De Oliveira, P.L., Aued-Pimentel, S., Da Silva, S.A., De Oliveira, M.A.L. 2013. An alternative method for rapid quantitative analysis of majority *cis–trans* fatty acids by CZE. *Food Res. Int.* 52, 33–41.

De Castro, P.C.M.A., Barra, M.M., Costa Ribeiro, M.C., Aued-Pimentel, S., Da Silva, S.A., De Oliveira, M.A.L. 2010. Total *trans* fatty acid analysis in Spreadable cheese by capillary zone electrophoresis. *J. Agric. Food Chem.* 58, 1403–1409.

De Freitas, E.V., Brandão, A.A., Pozzan, R., Magalhães, M.E., Fonseca, F., Pizzi, O., Campana, Brandão, A.P. 2011. Importance of high-density lipoprotein-cholesterol (HDL-C) levels to the incidence of cardiovascular disease (CVD) in the elderly. *Arch. Gerontol. Geriatr.* 52, 217–222.

De Oliveira, M.A.L., Solis, V.E.S., Gioielli, L.A., Polakiewicz, B., Tavares, M.F.M. 2003. Method development for the analysis of *trans*-fatty acids in hydrogenated oils by capillary electrophoresis. *Electrophoresis.* 24, 1641–1647.

De Roos, N.M., Bots, M.L., Katan, M.B. 2001. Replacement of dietary saturated fatty acids by *trans* fatty acids lowers serum HDL cholesterol and impairs endothelial function in healthy men and women. *Arterioscler. Thromb. Vasc. Biol.* 21, 1233–1237.

Delmonte, P., Rader, J. 2007. Evaluation of gas chromatographic methods for the determination of *trans* fat. *Anal. Bioanal. Chem.* 389, 77–85.

Delmonte, P., Fardin Kia, A.-R., Kramer, J.K.G., Mossoba, M.M., Sidisky, L., Rader, J.I. 2011. Separation characteristics of fatty acid methyl esters using SLB-IL111, a new ionic liquid coated capillary gas chromatographic column. *J. Chromatogr. A* 1218, 545–554.

Destaillats, F., Golay, P.-A., Joffre, F., De Wispelaere, M., Hug, B., Giuffrida, F., Fauconnot, L., Dionisi, F. 2007. Comparison of available analytical methods to measure *trans*-octadecenoic acid isomeric profile and content by gas-liquid chromatography in milk fat. *J. Chromatogr. A* 1145, 222–228.

Dionisi, F., Golay, P.A., Fay, L.B. 2002. Influence of milk fat presence on the determination of *trans* fatty acids in fats used for infant formulae. *Anal. Chim. Acta* 465, 395–407.

Dirix, C.E.H., Kester, A.D., Hornstra, G. 2009. Associations between term birth dimensions and prenatal exposure to essential and *trans* fatty acids. *Early Hum. Dev.* 85, 525–530.

FDA, Food and Drug Administration 21CFR101.9. Nutrition labeling of food (2010). [on-line]: http://www.accessdata.fda.gov/scripts/cdrh/cfdocs/cfCFR/CFRSearch.cfm?fr=101.9. Accessed 06.27.13.

Fritsche, J., Steinhart, H., Mossoba, M.M., Yurawecz, M.P., Sehat, N., Ku, Y. 1998. Rapid determination of *trans*-fatty acids in human adipose tissue: Comparison of attenuated total reflection infrared spectroscopy and gas chromatography. *J. Chromatogr. B Biomed. Sci. Appl.* 705, 177–182.

Glew, R.H., Herbein, J.H., Ma, I., Obadofin, M., Wark, W.A., Vanderjagt, D.J. 2006. The *trans* fatty acid and conjugated linoleic acid content of Fulani butter oil in Nigeria. *J. Food Compos. Anal.* 19, 704–710.

Grover-Páez, F., Zavalza-Gómez, A.B. 2009. Endothelial dysfunction and cardiovascular risk factors. *Diabetes Res. Clin. Pract.* 84, 1–10.

Gunawan, S., Melwita, E., Ju, Y.-H. 2010. Analysis of *trans–cis* fatty acids in fatty acid steryl esters isolated from soybean oil deodoriser distillate. *Food Chem.* 121, 752–757.

Han, S.N., Leka, L.S., Lichtenstein, A.H., Ausman, L.M., Schaefer, E.J., Meydani, S.N. 2002. Effect of hydrogenated and saturated, relative to polyunsaturated, fat on immune and inflammatory responses of adults with moderate hypercholesterolemia. *J. Lipid Res.* 43, 445–452.

Hayat, L., Al-Sughayer, M.A., Afzal, M. 1999. Fatty acid composition of human milk in Kuwaiti mothers. *Comp. Biochem. Physiol. B: Biochem. Mol. Biol.* 124, 261–267.

Hornstra, G., Van Eijsden, M., Dirix, C., Bonsel, G. 2006. *Trans* fatty acids and birth outcome: Some first results of the MEFAB and ABCD cohorts. *Atherosclerosis Suppl.* 7, 21–23.

Hou, J.-C., Wang, F., Wang, Y.-T., Xu, J., Zhang, C.-W. 2012. Assessment of *trans* fatty acids in edible oils in China. *Food Control.* 25, 211–215.

Huang, Z., Wang, B., Crenshaw, A.A. 2006. A simple method for the analysis of *trans* fatty acid with GC-MS and AT™-Silar-90 capillary column. *Food Chem.* 98, 593–598.

Innis, S.M. 2006. *Trans* fatty intakes during pregnancy, infancy and early childhood. *Atherosclerosis Suppl.* 7, 17–20.

Jakobsen, M.U., Bysted, A., Andersen, N.L., Heitmann, B.L., Hartkopp, H.B., Leth, T., Overvad, K., Dyerberg, J. 2006. Intake of ruminant *trans* fatty acids and risk of coronary heart disease—An overview. *Atherosclerosis Suppl.* 7, 9–11.

Jiménez, J.J., Bernal, J.L., Nozal, M.J., Toribio, L., Bernal, J. 2009. Profile and relative concentrations of fatty acids in corn and soybean seeds from transgenic and isogenic crops. *J. Chromatogr. A.* 1216, 7288–7295.

Juanéda, P. 2002. Utilisation of reversed-phase high-performance liquid chromatography as an alternative to silver-ion chromatography for the separation of *cis*- and *trans*-C18:1 fatty acid isomers. *J. Chromatogr. A.* 954, 285–289.

Kandhro, A., Sherazi, S.T.H., Mahesar, S.A., Bhanger, M.I., Younis Talpur, M., Rauf, A. 2008. GC-MS quantification of fatty acid profile including *trans* FA in the locally manufactured margarines of Pakistan. *Food Chem.* 109, 207–211.

Katan, M.B. 2006. Regulation of *trans* fats: The gap, the Polder, and McDonald's French fries. *Atherosclerosis Suppl.* 7, 63–66.

Kim, B.H., Lumor, S.E., Akoh, C.C. 2008. *Trans*-free margarines prepared with Canola oil/palm stearin/palm kernel oil-based structured lipids. *J. Agric. Food Chem.* 56, 8195–8205.

King, I.B., Kristal, A.R., Schaffer, S., Thornquist, M., Goodman, G.E. 2005. Serum *trans*-fatty acids are associated with risk of prostate cancer in β-carotene and retinol efficacy trial. *Cancer Epidemiol. Biomark. Prev.* 14, 988–992.

Koletzko, B., Decsi, T. 1997. Metabolic aspects of *trans* fatty acids. *Clin. Nutr.* 16, 229–237.

Kondoh, Y., Kawada, T., Urade, R. 2007. Activation of caspase 3 in HepG2 cells by elaidic acid (*t*18:1). *BBA-Mol. Cell. Biol. L.* 1771, 500–505.

Kramer, J.G., Blackadar, C.B., Zhou, J. 2002. Evaluation of two GC columns (60-m SUPELCOWAX 10 and 100-m CP sil 88) for analysis of milkfat with emphasis on CLA, 18:1, 18:2 and 18:3 isomers, and short- and long-chain FA. *Lipids* 37, 823–835.

Kramer, J.G., Hernandez, M., Cruz-Hernandez, C., Kraft, J., Dugan, M.R. 2008. Combining results of two GC separations partly achieves determination of all *cis* and *trans* 16:1, 18:1, 18:2 and 18:3 except CLA isomers of milk fat as demonstrated using Ag-ion SPE fractionation. *Lipids* 43, 259–273.

Larqué, E., Zamora, S., Gil, A. 2001. Dietary *trans* fatty acids in early life: A review. *Early Hum. Dev.* 65(Suppl. 2), 31–41.

Lee, J.H., Akoh, C.C., Himmelsbach, D.S., Lee, K.-T. 2008. Preparation of interesterified plastic fats from fats and oils free of *trans* fatty acid. *J. Agric. Food Chem.* 56, 4039–4046.

Leth, T., Jensen, H.G., Mikkelsen, A., Bysted, A. 2006. The effect of the regulation on *trans* fatty acid content in Danish food. *Atherosclerosis Suppl.* 7, 53–56.

Li, H., Fan, Y.-W., Li, J., Tang, L., Hu, J.-N., Deng, Z.-Y. 2013. Evaluating and predicting the oxidative stability of vegetable oils with different fatty acid compositions. *J. Food Sci.* 78, 633–641.

Lichtenstein, A.H., Erkkilä, A.T., Lamarche, B., Amp, X., Schwab, U.S., Jalbert, S.M., Ausman, L.M. 2003. Influence of hydrogenated fat and butter on CVD risk factors: Remnant-like particles, glucose and insulin, blood pressure and C-reactive protein. *Atherosclerosis* 171, 97–107.

Liu, W.H., Stephen Inbaraj, B., Chen, B.H. 2007a. Analysis and formation of trans fatty acids in hydrogenated soybean oil during heating. *Food Chem.* 104, 1740–1749.

Liu, X., Schumacher, F.R., Plummer, S.J., Jorgenson, E., Casey, G., Witte, J.S. 2007b. *Trans*-fatty acid intake and increased risk of advanced prostate cancer: Modification by RNASEL R462Q variant. *Carcinogenesis* 28, 1232–1236.

Liu, X., Deng, Z., Hu, J., Fan, Y., Liu, R., Li, J., Peng, J., Su, H., Peng, Q., Li, W. 2013. Erythrocyte membrane trans-fatty acid index is positively associated with a 10-year CHD risk probability. *Br. J. Nutr.* 109, 1695–1703.

Lopez-Garcia, E., Schulze, M.B., Meigs, J.B., Manson, J.E., Rifai, N., Stampfer, M.J., Willett, W.C., Hu, F.B. 2005. Consumption of *trans* fatty acids is related to plasma biomarkers of inflammation and endothelial dysfunction. *J. Nutr.* 135, 562–566.

Louheranta, A.M., Turpeinen, A.K., Vidgren, H.M., Schwab, U.S., Uusitupa, M.I.J. 1999. A high-*trans* fatty acid diet and insulin sensitivity in young healthy women. *Metabolism* 48, 870–875.

Martin, C.A., Milinsk, M.C., Visentainer, J.V., Matsushita, M., De-Souza, N.E. 2007. *Trans* fatty acid-forming processes in foods: A review. *An. Acad. Bras. Cienc.* 79, 343–350.

Massaro, M., Scoditti, E., Carluccio, M.A., Montinari, M.R., De Caterina, R. 2008. Omega-3 fatty acids, inflammation and angiogenesis: Nutrigenomic effects as an explanation for anti-atherogenic and anti-inflammatory effects of fish and fish oils. *Br. J. Nutr.* 1, 4–23.

Mckelvey, W., Greenland, S., Chen, M.-J., Longnecker, M.P., Frankl, H.D., Lee, E.R., Haile, R.W. 1999. A case–control study of colorectal adenomatous polyps and consumption of foods containing partially hydrogenated oils. *Cancer Epidemiol. Biomark. Prev.* 8, 519–524.

Micha, R., Mozaffarian, D. 2008. *Trans* fatty acids: Effects on cardiometabolic health and implications for policy. *Prostag. Leukotr. Ess.* 79, 147–152.

Mjøs, S.A. 2005. Properties of *trans* isomers of eicosapentaenoic acid and docosahexaenoic acid methyl esters on cyanopropyl stationary phases. *J. Chromatogr. A.* 1100, 185–192.

Molkentin, J., Precht, D. 1995. Optimized analysis of *trans*-octadecenoic acids in edible fats. *Chromatographia* 41, 267–272.

Momchilova, S., Nikolova-Damyanova, B., Christie, W.W. 1998. Silver ion high-performance liquid chromatography of isomeric *cis*- and *trans*-octadecenoic acids: Effect of the ester moiety and mobile phase composition. *J. Chromatogr. A.* 793, 275–282.

Morrison, J.A., Glueck, C.J., Wang, P. 2008. Dietary *trans* fatty acid intake is associated with increased fetal loss. *Fertil. Steril.* 90, 385–390.

Moss, J. 2006. Labeling of trans fatty acid content in food, regulations and limits—The FDA view. *Atherosclerosis Suppl.* 7, 57–59.

Mossoba, M., Yurawecz, M., Mcdonald, R. 1996. Rapid determination of the total trans content of neat hydrogenated oils by attenuated total reflection spectroscopy. *J. Am. Oil Chem. Soc.* 73, 1003–1009.

Mossoba, M.M., Milosevic, V., Milosevic, M., Kramer, J.K.G., Azizian, H. 2007. Determination of total *trans* fats and oils by infrared spectroscopy for regulatory compliance. *Anal. Bioanal. Chem.* 389, 87–92.

Mozaffarian, D., Pischon, T., Hankinson, S.E., Rifai, N., Joshipura, K., Willett, W.C., Rimm, E.B. 2004. Dietary intake of *trans* fatty acids and systemic inflammation in women. *Am. J. Clin. Nutr.* 79, 606–612.

Mozaffarian, D. 2006. *Trans* fatty acids—Effects on systemic inflammation and endothelial function. *Atherosclerosis Suppl.* 7, 29–32.

Mozaffarian, D., Katan, M.B., Ascherio, A., Stampfer, M.J., Willett, W.C. 2006. *Trans* fatty acids and cardiovascular disease. *N. Engl. J. Med.* 354, 1601–1613.

Mozaffarian, D., Aro, A., Willett, W.C. 2009. Health effects of *trans*-fatty acids: Experimental and observational evidence. *Eur. J. Clin. Nutr.* 63, S5–S21.

Mozaffarian, D., Clarke, R. 2009. Quantitative effects on cardiovascular risk factors and coronary heart disease risk of replacing partially hydrogenated vegetable oils with other fats and oils. *Eur. J. Clin. Nutr.* 63, S22–S33.

Nkondjock, A., Shatenstein, B., Maisonneuve, P., Ghadirian, P. 2003. Specific fatty acids and human colorectal cancer: An overview. *Cancer Detect. Prev.* 27, 55–66.

Oomen, C.M., Ocké, M.C., Feskens, E.J.M., Erp-Baart, M.-a.J.V., Kok, F.J., Kromhout, D. 2001. Association between trans fatty acid intake and 10-year risk of coronary heart disease in the Zutphen Elderly Study: A prospective population-based study. *Lancet* 357, 746–751.

Osso, F.S., Moreira, A.S.B., Teixeira, M.T., Pereira, R.O., Tavares Do Carmo, M.D.G., Moura, A.S. 2008. *Trans* fatty acids in maternal milk lead to cardiac insulin resistance in adult offspring. *Nutrition* 24, 727–732.

Pala, V., Krogh, V., Muti, P., Chajès, V., Riboli, E., Micheli, A., Saadatian, M., Sieri, S., Berrino, F. 2001. Erythrocyte membrane fatty acids and subsequent breast cancer: A prospective Italian study. *J. Natl. Cancer Inst.* 93, 1088–1095.

Pande, G., Akoh, C.C. 2013. Enzymatic synthesis of *trans*-free structured margarine fat analogs with high stearate soybean oil and palm stearin and their characterization. *LWT-Food Sci. Technol.* 50, 232–239.

Pande, G., Akoh, C.C., Shewfelt, R.L. 2013. Utilization of enzymatically interesterified cottonseed oil and palm stearin-based structured lipid in the production of *trans*-free margarine. *Biocat. Agric. Biotechnol.* 2, 76–84.

Precht, D., Molkentin, J. 1996. Rapid analysis of the isomers of *trans*-octadecenoic acid in milk fat. *Int. Dairy J.* 6, 791–809.

Priego-Capote, F., Ruiz-Jiménez, J., Garciá-Olmo, J., Luque De Castro, M.D. 2004. Fast method for the determination of total fat and *trans* fatty-acids content in bakery products based on microwave-assisted Soxhlet extraction and medium infrared spectroscopy detection. *Anal. Chim. Acta* 517, 13–20.

Qiu, B., Hu, J.-N., Liu, R., Fan, Y.-W., Li, J., Li, Y., Deng, Z.-Y. 2012. Caspase pathway of elaidic acid (9*t*-C18:1)-induced apoptosis in human umbilical vein endothelial cells. *Cell Biol. Int.* 36, 255–260.

Remig, V., Franklin, B., Margolis, S., Kostas, G., Nece, T., Street, J.C. 2010. *Trans* fats in America: A review of their use, consumption, health implications, and regulation. *J. Am. Diet. Assoc.* 110, 585–592.

Richter, E.K., Shawish, K.A., Scheeder, M.R.L., Colombani, P.C. 2009. *Trans* fatty acid content of selected Swiss foods: The *Trans* Swiss Pilot study. *J. Food Compos. Anal.* 22, 479–484.

Ridker, P.M., Hennekens, C.H., Buring, J.E., Rifai, N. 2000. C-Reactive protein and other markers of inflammation in the prediction of cardiovascular disease in women. *N. Engl. J. Med.* 342, 836–843.

Risérus, U. 2006. *Trans* fatty acids and insulin resistance. *Atherosclerosis Suppl.* 7, 37–39.

Robinson, J.E., Singh, R., Kays, S.E. 2008. Evaluation of an automated hydrolysis and extraction method for quantification of total fat, lipid classes and *trans* fat in cereal products. *Food Chem.* 107, 1144–1150.

Roe, M., Pinchen, H., Church, S., Elahi, S., Walker, M., Farron-Wilson, M., Buttriss, J., Finglas, P. 2013. *Trans* fatty acids in a range of UK processed foods. *Food Chem.* [online]: http://www.sciencedirect.com/science/article/pii/S0308814612013611

Ruiz-Jiménez, J., Priego-Capote, F., Castro, M.D.L.D. 2004. Identification and quantification of *trans* fatty acids in bakery products by gas chromatography–mass spectrometry after dynamic ultrasound-assisted extraction. *J. Chromatogr. A.* 1045, 203–210.

Santos, F.S., Chaves, C.R.M., Costa, R.S.S., Oliveira, O.R.C., Santana, M.G., Conceição, F.D., Sardinha, F.L.C., Veiga, G.V., Tavares Do Carmo, M.G. 2012. Status of *cis* and *trans* fatty ccids in Brazilian adolescent mothers and their newborns. *J. Pediatr. Adolesc. Gynecol.* 25, 270–276.

Saunders, D., Jones, S., Devane, G.J., Scholes, P., Lake, R.J., Paulin, S.M. 2008. *Trans* fatty acids in the New Zealand food supply. *J. Food Compos. Anal.* 21, 320–325.

Seppänen-Laakso, T., Laakso, I., Hiltunen, R. 2002. Analysis of fatty acids by gas chromatography, and its relevance to research on health and nutrition. *Anal. Chim. Acta.* 465, 39–62.

Sherazi, S.T.H., Kandhro, A., Mahesar, S.A., Bhanger, M.I., Talpur, M.Y., Arain, S. 2009a. Application of transmission FT-IR spectroscopy for the *trans* fat determination in the industrially processed edible oils. *Food Chem.* 114, 323–327.

Sherazi, S.T.H., Talpur, M.Y., Mahesar, S.A., Kandhro, A.A., Arain, S. 2009b. Main fatty acid classes in vegetable oils by SB-ATR-Fourier transform infrared (FTIR) spectroscopy. *Talanta* 80, 600–606.

Shin, J.-A., Akoh, C.C., Lee, K.-T. 2010. Enzymatic interesterification of anhydrous butterfat with flaxseed oil and palm stearin to produce low-trans spreadable fat. *Food Chem.* 120, 1–9.

Slattery, M.L., Benson, J., Ma, K.-N., Schaffer, D., Potter, J.D. 2001. *Trans*-fatty acids and colon cancer. *Nutr. Cancer.* 39, 170–175.

Smedes, F. 1999. Determination of total lipid using non-chlorinated solvents. *Analyst* 124, 1711–1718.

Smith, B.K., Robinson, L.E., Nam, R., Ma, D.W.L. 2009. *Trans*-fatty acids and cancer: A mini-review. *Br. J. Nutr.* 102, 1254–1266.

Stender, S., Dyerberg, J. 2004. Influence of *trans* fatty acids on health. *Ann. Nutr. Metab.* 48, 61–66.

Stender, S., Dyerberg, J., Astrup, A. 2006a. Consumer protection through a legislative ban on industrially produced *trans* fatty acids in foods in Denmark. *SSNU.* 50, 155–160.

Stender, S., Dyerberg, J., Bysted, A., Leth, T., Astrup, A. 2006b. A *trans* world journey. *Atherosclerosis Suppl.* 7, 47–52.

Sun, Q., Ma, J., Campos, H., Hankinson, S.E., Manson, J.E., Stampfer, M.J., Rexrode, K.M., Willett, W.C., Hu, F.B. 2007. A prospective study of *trans* fatty acids in erythrocytes and risk of coronary heart disease. *Circulation* 115, 1858–1865.

Tarrago-Trani, M.T., Phillips, K.M., Lemar, L.E., Holden, J.M. 2006. New and existing oils and fats used in products with reduced *trans*-fatty acid content. *J. Am. Diet. Assoc.* 106, 867–880.

Tavella, M., Peterson, G., Espeche, M., Cavallero, E., Cipolla, L., Perego, L., Caballero, B.N. 2000. *Trans* fatty acid content of a selection of foods in Argentina. *Food Chem.* 69, 209–213.

van Poppel, G. 1998. Intake of *trans* fatty acids in western Europe: The TRANSFAIR study. *Lancet* 351, 1099.

Vita, J.A., Keaney, J.F. 2002. Endothelial function: A barometer for cardiovascular risk? *Circulation* 106, 640–642.

Wang, Y., Jacome-Sosa, M.M., Proctor, S.D. 2012. The role of ruminant *trans* fat as a potential nutraceutical in the prevention of cardiovascular disease. *Food Res. Int.* 46, 460–468.

Willett, W.C., Stampfer, M.J., Manson, J.E., Colditz, G.A., Speizer, F.E., Rosner, B.A., Hennekens, C.H., Sampson, L.A. 1993. Intake of *trans* fatty acids and risk of coronary heart disease among women. *Lancet* 341, 581–585.

Willett, W.C. 2006. *Trans* fatty acids and cardiovascular disease-epidemiological data. *Atherosclerosis Suppl.* 7, 5–8.

Wolff, R., Combe, N., Destaillats, F., Boué, C., Precht, D., Molkentin, J., Entressangles, B. 2000. Follow-up of the Δ4 to Δ16 *trans*-18:1 isomer profile and content in French processed foods containing partially hydrogenated vegetable oils during the period 1995–1999. Analytical and nutritional implications. *Lipids* 35, 815–825.

Żbikowska, A. 2010. Formation and properties of *trans* fatty acids—A review. *Pol. J. Food Nutr. Sci.* 60, 107–114.

9 Application of Vibrational Spectroscopy, Portable Detectors, and Hyphenated Techniques to the Analysis of Heat-Induced Changes in Oil and Food Constituents

Magdi M. Mossoba, Didem Peren Aykas, and Luis Rodriguez-Saona

CONTENTS

9.1   INTRODUCTION

Deep fat frying is a food preparation method during which physical and chemical transformations may occur (Velasko et al. 2008). The resulting changes in fried foods impart crisp texture, and desirable color and flavor. However, these desirable sensory characteristics are accompanied by the formation of new volatile and nonvolatile decomposition products that can alter the nutritional quality and functionality of frying oils. Starting from approximately 190°C, frying oils decompose due to heat and oxidation. Heat gives rise to *cis*-to-*trans* geometric isomerization and the formation of cyclic fatty acid monomers (CFAM), as well as nonpolar dimers and oligomers, while oxidation leads to products such as oxidized triacylglycerol monomers, dimers, and oligomers, in addition to volatiles that include ketones, aldehydes, and hydrocarbons. The formation of nonvolatile decomposition products is due to thermal oxidation of fats during frying. The well-documented complex patterns of oxidative and thermolytic reactions that occur during the heating of fats and oils include polymerization, hydrolysis, isomerization, and cyclization (White 1991). These processes affect viscosity, color, and foaming characteristics, and lead to the breakdown of triacylglycerols, the formation of free fatty acids, an increase in carbonyl value, hydroxyl content, and saponification value, and a decrease in unsaturation level. The nature of the chemical products of these reactions is of interest because they affect the sensory quality and the nutritive value of the frying oil and the fried food. Moreover, published reports have linked the ingestion of heat-abused lipids to harmful physiological effects (Sebedio and Grandgirard 1989).

Although advanced instrumentation such as gas chromatography (GC) and high-performance liquid chromatography are in widespread use in the study of lipid chemistry, the cost and complexity of these techniques are beyond the grasp of many producers and processors. Emphasis on maintaining oil quality has opened the door for more rapid and accessible analytical techniques. Vibrational spectroscopy, especially in combination with pattern recognition techniques, offers opportunities for the rapid determination of oil degradation products by analysts who have limited experience using these techniques. These analytical tools allow precise measurement of a few microliters of an edible oil without requiring any sample preparation.

In this chapter, we present a review of the application of novel vibrational spectroscopic methods in monitoring the oxidative stability of oil and in studying heat-induced changes in food constituents. The advent of handheld and portable infrared (IR) units with spectral resolution equivalent to those of benchtop instruments has allowed the extension of the convenience, speed, and analytical precision of vibrational spectroscopy to a number of field applications.

We also present the unique role of IR spectroscopy in the structural elucidation of complex mixtures of CFAM produced in heated oils. To confirm the configuration of a double bond along the fatty acid chain and in monounsaturated five- and six-membered rings, as well as the identity of functional groups for individual components of complex diunsaturated CFAM mixtures produced during deep frying, we discuss the highly sensitive hyphenated GC-IR spectroscopic technique with interfaces operating at cryogenic temperatures, which were applied for the first time to the structural elucidation of the potentially toxic and readily absorbable CFAM class of chemical contaminants in heat-abused oils.

9.2 THERMAL DEGRADATION PRODUCTS

Thermal reactions of lipids occur without the participation of oxygen and result in products with lower polarity than those produced by oxidation (Velasco et al. 2008). The lower solubility of oxygen at high frying temperatures promotes thermal reactions leading to the formation of isomerization products, containing cyclic or *trans* fatty acyl groups, and triacylglycerol dimers (Velasco et al. 2008). Isomerization of *cis* C=C bonds to the *trans* form is a by-product of incomplete hydrogenation of oils. Concerns over the health implications of *trans* fats with isolated double bonds have resulted in labeling requirements and, in some countries, acceptable upper limits in foods. Acyl chains from linoleic and linolenic acids are particularly susceptible to isomerization reactions at frying temperatures (Lambelet et al. 2003, Wolff 1992). A decrease in the concentrations of unsaturated fatty acids attained from isomerization during frying is very low, and the quality of the final products is more likely to be dependent on the nature of frying oils (Sebedio et al. 1996b). Cyclic thermal degradation products are thought to be toxic, but are found at very low concentrations in degraded oils (Velasco et al. 2008). Free fatty acids in oils are a result of triacylglycerol hydrolysis, driven by enzymatic or thermal factors. Although the solubility of water in lipid systems is low, the introduction of fried foods with high water contents can promote hydrolysis (McClements and Decker 2007). The oxidative stability of free fatty acids, monoacylglycerols, and diacylglycerols is lower than the intact triacylglycerol (Guillén and Cabo 1997). These hydrolyzed compounds directly affect oil quality by producing off-flavors, reducing the smoke point of the oil, and accelerating further hydrolysis reactions (Frega et al. 1999).

9.3 LIPID OXIDATION PRODUCTS

Lipid oxidation refers to the interaction of fatty acids with oxygen. For processors and consumers, oxidation is the primary degradation reaction of concern in edible oils. Thermal oxidation is a complex series of chemical reactions initiated by free-radical species, promoted by high temperature and oxygen concentration (Choe and Min 2007). The mechanism of lipid and thermal oxidation is given in Figure 9.1. The free-radical mechanism occurs in three phases: initiation, propagation, and termination. Alkyl radicals are formed initially with the loss of hydrogen radical on the fatty acid aliphatic chain. The carbon–hydrogen bond that is *alpha* to a double bond on an unsaturated fatty acid backbone has lower bond energy as a result of localization of electrons on the double bond. In polyunsaturated fatty acids, where a carbon–hydrogen bond is *alpha* to two double bonds, the bond energy is even lower. This property is responsible for the differences in oxidation rates of fatty acids with varying degrees of unsaturation. Radical formation in oil is a process promoted by heat, light, metals, and reactive oxygen species (Choe and Min 2007). Alkyl radicals from fatty acids can react with oxygen, forming unstable peroxy radicals, which degrade further, yielding alkoxy and hydroxyl radicals. These radicals propagate, removing hydrogen atoms from other fatty acyl chains and creating new radical species. The large range of radicals that can be formed at different positions along the fatty acid aliphatic chain creates a complex mixture of radicals. The end of the oxidative

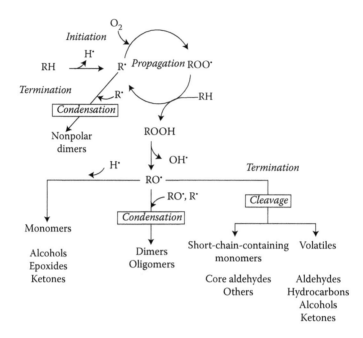

FIGURE 9.1 Mechanism of lipid and thermal oxidation. (Adapted from Velasco, J. et al. 2008. Chemistry of frying. In *Advances in Deep-Fat Frying of Foods*, edited by S. G. Sumnu and S. Sahin, 33–56. CRC Press (Taylor & Francis).)

process is known as the termination step, in which the formation of nonradical compounds halts the reaction.

Hydroperoxides are primary oxidation products; they are highly unstable and decompose to form a variety of volatile and nonvolatile secondary oxidation products (Melton et al. 1994). Decomposition of alkoxy and hydroxyl radicals eventually yields nonradical compounds, which can undergo further reactions to form secondary oxidation products. Common examples of these secondary oxidation products include alcohols, acids, aldehydes, epoxides, and hydrocarbons (Frankel 1987, Paul and Mittal 1997).

Although these compounds significantly contribute to the quality of flavor of frying oils and foods, some may have toxic or carcinogenic properties (Marnett 1999). Due to their small size and relatively nonpolar nature, many secondary oxidation products are highly volatile. During frying, evaporation and reaction with other food components reduce the concentration of volatile compounds (Nawar and Witchwoot 1985). Dimerization and polymerization of radicals generate large macromolecules, which may increase oil viscosity and color (Melton et al. 1994). Studies have shown that triacylglycerol polymers are the most prevalent and complex group of degradation products formed during frying, constituting 12–15 wt% of oil (Velasco et al. 2008). Highly polymerized oils are also more able to interact with the matrix of fried foods, resulting in higher oil absorption in the fried product (Paul and Mittal 1997). Unlike volatile oxidation products, polymerized compounds build up in the oil and may eventually produce a brown, resin-like residue on the sides of the fryer (Choe and Min 2007). Research into the chemistry of edible oils and fats has led to the development of techniques for controlling reactions that lead to undesirable products. Although the nature of lipid oxidation makes it impossible to completely prevent lipid breakdown, slowing these processes has been the focus of much research.

9.4 CYCLIC FATTY ACID MONOMERS

Monocyclic (and some minor bicyclic) fatty acid monomers are among the complex products formed during thermal abuse (Crampton et al. 1956, Firestone 2007, Sebedio and Grandgirard 1989). These toxic CFAM products (Combe et al. 1978, Crampton et al. 1956, Iwaoka and Perkins 1978, Saito and Kaneda 1976) are, reportedly, readily absorbed by the digestive system (Combe et al. 1978, Saito and Kaneda 1976). To assess the physiological effects of heated oils, enriched or pure fractions of well-characterized five- and six-membered CFAM were isolated from heated oils and used in nutritional studies. A discussion of the occurrence, toxicity, and structural elucidation of CFAM is provided in the next section.

9.4.1 CONCENTRATION OF CFAM IN COMMERCIAL FRYING OILS

Most published studies were carried out under simulated frying conditions in the laboratory, and had investigated the formation of CFAM as a function of the nature of oil used, frying temperature, and duration of heat treatment. Only a handful reports on the occurrence of CFAM in frying oils from industrial operations and restaurants were found in the literature. Table 9.1 summarizes the concentration of CFAM found in the heated oils used in various frying operations reported for several countries between 1984 and 2006. All the total CFAM content values consistently fell within a relatively small range, 0.01–0.7% of total oil.

Commercially produced pre-fried French fries and crisps were reported to contain up to 0.1% CFAM (Sebedio et al. 1991, 1996b). In order to determine whether or when

TABLE 9.1
CFAM Levels (as% of Total Oil) in Heated Oils

Nature of Heated Oil Used	CFAM, % of Total Oil	References
Vegetable shortening, animal–vegetable shortening, partially hydrogenated vegetable oil, cottonseed oil, and soybean oil	0.02–0.50 (USA) 0.17–0.66 (Middle East)	Frankel et al. (1984)
Sunflower oil and lard	0.02–0.16 (Hungary)	Gere et al. (1985)
Peanut, sunflower, and soybean oils	0.01–0.25 (France)	Sebedio and Grandgirard (1987)
Palm and peanut oils	0.01–0.09 (France)	Poumeyrol (1987)
Sunflower, peanut, and rapeseed oils	0.15–0.17 (France)	Juaneda et al. (2001)
Sunflower oil	0.09 (Spain)	Romero et al. (2006)
High oleic sunflower oil	0.07 (Spain)	Romero et al. (2006)

frying oils should be discarded, Romero et al. (2006) compared the deterioration of sunflower oil (SO) to that of high-oleic sunflower oil (HOSO) after 20 separate deep fat frying operations without oil replenishment. The reported total CFAM content before and after 20 frying cycles was 64 and 706 mg/kg HOSO oil, respectively, and 71 and 855 mg/kg SO oil, respectively. The low concentration of total CFAM found in oils prior to frying was attributed to the heating process required for deodorization of refined oils. The reported concentrations of CFAM were directly proportional to the number of frying cycles. The vastly different concentration of oleic acid content of SO and HOSO was responsible for the difference in total CFAM levels in the two frying oils. This result led the authors to suggest that oils high in monounsaturated fatty acids, such as HOSO, should be recommended for frying operations.

9.4.2 Toxicity of CFAM

A limited number of studies on the biological effects of CFAM were published between 1951 and 1998 and reviewed by Sebedio et al. (2007), Perkins (1976), and Artman (1969). Three more reports on this subject appeared (Bretillon et al. 2003, Joffre 2001, Martin et al. 2000). These investigations were carried out on laboratory animals which were fed diets that included either heated oils or CFAM fractions isolated from heated oils; such studies indicated that CFAM may be harmful. To date, no human studies have been reported.

In reproduction experiments on rats fed diets that included CFAM, pups had lower weights at birth and their mortality rate was high (Iwaoka and Perkins 1978, Perkins 1976, Saito and Kaneda 1976, Sebedio and Grandgirard 1989); the number of pups per litter was also decreased (Sebedio et al. 1996a). Joffre (2001) further investigated the role of CFAM (in the form of triacylglycerols) in inducing cellular lesions in the livers of rat fetuses during gestation. The diet of pregnant rats was spiked with between 0.3% and 0.9% CFAM; at the 0.9% concentration, the rats exhibited reduced body weight and food intake. The livers of their fetal pups were also lower relative to those of control rats. However, no significant differences were found in the average number or weight of fetuses at 20 days. A higher ratio of the five-membered to the six-membered ring CFAM was found in the fatty acid profile of the liver of fetuses associated with the 0.3% CFAM than that with the 0.9% CFAM.

The effect of CFAM on biochemical pathways of the intermediary metabolism has also been investigated. To determine the impact of CFAM on the activity of liver enzymes of rats, cyclic monomers, isolated from partially hydrogenated soybean oil that had been continuously used in commercial frying operations for a week, were used in feeding studies (Lamboni et al. 1998). The large increase observed in the activity of NADPH cytochrome P450 reductase in the rat livers was consistent with active detoxification. The authors limited their evaluation to enzymes involved in lipid metabolism. Reduced activities were also noted, in experimental groups relative to control, for carnitine palmitoyl transferase-1, isocitrate dehydrogenase, and glucose 6-phosphate dehydrogenase. To determine the ability of CFAM to modify major biochemical pathways involving lipids, Martin et al. (2000) carried out a feeding study on rats using a pure fraction of CFAM, isolated from heated linseed oil, at dose levels of 0.1 and 1 g/100 g diet. They observed an increase in the activity of peroxisomal acyl-CoA oxidase (ACO), microsomal ω- and (ω-1)-laurate hydroxylase (CYP4A1 and CYP2E1, respectively), thus providing indirect evidence that cyclic monomers exhibit a proliferator-like effect. In addition, these authors reported that a coordinated regulation was induced by CFAM between peroxisomal oxidation and the activities of Δ9-desaturase and phosphatidate phosphohydrolase. CFAM was responsible for a reduction, as a function of dose, in Δ-9 desaturase activity and the concentration of monounsaturated fatty acids in the liver. Intake of CFAM also led to an increase in the concentration of γ-linolenic acid that was consistent with an increase in the Δ-6 desaturase activity. The peroxisomal retroconversion process was reportedly responsible for changes in the concentrations of eicosapentaenoic and arachidonic acids in the liver of rats fed with CFAM. These results indicated that CFAM could activate the peroxisome proliferator-activated receptor α (PPARα).

To evaluate the role of PPARα in mediating the induction activity of ACO and CYP4A, Bretillon et al. (2003) fed male and female wild-type and PPARα-null mice a diet of CFAM isolated from heated α-linolenic acid, and determined the activity of ACO, CYP4A, CYP2E1, and stearoyl-coA desaturase at the end of the feeding period. The results indicated that PPARα plays an important role in the liver metabolism of CFAM. However, these authors concluded that PPARα is not the sole mediator of the effects of cyclic monomers in the lipid metabolism of mice. Since the expression of PPARα in human is significantly lower than that in mice (Gonzalez 1997, Roberts 1999), humans may not be able to adequately metabolize CFAM. The effect of CFAM on humans has yet to be determined.

In a recent toxicity study on CFAM produced from linolenic acid (Bretillon et al. 2008), pregnant rats were fed a diet containing 0.7% CFAM, as percent of total diet. A 20% reduction of food intake by female rats was noted, while the fetus numbers and weight were unaffected. However, their one-day-old litter was 20% lighter than those for the control group. Glycemia was reduced in both the female rats and the litter. A significant reduction in insulinemia was also observed in female rats. The authors associated the adverse effects of high CFAM doses with hypoinsulinemia as well as the low growth of the pups.

9.5 VIBRATIONAL SPECTROSCOPY AS A TOOL FOR MONITORING OIL DEGRADATION PRODUCTS

Spectroscopy is the study of the interaction between light and matter and it explores the production, measurement, and interpretation of spectra arising from this interaction (Penner 2010). Electromagnetic radiation interacts with matter in predictable and useful ways; these interactions are the basis of spectroscopy. IR spectroscopy is based on the absorption of IR radiation by chemical bonds and bond structures. Methods for qualitative and quantitative analyses by means of IR spectroscopy have been widely used for moisture, lipid, protein, and carbohydrate determination in food products (Ismail et al. 1999). Since the middle of the 20th century, IR spectroscopy has been an important part of fundamental research on lipid systems (Chapman 1965). Table 9.2 summarizes the different applications of near-IR (NIR) and mid-IR (MIR) for monitoring oil degradation parameters. The IR spectrum is a subset of the electromagnetic spectrum with wavelengths (λ) shorter than that of

TABLE 9.2

Performances and Statistics of Degradation Products in Heated Edible Oils Analyzed by MIR and NIR Techniques[a]

Degradation Oil Analysis	Method	Multivariate Analysis	Performance of Models	Source
FFA[a]	ATR-FTIR[a]	PCA[a]	R[a] from 0.84 to 0.94 for three different oils	Innawong et al. (2004)
		PLS[a]	$R^{2a} = 0.954$, SEP[a] $= 0.14$	Du et al. (2012)
PV[a]	ATR-FTIR	PCA	R from 0.90 to 0.97 for three different oils	Innawong et al. (2004)
		PLS	$R^2 = 0.893$, SEP $= 6.17$	Du et al. (2012)
PTG[a]	ATR-FTIR	PLS	$R^2 = 0.991$, RMSECV[a] $= 1.21\%$, RMSEP[a] $= 1.40\%$	Kuligowski et al. (2010)
			$R^2 = 0.986$, RMSEP $= 1.59\%$ for olive oil	Kuligowski et al. (2011)
			$R^2 = 0.978$, RMSEP $= 2.91\%$ for corn oil	
			$R^2 = 0.995$, RMSEP $= 1.06\%$ for sunflower oil	
TPM[a]	ATR-FTIR	SMLRA[a]	$R^2 = 0.98$, STD[a] $= 0.012$	Tena et al. (2009)
trans (TFA)[a]	ATR-FTIR	PLS	$R^2 = 0.999$, RMSEP $= 0.385$	Talpur (2012)
			$R^2 = 0.999$, RMSEP $= 0.625$, RMSEC[a] $= 0.252$	Sherazi et al. (2009)
PTG	FT-NIR[a]	PLS	$R^2 = 96.47$, RMSEP $= 1.38$	Gertz et al. (2013)
	Transmission, FT-NIR		$R^2 = 0.984$; RMSECV $= 1.85\%$	Kuligowski et al. (2012)
TPM	Transmission, NIR	PLS and FSMLR[a]	PLS $R = 0.999$, rmsd[a] $= 0.731$ FSMLR $R = 0.999$, rmsd $= 0.719$	Ng et al. (2007)
			PLS $R = 0.994$, rmsd $= 0.78$ FSMLR $R = 0.992$, rmsd $= 0.91$	Ng et al. (2011)
	FT-NIR	PLS	$R^2 = 96.86$, RMSEP $= 1.97$	Gertz et al. (2013)
			$R = 93.61$, RMSEP $= 5.78$	Ogutcu et al. (2012)
	Reflectance, NIR		$R^2 = 0.984$, SECV[a] $= 1.068$	Gerde et al. (2007)
FFA	Transmission, NIR	PLS and FSMLR	PLS $R = 0.987$, rmsd $= 0.019$ FSMLR $R = 0.978$, rmsd $= 0.026$	Ng et al. (2007)
			PLS $R = 0.981$, rmsd $= 0.018$ FSMLR $R = 0.963$, rmsd $= 0.025$	Ng et al. (2011)
	Transflectance, NIR	PLS	$R = 0.98$, SEP $= 0.50$	Cozzolino et al. (2005)
	vis-NIR[a]		$R^2 = 0.86$, SEP $= 0.35$	Sanchez et al. (2013)

TABLE 9.2 (continued)
Performances and Statistics of Degradation Products in Heated Edible Oils Analyzed by MIR and NIR Techniques[a]

Degradation Oil Analysis	Method	Multivariate Analysis	Performance of Models	Source
	Reflectance, NIR		$R^2 = 0.973$, SECV $= 0.232$	Gerde et al. (2007)
	Transmission, FT-NIR		$R^2 = 0.948$, SEP $= 0.14$	Du et al. (2012)
	FT-NIR		$R^2 = 96.42$, RMSEP $= 0.259$	Gertz et al. (2013)
			$R = 92.58$, RMSEP $= 0.121$	Ogutcu et al. (2012)
AV[a]	FT-NIR	PLS	$R^2 = 95.93$, RMSEP $= 6.23$	Gertz et al. (2013)
	Transmission, NIR		$R = 0.938$, SEP $= 0.328$	Yildiz et al. (2001)
	Transflectance, NIR		$R = 0.77$, SEP $= 6.2$	Cozzolino et al. (2005)
PV	Transmission, FT-NIR	PLS	$R^2 = 0.953$, SEP $= 4.15$	Du et al. (2012)
	vis-NIR		$R^2 = 0.87$, SEP $= 3.82$	Sanchez et al. (2013)
	Transmission, NIR		$R = 0.994$, SEP $= 0.720$	Yildiz et al. (2001)
	Transflectance, NIR		$R = 0.40$, SEP $= 3.9$	Cozzolino et al. (2005)
	Transmission, NIR	PLS and FSML[a]	PLS $R = 0.991$, SEP $= 0.75$	Yildiz et al. (2002)
			FSML $R = 0.991$, SEP $= 0.76$	
CD[a]	Transmission, NIR	PLS	$R = 0.945$, SEP $= 0.020$	Yildiz et al. (2001)
	Reflectance, NIR		$R^2 = 0.902$, SECV $= 0.088$	Gerde et al. (2007)
	vis-NIR		$R^2 = 0.82$, SEP $= 0.32$	Sanchez et al. (2013)
Viscosity	FT-NIR	PLS	$R = 81.95$, RMSEP $= 22.30$	Ogutcu et al. (2012)
Smoke point	FT-NIR	PLS	$R = 84.07$, RMSEP $= 8.74$	Ogutcu et al, (2012)
OSI[a]	vis-NIR	PLS	$R^2 = 0.93$, SEP $= 6.68$	Sanchez et al. (2013)
PV	Handheld ATR-FTIR	PLS	$R = 0.98$, SECV $= 1.01$	Allendorf et al. (2012)
FFA			$R = 0.96$, SECV $= 0.09$	

[a] FSMLR: Forward stepwise multiple linear regression, SMLRA: Stepwise multiple linear regression analysis, FSML: Forward stepwise multiple linear regression, PTG: Polymerized triacylglyceride, TPM: Total polar materials, FFA: Free fatty acids, AV: Anisidine value, PV: Peroxide value, OSI: Oil stability index, CD: Conjugated diene, rmsd: Root-mean-square of the differences, ATR-FTIR: Attenuated total reflectance-Fourier transform infrared spectroscopy, RMSEC: Root mean square error of calibration, RMSECV: Root mean square error of cross validation, RMSEP: Root mean square error of prediction, PLS: Partial least squares, SEP: Standard error of prediction, PCA: Principal component analysis, SECV: Standard error of cross-validation, FT-NIR: Fourier transform-near infrared spectroscopy, TFA: *trans* fatty acid, STD: Standard deviation, R: Correlation coefficient, R^2: Coefficient of determination, vis–NIR: Visible–near-infrared spectroscopy.

microwave radiation and longer than visible light (Wehling 2010). The IR spectrum is divided into three parts: the far-IR (40–400 cm^{-1}), MIR (400–4000 cm^{-1}), and NIR (4000–14,000 cm^{-1}) (Guillen and Cabo 1997). Absorption of IR radiation by a molecule causes a shift in the dipole moment as a result of molecular vibrations. Vibrational energy is directly proportional to the strength of the bond, and the unique connectivity and environment of each molecule gives it slightly different vibrational modes (Griffiths and de Haseth 2007). The sensitivity of this spectral region to slight changes in structure and measurement conditions makes it a powerful tool for analysis of components in a complex matrix. IR spectroscopy has been widely used for characterization (Arnold and Hartung 1971, Guillén and Cabo 1997, Safar et al. 1994), authentication (de la Mata et al. 2012, Maurer et al. 2012, Ozen and Mauer 2002), and classification (De Luca et al. 2011, Tapp et al. 2003, Yang et al. 2005) of edible oils. NIR and MIR spectroscopies have both been the focus of a large amount of research into the determination of oil quality (Allendorf et al. 2012, Du et al. 2012, Innawong et al. 2004, Ismail et al. 1993, van de Voort et al. 2001).

9.5.1 MID-IR SPECTROSCOPY

MIR is very useful in the study of organic compounds because the observed absorption bands (Figure 9.2) can be attributed to vibrational modes of specific functional groups (Guillén and Cabo 1997). The positioning of the band (within a few wavenumbers) and its intensity are correlated with the energy of the band and its concentration in the matrix (Figure 9.2). These characteristics and the presence of many well-resolved bands in a spectrum make MIR spectroscopy ideal for both qualitative and quantitative applications (2010).

Attenuated total reflectance (ATR, also known as internal reflection) represents an important advance in MIR measurement mode technology (Figures 9.3 and 9.4), overcoming many of the issues that had plagued MIR transmission or external reflection modes (Griffiths and de Haseth 2007). Briefly, placed in the signal path between the beamsplitter and the detector, an ATR accessory consists of a high refractive-index crystal that is in direct contact with the test sample. When light passing through this crystal hits an interface with a medium of lower refractive index at an angle equal to or greater than the critical angle, the light will totally internally reflect in the crystal. This process results in the formation of an evanescent wave in the medium of lower refractive index (e.g., the fat or oil test sample) whose amplitude decays exponentially with distance from the interface (Günzler and Heise 2000, Ismail et al. 1999, Sherazi et al. 2009). Although the wave penetrates minimally into the fat or oil test sample (between 1 and 4 µm depending on wavelength), multiple bounce ATR accessories allow for many points of contact with the test sample, increasing the effective pathlength by allowing the IR light to bounce through the crystal undergoing multiple internal reflections (Ismail et al. 1999). A wide array of high refractive-index materials can be used when constructing an ATR accessory; ZnSe, Ge, and diamond crystals all provide unique properties for a variety of sample types. As stated above, while the depth of penetration is very short, higher sensitivity can be obtained by using multiple-reflection ATR devices by extending the

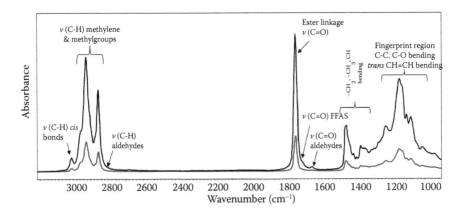

FIGURE 9.2 Increased absorbance is observed for a 5-reflection ATR crystal relative to that of a single-reflection one. (The data for MIR band assignments for typical oil spectra is reported by van de Voort et al. 2001. *European Journal of Lipid Science and Technology* 103 (12):815–826. doi: 10.1002/1438-9312(200112)103:12<815::AID-EJLT1111815>3.0.CO;2-P.; Guillen, M. D., and Cabo, N. 1997. *Journal of the Science of Food and Agriculture* 75(1):1–11. doi: 10.1002/(Sici)1097-0010(199709)75:1<1::Aid-Jsfa842>3.0.Co;2-R. and Sinelli, N. et al. 2010. *Food Research International* 43 (8):2126–2131. doi: 10.1016/j.foodres.2010.07.019.)

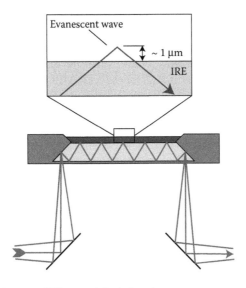

FIGURE 9.3 Five-bounce ATR crystal depicting the evanescent wave (see arrow in inset) and depth of penetration (1 μm) from the yellow internal reflection element (IRE) into the test sample (white area). (Adapted from Martín-Gil, J. et al. 2007. *Journal of Interdisciplinary Celtic Studies*, 563–576.)

FIGURE 9.4 Portable (Cary 630, Agilent Technologies), heated, 5-bounce ATR-MIR spectrometer (far right) equipped with a temperature-controller unit (far left) needed for melting fat.

effective pathlength, as shown in Figure 9.3. Figure 9.2 shows the marked increase in test sample absorbance by comparing a 5-reflection to a single-reflection ZnSe crystal, thus improving the signal-to-noise ratio and enhancing the response of low-concentration components. The theoretical increase in sample absorbance (with all other ATR experimental parameters being equivalent) is equal to the multiple of the effective pathlength and the number of internal reflections of the IR beam from the ATR element surface (PIKE Technologies, Madison, WI).

MIR has been used to monitor degradation reactions in edible oils, providing valuable information about the spectral changes during frying (Du et al. 2012, Guillén and Cabo 1997, Ismail et al. 1993, Pinto et al. 2010, Voort et al. 1994). Figure 9.5 shows the corn oil spectral changes for the control (no thermal treatment) and after 5 days of frying cycle (8 h/day at 185°C) for the ATR-MIR bands associated with C=O stretching of carboxylic acids (1710 cm^{-1}) and C=C–H bending (988 cm^{-1}) which increased during frying, while the band at 3010 cm^{-1} (associated with degree of *cis* unsaturation) simultaneously decreased.

Moros et al. (2009) used ATR-MIR to determine the changes during oil degradation by determining the level of *cis* unsaturation and the formation of *trans* fatty acids as a function of time and heating temperature. In this study, small differences were found between before and after heating; these differences correspond to the variations in the degree and form of unsaturation of acyl groups and their chain length, and they are determined by monitoring the frequencies of the characteristic bands assigned to C=C–H, CH$_2$, and CH$_3$ bands. These MIR spectra showed that increasing heating time resulted in a decrease in *cis* unsaturation, and increase in *trans* fatty acids and FFA content.

Innawong et al. (2004) studied oil degradation by collecting fresh, used, and discarded oils from several fast food restaurants. ATR-MIR spectra were collected and frying oil samples showed strong C–H absorption between 3000 and 2850 cm^{-1}, and

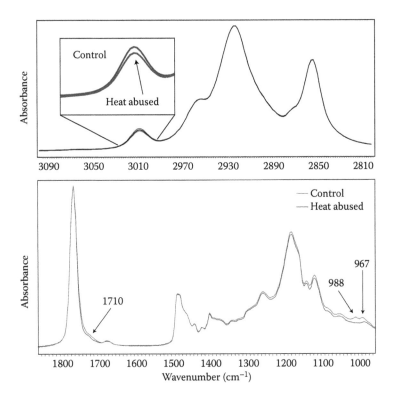

FIGURE 9.5 Comparison of ATR-MIR partial spectral regions for corn oil test portions heated at 185°C for 5 days (8 h/day cycles) and control (unheated) oil.

strong bands at 1749, 1464, and 1165 cm^{-1} that correspond to C=O (ester) stretching, C–H bending (scissoring), C–O (stretching), and CH$_2$ bending, respectively.

The critical absorption bands corresponding to common oxidation end products from the frying process can be observed in the range of 3800–3200 cm^{-1} arising from OH stretching vibrations. The band near 3471 cm^{-1} corresponds with OH stretching vibration of hydroperoxides and a smaller band near 3300 cm^{-1} is attributed to the formation of FFA (Innawong et al. 2004). Pinto et al. (2010) carried out a study by ATR-MIR and showed that aldehydes exhibited a peak at around 1728 cm^{-1} and a peak around 985 cm^{-1} assigned to *trans* conjugated double bonds present in either isomerization products such as conjugated linoleic acid (CLA) or oxidation products such as 2,4-decadienal. The peak at 970 cm^{-1} can be related to isolated (nonconjugated) *trans* double bonds present in either isomerization products such as elaidic acid, or oxidation products such as 3-nonenal and 2-heptenal.

9.5.2 NEAR-IR SPECTROSCOPY

NIR is a well-established branch of spectroscopy that correlates functional groups with overtone and combination bands in the spectrum from 4000 to 14,000 cm^{-1}.

Although NIR bands are one or two orders of magnitude lower in intensity than corresponding bands in the MIR region, the NIR region has some distinct advantages over MIR such as enabling direct analysis of samples that are highly absorbing and strongly light scattering without dilution or extensive sample preparation. Unlike MIR, it also allows measuring test samples contained in glass or plastic containers (Rodriguez-Saona et al. 2001). In contrast to the well-separated, distinctly identifiable bands of MIR, the overlapping bands in the NIR region do not allow for determination of functional groups or chemical structure from the observed spectra alone (Ismail et al. 1999). The wide, overlapping bands of NIR are much more effective at determining major components in complex matrices. Differentiating major components in food matrices by the type of molecule bound to hydrogen allows for quantitative determination of proteins (primarily N–H bonds), lipids (C–H), and carbohydrates (O–H) (2010).

Unlike MIR, the higher-energy NIR radiation can be used to directly analyze solids with diffuse reflectance spectroscopy. In this technique, a small amount of incident radiation penetrates the sample surface and is reflected several times before exiting and captured by an NIR detector (Ismail et al. 1999). More IR radiation is absorbed as a result of the multiple diffusion reflections within the test sample, allowing for analyses which require minimal test sample preparation (Günzler and Heise 2000). For the analysis of liquid samples such as edible oils, the transmittance or transflectance measurement modes are used. In transflectance, the test sample is placed in front of a reflecting surface (reflector) allowing the light to be reflected back to the NIR detector, thus doubling the optical pathlength as the radiation beam passes twice through the test sample.

These properties have led to the development of NIR methods for a variety of lipid applications (Du et al. 2012, Gonzaga and Pasquini 2006) such as the determination of major and minor components and chemical indexes related to oil quality. The NIR spectra of oils (Figure 9.6) show major bands in the 8100–8700 cm^{-1} region arising from second overtones of C–H stretching vibrations, the 7000–7200 cm^{-1} region associated with the combination bands of C–H vibrations, the 5500–6000 cm^{-1} region due to the first overtone of C–H stretching vibrations of methyl, methylene, and ethylene groups (Cozzolino et al. 2005, Christy et al. 2004, Du et al. 2012), and the 4500–4800 cm^{-1} region attributed to combination bands of C–H and C–O stretching vibrations (Downey et al. 2003). It has been reported that oleic acid shows bands at 1725 nm (5797 cm^{-1}), while saturated and *trans* unsaturated triglycerides absorb at, respectively, 1725 and 1760 nm (5797 and 5681 cm^{-1}) (Sinelli et al. 2010).

Oxidation levels in soybean oil have been determined by NIR spectroscopy. Peroxide value (PV), conjugated diene value (CD), and anisidine value (AV) in soybean oil samples have been qualitatively determined by NIR (Yildiz et al. 2001). The NIR technique has been used to analyze degradation products in frying oils, total polar materials (TPMs), and FFAs (Ng et al. 2011). Hein et al. (1998) applied Fourier-transformed near-infrared (FT-NIR) spectroscopy and partial least squares (PLS) as the chemometric method for monitoring polar components in stressed frying oils and fats, monitoring wavenumbers specific for aldehydes, ketones, and hydroperoxides during deep fat frying. Fat oxidation in cereal-based products has

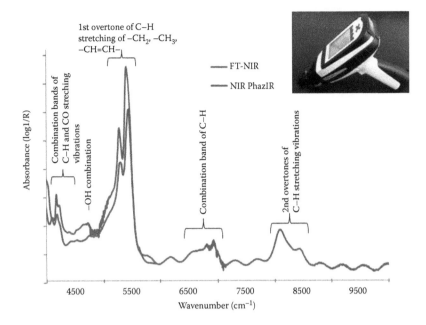

FIGURE 9.6 Observed typical benchtop FT-NIR and dispersive hand-held NIR (micro-Phazir from Thermo Fisher Scientific, see inset) spectra and band assignments for typical oil spectra. (The data for band assignments are reported from Christy, A. A. et al. 2004. *Analytical Sciences* 20 (6):935–940; Cozzolino, D. et al. 2005. *LWT—Food Science and Technology* 38 (8):821–828; and Downey, G. et al. 2003. *Applied Spectroscopy* 57 (2): 158–163.)

been studied and significant differences have been found in the NIR spectra of fresh and stored products; with increasing oxidation, decrease in absorption values was observed arising from the –CH$_2$ groups, at 1400, 1775, 2270, and 2445 nm. In the same study, the authors reported that NIR also enabled the simultaneous measurement of the chemical changes associated with oxidation, such as the formation of peroxides and aldehydes (Kaddour et al. 2006). Using NIR, the 350–2500 nm region allowed a good comparison with official methods for the determination of FFA, conjugated dienoic acids (CDA), and polar substances content (e.g., TPM) (Kleinova and Cventros 2009).

NIR could be used as an alternative technique to preparative column chromatography to detect heat abuse of frying fats and oils (Hein et al. 1998).

9.5.3 Portable and Handheld IR Systems

State-of-the-art benchtop spectrometer systems with specific accessories for measuring liquids or solids provide food, pharmaceutical, and other analytical industries with rapid and sensitive equipment for routine, in-laboratory research analysis. However, the test sample has to be brought to the benchtop spectrometer.

Benchtop and portable
FTIR

FIGURE 9.7 Examples of commercially available portable MIR (two systems on left, Agilent Technologies) and handheld MIR and NIR spectrometers (systems on right, respectively, Thermo Fisher Scientific, Inc.).

By contrast, portable and handheld IR systems (Figure 9.7) have overcome this limitation by providing in many cases reliability and sensitivity equal to those of benchtop systems, but allowing for more flexibility since these systems can be easily transported to the test sample particularly for field applications. Advantages of these approaches potentially include low cost, small size, compactness, robustness, high-throughput, and ease of operation for *in-field* routine analysis. Since minimum background training is required to operate, optically based fingerprinting approaches can enable food producers to obtain real-time information for routine in-process or onsite measurements for quality control (Birkel and Rodriguez-Saona 2011).

The first portable Fourier transform (FT)-MIR systems were developed in 2000 and were originally employed by the US government for the identification of illegal drugs, weapon materials, and hazardous substances (Rein 2008). Development of handheld FTIR technology has been further implemented for identification of surface metals, polymers, contaminants, coatings, and components in chemical manufacturing (Rein 2008). To be useful, a handheld system must operate at varying temperature and humidity settings, in any orientation, and be stable to vibration. Systems were designed for use with either an external reflectance probe for hard, reflective surfaces or a diamond internal reflectance probe for soft, nonreflective samples (Rein 2008).

Allendorf et al. (2012) reported a study which evaluated the capabilities of a handheld MIR spectrometer combined with multivariate analysis to monitor oil oxidation. PLS regression models were developed using various vegetable frying oils (corn, peanut, sunflower, safflower, cottonseed, and canola) stored at 65°C for 30 days to accelerate oxidation reactions. Models developed from reference benchtop tests and handheld systems showed strong correlation with satisfactory R value ($R > 0.9$; Figure 9.8) and prediction errors (SECV) of 1 meq/kg for peroxide value and 0.09% for acid value. These measurements were based on bands in the spectral region of 3012–2850 cm^{-1} (C–H stretching bands/shoulders of fatty acids), at 1740 cm^{-1} (C=O stretching of esters), and 1114 cm^{-1} (–C–O stretching) which were found to be important for prediction of unknown test samples.

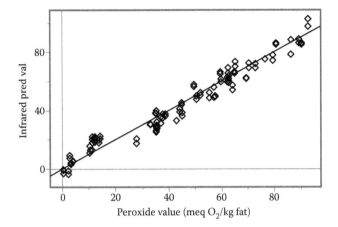

FIGURE 9.8 Application of a partial least-square regression (PLSR) model to the determination of peroxide value for vegetable (corn, cottonseed, sunflower, safflower, canola, and peanut) oils using spectra from a handheld ATR-MIR spectrometer. (Adapted from Allendorf, M. et al. 2012. *Journal of the American Oil Chemists Society* 89 (1):79–88.)

9.6 ROLE OF CRYOGENIC TRAPPING GC-FTIR IN THE STRUCTURAL ELUCIDATION OF CFAM ISOLATED FROM FRYING OILS

While handheld systems are attractive, hyphenated IR instrumentation proved to be sensitive analytical tools with unique advantages for analyzing complex mixtures of CFAM. The heat-induced formation of CFAM has been well documented in the literature (Berdeaux et al. 2007, Christie and Dobson 2000, Sebedio and Juaneda 2007, Sebedio and Grandgirard 1989). CFAM, produced as a result of frying or deodorization of edible oils, are usually generated from oleic, linoleic, and linolenic acids. Oleic acid leads to the formation of eight C18 saturated CFAM with cyclopentyl or cyclohexyl rings along the fatty acid chains (Dobson et al. 1995). Linoleic acid gives rise to 13 monounsaturated cyclic fatty acids mostly with a five-membered ring (Sebedio and Grandgirard 1989). The trienoic α-linolenic acid is in an order of magnitude more susceptible to heat than the dienoic linoleic acid, and produces 16 diunsaturated CFAM having mainly five- and six-membered rings and a few minor bicyclic fatty acid structures (Dobson et al. 1995, 1997, Mossoba et al. 1994, 1995a, b, 1996a, b, Sebedio and Grandgirard 1989, Sebedio and Juaneda 2007). The most common cyclization products are CFAM with five- or six-membered rings in different positions along the alkyl (straight) chain of the parent fatty acids. Each ring has two ortho-substituents, which could be *cis* or *trans* relative to the ring, thus leading to the formation of stereoisomers. Double bonds can be located either within the CFAM ring or on one of the two adjacent alkyl substituents of the ring; the latter can have either a *cis* or a *trans* configuration. For a detailed account on the isolation of CFAM fractions from heated oils and their chromatographic separation, see the report by Sebedio and Juaneda (2007) and references therein.

9.6.1 HYPHENATED GC-SPECTRAL METHODS

Since the 1950s, several studies on the analysis of CFAM by a number of different techniques have been reported (Christie and Dobson 2000, Dobson et al. 1995, Sebedio and Grandgirard 1989, Sebedio and Juaneda 2007). The following section describes a uniquely suitable and sensitive hyphenated IR spectroscopic tool known as gas chromatography-matrix isolation-Fourier transform infrared spectroscopy (GC-MI-FTIR) (Figure 9.9) and its application for the first time to the structural elucidation of CFAM in our laboratory (Mossoba et al. 1994, 1995a, b, 1996a, b). As discussed below, due to its unique sensitivity, GC-MI-FTIR provides information about double-bond configuration for diunsaturated five- and six-membered CFAM structures (Figure 9.10 and Table 9.3). The complementary GC-mass spectrometry (GC-electron ionization (EI)-MS) technique was also essential in these studies (see below). A comprehensive review on the formation and measurement of CFAM by MS has been recently published (Christie and Dobson 2000).

Mossoba et al. (1994, 1995a, b, 1996a, b) reported the separation (Figure 9.10) and determination of the *cis* or *trans* double-bond configuration (Table 9.3) as well as the location of unsaturation sites for a purified, complex mixture of diunsaturated cyclized monomers isolated from heated flaxseed (linseed) oil used as a model system. GC-FTIR interfaces (Bourne et al. 1984, Mossoba et al. 1990, Reedy et al. 1985), such as MI and direct deposition (DD), operate under high vacuum at cryogenic temperatures. With these interfaces, the GC effluent is condensed as a frozen solid for either subsequent or real-time measurement by FTIR. With MI or DD, the effluent is sprayed in real time during a GC run onto the outer rim of a slowly rotating gold-plated disk (Figure 9.9) or a slowly moving IR-transparent slide held at 12 K or 77 K, respectively, under 10^{-5} Torr vacuum.

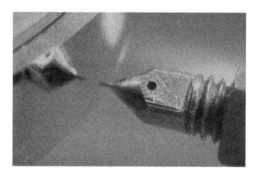

FIGURE 9.9   Picture of the GC-FTIR matrix isolation interface. The hot (250°C or 523 K) tip of the GC transfer line is carefully placed only a few micrometers away from a cryogenically cooled (12 K or −261°C) gold mirror. During a GC-MI-FTIR run, the hot GC effluent is sprayed in real time onto the outer rim of a slowly rotating gold-plated disk (gold mirror) held at 12 K under vacuum. Individual GC components are instantly frozen for subsequent and sequential measurement of each solid "peak" by FTIR spectroscopy.

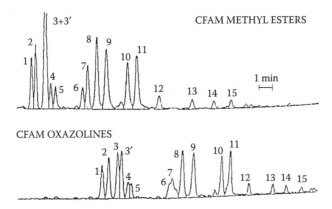

FIGURE 9.10 GC-FID profiles for mixtures of diunsaturated cyclic fatty acid monomer (CFAM) methyl ester (top) and CFAM DMOX (bottom) derivatives. A fairly similar chromatographic profile was observed for methyl ester and oxazoline derivatives.

Hence, mixture components are individually trapped as microscopic "peaks" in the solid state on a slowly moving IR substrate (gold mirror or zinc selenide slide). In MI, a small quantity (1.5%) of IR-transparent argon gas is pre-mixed with the helium GC carrier gas. At 12 K, helium is pulled away by the vacuum, while the argon atoms freeze with the separated mixture components (CFAM derivatives) on the cryogenic gold disk. Each individual CFAM (or other analyte) *molecule* is surrounded by a large excess of argon atoms, hence isolating it in a solid matrix of argon. After the gas chromatographic separation run is completed, each frozen "peak" is sequentially placed in the path of the IR beam for measurement in the transflection (MI) or transmission (DD) mode. The higher sensitivity of these hyphenated IR techniques results partly from signal averaging, which significantly improves signal-to-noise ratios. For instance, with the MI interface, signal averaging time can be long (several minutes) since the separated GC components or "peaks" are trapped as solids for as long as the interface is maintained at cryogenic temperatures under vacuum.

CFAM derivatives other than methyl esters are more appropriate for analysis by GC-EIMS because double-bond migration in CFAM ME derivatives had been found to obscure the location of double bonds in the observed EI mass spectra (Christie and Dobson 2000, Sebedio and Grandgirard 1989). Therefore, the ME mixture of CFAM derivatives are converted into one of the more appropriate derivatives. One derivative in particular, namely the 2-alkenyl-4,4-dimethyloxazoline (DMOX) derivative (Figure 9.10), first reported by Zhang et al. (1988), was found to be the most suitable and was used to derivatize CFAM in our laboratory for subsequent analysis by GC-EIMS (Mossoba et al. 1994, 1995a, b, 1996a, b); some of the other derivatives that are potentially useful for MS analysis of CFAM degraded the gas chromatographic resolution, and therefore were avoided. Using a few selected examples, we demonstrate in the following section the advantage of applying complementary IR and mass spectral detection methods to the elucidation of CFAM components eluting from a gas chromatographic column.

TABLE 9.3
GC-MI-IR Bands for Diunsaturated CFAM Methyl Ester Components That Gave Rise to the GC Profile in Figure 9.10

Peak	Ring (Five-membered)	Chain (trans)	Ring (Six-membered)	Infrared Bands for Unsaturated CFAM					
				Chain (cis)	Chain (trans)	Chain (trans)	Ring (Six-membered)	Ring (Five-membered)	Ring (Six-membered)
1	3061	3035			3003	970		719	
2	3061	3035			3003	970		719	
3 + 3	3061	3035		3005				716	
4	3063	3035			3005	979		716	
6	3063			3006				711	
7	3063			3006				711	
8			3032		3000	976			663
9			3032		3005	972			664
10			3032		3005	972			664
11			3032		3004	975			663
12			3031				723		664
13			3031				725		662
14			3025						
15			3021						

The hyphenated GC-MI-FTIR techniques proved to be essential for observing minor yet highly characteristic IR bands (Mossoba et al. 1994, 1995a, b, 1996a, b) for individual CFAM mixture components (Table 9.3). The complementary GC-MI-FTIR and GC-EIMS were used to establish the *cis* or *trans* double-bond configuration, the position of both double bonds, and five- or six-membered rings along the hydrocarbon chain in CFAM, the molecular mass, the size of the cyclic ring, and the number of double bonds in a CFAM (Figures 9.11 through 9.14). For comparison, the mixtures of CFAM DMOX and CFAM ME derivatives were both measured by GC-MI-FTIR spectroscopy and yielded similar spectral information.

Mass spectra for DMOX derivatives of saturated fatty acids were found to exhibit (a) the molecular ion, (b) a characteristic peak at m/z 113 attributed to McLafferty rearrangement (Yu et al. 1989, Zhang et al. 1988, 1989), and (c) an even-mass homologous series, in which the most abundant peak in each cluster is at m/z ($126 + 14n$, where $n = 0, 1, 2,...$). The first ion at m/z 126, the low-mass end of this series, is reportedly (Zhang et al. 1988, 1989) formed via a cyclization–displacement reaction induced by the nucleophilic center. The homologous ion series with a pattern of peaks separated by 14 Daltons (Da) is due to the sequential cleavage of one methylene group at a time. For CFAM DMOX derivatives, this fragmentation at each skeletal C–C bond is interrupted when a double bond and/or a ring is present along the fatty acid hydrocarbon chain. A mass interval of 68 Da or 82 Da was consistent with the identity and location of a cyclopentyl (Figure 9.13a) or cyclohexyl ring, respectively, in a CFAM. The ring 1,2-disubstitution pattern had been established for CFAM structures with five- (Gast et al. 1963, Le Quere et al. 1991) or six-membered (Hutchison and Alexander 1963, Macdonald 1956, McInnes et al.1961, Saito and

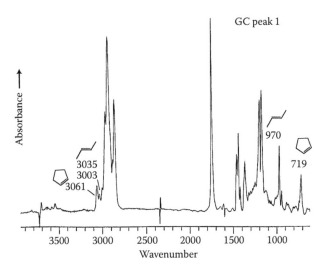

FIGURE 9.11 Unique MI-FTIR spectral features observed at 4 cm⁻¹ resolution for the CFAM methyl ester mixture components that gave rise to GC peak 1 (GC trace shown in Figure 9.10). IR spectral data are consistent with a cyclopentenyl ring and a *trans* double bond along the fatty acid chain (Table 9.3).

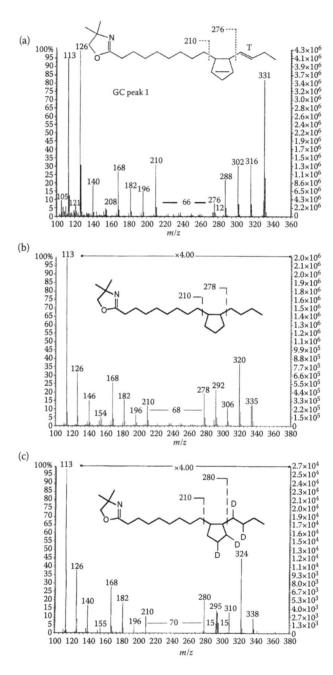

FIGURE 9.12 EIMS observed for the CFAM oxazoline mixture components that gave rise to GC peak 1 (shown in Figure 9.10). The position of the ring double bond may be at the 3,4- or 4,5-ring carbon positions. EIMS and structures for CFAM oxazoline mixture components that gave rise to GC peak 1 (shown in Figure 9.10) after (b) hydrogenation and (c) deuteration. The ring deuterium atoms may be at the 3,4- or 4,5-ring carbon positions.

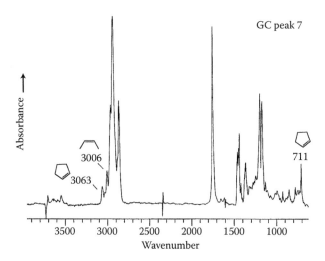

FIGURE 9.13 Unique MI-FTIR spectral features observed at $4\ cm^{-1}$ resolution for the CFAM methyl ester mixture components that gave rise to GC peak 7 (GC profile shown in Figure 9.10). IR spectral data are consistent with a cyclopentenyl ring and a *cis* double bond along the fatty acid chain (Table 9.3).

Kaneda 1976) rings isolated from heated linseed oil, or from synthesized standards (Rojo and Perkins 1989, Vatèla et al. 1988).

The GC chromatograms for the ME and DMOX derivatives of the unsaturated CFAM mixture isolated from heated flaxseed oil were qualitatively similar (Figure 9.10). Typical examples of CFAM mass spectra for DMOX derivatives are shown in Figures 9.12 and 9.14. The mass spectrum for the analyte that gave rise to GC peak 1 exhibited successive mass intervals of 14 Da with interruptions of 66 and 12 Da (Figure 9.12). This spectrum is due to a C18 CFAM structure with a cyclopetenyl ring having a 2″-*n*-butene substituent.

The double bond is on C1 of the butene group based on empirical rules by Zhang et al. (1988, 1989). Namely, if two consecutive even-mass homologous fragments separated by a 12-Da interval are observed, and these fragments contain $n - 1$ and n carbons of the original fatty acid moiety, then these spectral features would be attributed to the presence of a double bond between carbon atoms n and $n + 1$. The double-bond configuration of this butene group was found by GC-MI-IR to be *trans* (Figure 9.11 and Table 9.3). A disubstituted double bond in five- or six-membered ring was always found to have a *cis* configuration. We note that more than one position for a double bond within a 1,2-disubstituted ring may be possible. The double-bond configuration (Table 9.3) obtained by GC-MI-IR (Mossoba et al. 1995a, b) for the different GC peaks observed for the heated linseed oil (Figure 9.10) was consistent with those previously reported in the literature (Sebedio et al. 1987a, b).

To verify the MS assignments of double-bond location, two portions of the CFAM ME mixture were hydrogenated and deuterated, converted into the DMOX derivative, and analyzed by GC-EIMS. In the mass spectra for hydrogenated or deuterated

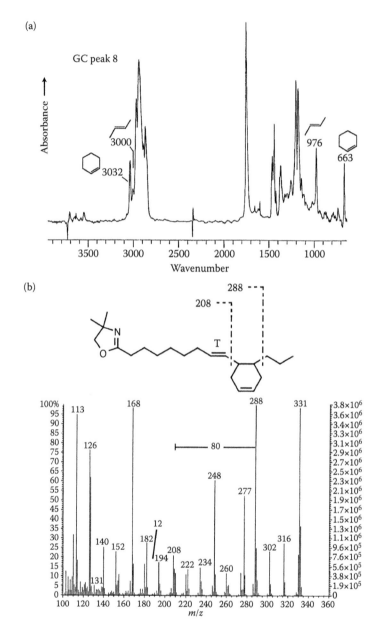

FIGURE 9.14 (a) Unique MI-FTIR spectral features observed at 4 cm⁻¹ resolution, and (b) EIMS observed for the CFAM methyl ester mixture components that gave rise to GC peak 8 (GC trace shown in Figure 9.10). IR and MS spectral data are consistent with a cyclohexenyl ring, and IR data are consistent with a *trans* double bond along the fatty acid chain (Table 9.3).

CFAM (Figure 9.12b, c), consecutive fragments separated by an interval of 14 or 15 Da each were observed and attributed to the presence of a –CH$_2$–CH$_2$– or –CHD–CHD– saturated moieties, respectively, along the fatty acid hydrocarbon chain. Similar mass increases were found for five- and six-membered rings upon hydrogenation and deuteration and confirmed the saturation of a double bond within the ring. Two GC-MI-FTIR spectra representative of other CFAM components are shown in Figures 9.13 and 9.14a. Unique IR features (Table 9.3) not observed by other less sensitive techniques, such as light-pipe GC-FTIR (Mossoba et al. 1995a), demonstrate the essential role of cryogenic trapping GC-FTIR in elucidating the structures of complex mixtures of CFAM isolated from frying oils.

9.7 CONCLUSION

Vibrational spectroscopy combined with multivariate statistical techniques is a powerful technology for measuring oxidative parameters during the frying process. Integration of robust and reliable prediction models into portable IR spectrometers can provide a rapid, simple, cost-effective, and convenient quality control tool for the edible oil industry.

The reported formation of appreciable amounts (up to 0.7% of total oil) of CFAM in frying oils is critically dependent on the temperature of frying operations particularly above 190°C. Oleic, linoleic, and linolenic acid constituents of frying oils are responsible for the formation of CFAM. GC-cryogenic trapping-FTIR spectroscopic techniques uniquely provide the required sensitivity for confirming the double-bond configuration and the presence of a double bond in five- and six-membered rings and along the fatty acid chain for individual components of complex CFAM mixtures. Hyphenated GC-IR and GC-MS techniques are essential complementary methods for the structural elucidation of CFAM. Hydrogenation and deuteration can help confirm MS structural assignments for CFAM. Experiments on laboratory animals suggest that CFAM are incorporated into membrane phospholipids and may have an adverse impact on fatty acid metabolism and physiological functions. No human studies on CFAM toxicity have been reported. Frying oils high in monounsaturated fatty acids produce relatively low quantities of CFAM.

REFERENCES

Allendorf, M., Subramanian, A., and Rodriguez-Saona, L. 2012. Application of a handheld portable mid-infrared sensor for monitoring oil oxidative stability. *Journal of the American Oil Chemists Society* 89 (1):79–88. doi: 10.1007/S11746-011-1894-9.

Arnold, R. G., and Hartung T. E. 1971. Infrared spectroscopic determination of degree of unsaturation of fats and oils. *Journal of Food Science* 36 (1):166–168. doi: 10.1111/J.1365-2621.1971.Tb02061.X.

Artman, N.R. 1969. The chemical and biological properties of heated and oxidized fat. *Advances in Applied Lipid Research* 7:245–330.

Berdeaux, O., Fournier, V., Lambelet, P., Dionisi, F., Sebedio, J. L., and Destaillats, F. 2007. Isolation and structural analysis of the cyclic fatty acid monomers formed from eicosapentaenoic and docosahexaenoic acids during fish oil deodorization. *Journal of Chromatography A* 1138 (1–2):216–224. doi: 10.1016/J.Chroma.2006.10.061.

Birkel, E., and Rodriguez-Saona, L. 2011. Application of a portable handheld infrared spectrometer for quantitation of trans fat in edible oils. *Journal of the American Oil Chemists Society* 88 (10):1477–1483. doi: 10.1007/s11746-011-1814-z.

Bourne, S., Reedy, G., Coffey, P., and Mattson, D. 1984. Matrix isolation GC/FTIR. *American Laboratory (Shelton, CT)* 16:90–101.

Bretillon, L., Roy, A., Pasquis, B., and Sebedio, J. L. 2008. Dietary cyclic fatty acids derived from linolenic acid do not exhibit intrinsic toxicity in the rat during gestation. *Animal* 2 (10):1534–1537. doi: 10.1017/S1751731108002668.

Bretillon, L., Alexson, S. E. H., Joffre, F., Pasquis, B., and Sébédio, J. L. 2003. Peroxisome proliferator-activated receptor α is not the exclusive mediator of the effects of dietary cyclic FA in mice. *Lipids* 38 (9):957–963. doi: 10.1007/s11745-003-1149-y.

Chapman, D. 1965. Infrared spectroscopy of lipids. *Journal of the American Oil Chemists Society* 42 (5):353–371. doi: 10.1007/BF02635571.

Choe, E., and Min, D. B. 2007. Chemistry of deep-fat frying oils. *Journal of Food Science* 72 (5):R77–R86. doi: 10.1111/j.1750–3841.2007.00352.x.

Christie, W. W., and Dobson, G. 2000. Formation of cyclic fatty acids during the frying process. *European Journal of Lipid Science and Technology* 102 (8–9):515–520. doi: 10.1002/1438-9312(200009)102:8/9<515::Aid-Ejlt515>3.3.Co;2-Q.

Christy, A. A., Kasemsumran, S., Du, Y., and Ozaki, Y. 2004. The detection and quantification of adulteration in olive oil by near-infrared spectroscopy and chemometrics. *Analytical Sciences* 20 (6):935–940.

Combe, N., Constantin, M.J., and Entressangles, B. 1978. Etude sur les huiles chauffees IV. Absorption intestinale des especes chimiques nouvelles (E. C. N.) formees lors du chauffage des huiles. (Study on heated oils. IV. Intestinal absorption of new chemical species formed during the heating of oils.) *Revue Francaise des Corps Gras* 25:27–28.

Cozzolino, D., Murray, I., Chree A., and Scaife, J. R. 2005. Multivariate determination of free fatty acids and moisture in fish oils by partial least-squares regression and near-infrared spectroscopy. *LWT-Food Science and Technology* 38 (8):821–828. doi: 10.1016/j.lwt.2004.10.007.

Crampton, E. W., Common, R. H., Pritchard, E. T., and Farmer, F. A. 1956. Studies to determine the nature of the damage to the nutritive value of some vegetable oils from heat treatment IV. Ethyl esters of heat-polymerized linseed, soybean and sunflower seed oils. *Journal of Nutrition* 60:13–24.

de la Mata, P., Dominguez-Vidal, A., Bosque-Sendra, Ruiz-Medina, A., Cuadros-Rodríguez, L., and Ayora-Cañada, M. J., 2012. Olive oil assessment in edible oil blends by means of ATR-FTIR and chemometrics. *Food Control* 23 (2):449–455. doi: 10.1016/j.foodcont.2011.08.013.

De Luca, M., Terouzi, W., Ioele, G., Kzaiber, F., Oussama, A., Oliverio, F., Tauler, R., and Ragno, G. 2011. Derivative FTIR spectroscopy for cluster analysis and classification of morocco olive oils. *Food Chemistry* 124 (3):1113–1118.

Dobson, G., Christie, W. W., Brechany, E. Y., Sebedio, J. L., and Lequere, J. L. 1995. Silver ion chromatography and gas-chromatography mass-spectrometry in the structural-analysis of cyclic dienoic acids formed in frying oils. *Chemistry and Physics of Lipids* 75 (2):171–182. doi: 10.1016/0009-3084(95)02420-N.

Dobson, G., Christie, W. W., and Sebedio, J. L. 1997. Saturated bicyclic fatty acids formed in heated sunflower oils. *Chemistry and Physics of Lipids* 87 (2):137–147. doi: 10.1016/S0009-3084(97)00036-4.

Downey, G., McIntyre, P., and Davies, A. N. 2003. Geographic classification of extra virgin olive oils from the eastern Mediterranean by chemometric analysis of visible and near-infrared spectroscopic data. *Applied Spectroscopy* 57 (2):158–163.

Du, R., Lai, K., Xiao, Z., Shen, Y., Wang, X., and Huang, Y. 2012. Evaluation of the quality of deep frying oils with Fourier transform near-infrared and mid-infrared spectroscopy. *Journal of Food Science* 77 (2):C261-6. doi: 10.1111/j.1750-3841.2011.02551.x.

Firestone, D. 2007. Regulation of frying fat and oil. In *Deep Frying: Chemistry, Nutrition and Practical Applications*, edited by M.D. Erickson, 373–386. Urbana, IL: AOCS Press.

Frankel, E. N. 1987. Secondary products of lipid oxidation. *Chemistry and Physics of Lipids* 44 (2–4):73–85.

Frega, N., Mozzon, M., and Lercker, G. 1999. Effects of free fatty acids on oxidative stability of vegetable oil. *Journal of the American Oil Chemists Society* 76 (3):325–329. doi: 10.1007/s11746-999-0239-4.

Gast, L. E., Schneider, W. J., Forest, C. A., and Cowan, J. C. 1963. Composition of methyl esters from heat-bodied linseed oils. *Journal of the American Oil Chemists Society* 40 (7):287–289.

Gonzaga, F. Barbieri, and Pasquini, C. 2006. A new method for determination of the oxidative stability of edible oils at frying temperatures using near infrared emission spectroscopy. *Analytica Chimica Acta* 570 (1):129–135. doi: 10.1016/j.aca.2006.03.109.

Gonzalez, F. J. 1997. Recent update on the PPAR alpha-null mouse. *Biochimie* 79 (2–3):139–44.

Griffiths, P. R., and de Haseth J. A. 2007. *Fourier Transform Infrared Spectrometry*. 2nd ed. Hoboken, NJ: John Wiley and Sons.

Guillen, M. D., and Cabo, N. 1997. Infrared spectroscopy in the study of edible oils and fats. *Journal of the Science of Food and Agriculture* 75 (1):1–11. doi: 10.1002/(Sici)1097-0010(199709)75:1<1::Aid-Jsfa842>3.0.Co;2-R.

Guillén, M. D., and Cabo, N. 1997. Characterization of edible oils and lard by fourier transform infrared spectroscopy. Relationships between composition and frequency of concrete bands in the fingerprint region. *Journal of the American Oil Chemists Society* 74 (10):1281–1286. doi: 10.1007/s11746-997-0058-4.

Günzler, H., and Heise, H. 2000. *Infrared Spectroscopy: An Introduction*. Weinheim: Wiley-VCH.

Hein, M., Henning, H., and Isengard, H. D. 1998. Determination of total polar parts with new methods for the quality survey of frying fats and oils. *Talanta* 47 (2):447–454. doi: 10.1016/S0039-9140(98)00148-9.

Hutchison, R. B., and Alexander J. C. 1963. The structure of a cyclic C18 acid from heated linseed oil. *The Journal of Organic Chemistry* 28 (10):2522–2526. doi: 10.1021/jo01045a007.

Innawong, B, Mallikarjunan, P.,Irudayaraj, J., and Marcy, J. E. 2004. The determination of frying oil quality using Fourier transform infrared attenuated total reflectance. *LWT-Food Science and Technology* 37 (1):23–28. doi: 10.1016/S0023-6438(03)00120-8.

Ismail, A. A., Nicodemo, A., Sedman, J.,van de Voort, F. R., and Holzbauer, I. E. 1999. Infrared spectroscopy of lipids: Principles and applications. In *Spectral Properties of Lipids*, edited by R. Hamilton and J. Cast, 262–269. Sheffield, UK: Sheffield Academic Press.

Ismail, A. A., Voort, F. R., Emo, G., and Sedman, J. 1993. Rapid quantitative determination of free fatty acids in fats and oils by fourier transform infrared spectroscopy. *Journal of the American Oil Chemists Society* 70 (4):335–341. doi: 10.1007/BF02552703.

Iwaoka, W. T., and Perkins, E. G. 1978. Metabolism and lipogenic effects of the cyclic monomers of methyl linolenate in the rat. *Journal of the American Oil Chemists Society* 55 (10):734–738.

Joffre, F. 2001. Effects nutritionnels des monomeres cycliques issus de l'acide α-linolenique chez l'animal. Dijon, France: Universite de Bourgogne.

Kaddour, A. A., Grand, E., Barouh, N., Barea, B., Villeneuve, P., and Cuq, B. 2006. Near-infrared spectroscopy for the determination of lipid oxidation in cereal food products. *European Journal of Lipid Science and Technology* 108 (12):1037–1046. doi: 10.1002/Ejlt.200600132.

Kleinova, A., and Cventros, J. 2009. Utilization of frying oils/fats in FAME production. In *44th International Petroleum Conference*. Bratislava, Slovak Republic.

Lambelet, P., Grandgirard, A., Gregoire, S., Juaneda, P., Sebedio, J. L., and Bertoli, C. 2003. Formation of modified fatty acids and oxyphytosterols during refining of low erucic acid rapeseed oil. *Journal of Agricultural and Food Chemistry* 51 (15):4284–4290. doi: 10.1021/jf030091u.

Lamboni, C., Sebedio, J. L., and Perkins, E. G. 1998. Cyclic fatty acid monomers from dietary heated fats affect rat liver enzyme activity. *Lipids* 33 (7):675–81.

Le Quere, J. L., Sebedio, J. L., Henry, R., Couderc, F., Demont, N., and Prome, J. C. 1991. Gas chromatography-mass spectrometry and gas chromatography-tandem mass spectrometry of cyclic fatty acid monomers isolated from heated fats. *Journal of Chromatography* 562 (1–2):659–672.

Li-Chan, E., Griffiths, P. R., and Chalmers, J. M. Editors. 2010. *Applications of Vibrational Spectroscopy in Food Science.* 2 vols. Vol. 1. Chichester, UK: John Wiley and Sons.

Macdonald, J. A. 1956. Evidence for cyclic monomers in heated linseed oil. *Journal of the American Oil Chemists Society* 33 (9):394–396. doi: 10.1007/BF02630764.

Marnett, L. J. 1999. Lipid peroxidation—DNA damage by malondialdehyde. *Mutation Research-Fundamental and Molecular Mechanisms of Mutagenesis* 424 (1–2):83–95. doi: 10.1016/S0027-5107(99)00010-X.

Martín-Gil, J., Palacios-Leblé, G., Ramos, P.M., and Martín-Gil, F. J. 2007. Analysis of a Celtiberian protective paste and its possible use by Arevaci warriors. *Journal of Interdisciplinary Celtic Studies*, 5:563–576.

Martin, J. C., Joffre, F., Siess, M. H., Vernevaut, M. F., Collenot, P., Genty, M., and Sebedio, J. L. 2000. Cyclic fatty acid monomers from heated oil modify the activities of lipid synthesizing and oxidizing enzymes in rat liver. *Journal of Nutrition* 130 (6):1524–1530.

Maurer, N. E., Hatta-Sakoda, B., Pascual-Chagman, G., and Rodriguez-Saona, L. E. 2012. Characterization and authentication of a novel vegetable source of omega-3 fatty acids, sacha inchi (*Plukenetia volubilis* L.) oil. *Food Chemistry* 134 (2):1173–80. doi: 10.1016/j.foodchem.2012.02.143.

McClements, D. J., and Decker, E.A. 2007. Lipids. In *Fennema's Food Chemistry*, edited by S. Damodaran, K. L. Parkin and O. R. Fennema. New York: CRC Press/Taylor & Francis.

McInnes, A. G., Cooper, F. P., and MacDonald, J. A. 1961. Further evidence for cyclic monomers in heated linseed oil. *Canadian Journal of Chemistry* 39 (10):1906–1914. doi: 10.1139/v61-256.

Melton, S. L., Jafar, S., Sykes, D., and Trigiano, M. K. 1994. Review of stability measurements for frying oils and fried food flavor. *Journal of the American Oil Chemists Society* 71 (12):1301–1308. doi: 10.1007/Bf02541345.

Moros, J., Roth, M., Garrigues, S., and de la Guardia, M. 2009. Preliminary studies about thermal degradation of edible oils through attenuated total reflectance mid-infrared spectrometry. *Food Chemistry* 114 (4):1529–1536. doi: 10.1016/J.Foodchem.2008.11.040.

Mossoba, M. M., McDonald, R. E., Chen, J. Y. T., Armstrong, D. J., and Page, S. W. 1990. Identification and quantitation of *trans-9,trans*-12-octadecadienoic acid methyl ester and related compounds in hydrogenated soybean oil and margarines by capillary gas chromatography/matrix isolation/Fourier transform infrared spectroscopy. *Journal of Agricultural and Food Chemistry* 38 (1):86–92.

Mossoba, M. M., Yurawecz, M. P., Roach, J. A. G., Lin, H. S., Mcdonald, R. E., Flickinger, B. D., and Perkins, E. G. 1994. Rapid-determination of double-bond configuration and position along the hydrocarbon chain in cyclic fatty-acid monomers. *Lipids* 29 (12):893–896. doi: 10.1007/Bf02536259.

Mossoba, M. M., Yurawecz, M. P., Roach, J. A. G., Lin, H. S., McDonald, R. E., Flickinger, B. D., and Perkins, E. G. 1995a. Application of gas chromatography-matrix isolation-Fourier transform infrared spectroscopy to structural elucidation of cyclic fatty acid monomers. *American Laboratory (Shelton, CT)* 27 (14):16 K-16O.

Mossoba, M. M., Yurawecz, M. P., Roach, J. A. G., Lin, H. S., Mcdonald, R. E., Flickinger, B. D., and Perkins, E. G. 1995b. Elucidation of cyclic fatty-acid monomer structures—Cyclic and bicyclic ring sizes and double-bond position and configuration. *Journal of the American Oil Chemists Society* 72 (6):721–727. doi: 10.1007/Bf02635662.

Mossoba, M. M., Yurawecz, P. M., Roach, J. A. G., McDonald, R. E., and Perkins, E. G. 1996a. Confirmatory mass-spectral data for cyclic fatty acid monomers. *Journal of the American Oil Chemists Society* 73 (10):1317–1321. doi: 10.1007/BF02525462.

Mossoba, M. M., Yurawecz, M. P., Roach, J. A. G., McDonald, R, E., Flickinger, and Perkins, E. G. 1996b. Analysis of cyclic fatty acid monomer 2-alkenyl-4,4-dimethyloxazoline derivatives by gas chromatography–matrix isolation–Fourier transform infrared spectroscopy. *Journal of Agricultural and Food Chemistry* 44 (10):3193–3196. doi: 10.1021/jf960002w.

Nawar, W. W., and Witchwoot, A. 1985. Thermal interaction of lipids with amino acids. *Abstracts of Papers of the American Chemical Society* 189 (April):26-Agfd.

Ng, C. L., Wehling, R. L., and Cuppett, S. L. 2011. Near-infrared spectroscopic determination of degradation in vegetable oils used to fry various foods. *Journal of Agricultural and Food Chemistry* 59 (23):12286–12290. doi: 10.1021/Jf202740e.

Ozen, B. F., and Mauer, L. J. 2002. Detection of hazelnut oil adulteration using FT-IR spectroscopy. *Journal of Agricultural and Food Chemistry* 50 (14):3898–3901. doi: 10.1021/Jf0201834.

Paul, S., and Mittal G. S. 1997. Regulating the use of degraded oil/fat in deep-fat/oil food frying. *Critical Reviews in Food Science and Nutrition* 37 (7):635–662.

Penner, M.H. 2010. Basic principles of spectroscopy. In *Food Analysis*, edited by S.S. Nielsen, 375–386. New York: Springer.

Perkins, E. G. 1976. Chemical, nutritional, and metabolic effects of heated fats. II. Nutritional aspects. *Revue Francaise des Corps Gras* 23:313–322.

Pinto, R. C., Locquet, N., Eveleigh, L., and Rutledge, D. N. 2010. Preliminary studies on the mid-infrared analysis of edible oils by direct heating on an ATR diamond crystal. *Food Chemistry* 120 (4):1170–1177. doi: 10.1016/j.foodchem.2009.11.053.

Reedy, G. T., Ettinger, D. G., Schneider, J. F., and Sidney Bourne. 1985. High-resolution gas chromatography/matrix isolation infrared spectrometry. *Analytical Chemistry* 57 (8):1602–1609. doi: 10.1021/ac00285a024.

Rein, A. 2008. Handheld FT-IR spectrometers: Bringing the spectrometer to the sample. *Spectroscopy* August: 44–50.

Roberts, R. A. 1999. Peroxisome proliferators: mechanisms of adverse effects in rodents and molecular basis for species differences. *Archives of Toxicology* 73 (8–9):413–418. doi: 10.1007/s002040050629.

Rodriguez-Saona, L. E., Fry, F. S., McLaughlin, M. A., and Calvey, E. M. 2001. Rapid analysis of sugars in fruit juices by FT-NIR spectroscopy. *Carbohydrate Research* 336 (1):63–74. doi: 10.1016/S0008-6215(01)00244-0.

Rojo, J. A., and Perkins, E. G. 1989. Cyclic fatty-acid monomer—Isolation and purification with solid-phase extraction. *Journal of the American Oil Chemists Society* 66 (11):1593–1595. doi: 10.1007/Bf02636183.

Romero, A., Bastida S., and Sanchez-Muniz F. J. 2006. Cyclic fatty acid monomer formation in domestic frying of frozen foods in sunflower oil and high oleic acid sunflower oil without oil replenishment. *Food and Chemical Toxicology* 44 (10):1674–1681. doi: 10.1016/J.Fct.2006.05.003.

Safar, M., Bertrand D., Robert, P., Devaux, M. F., and Genot, C. 1994. Characterization of edible oils, butters and margarines by Fourier-transform infrared-spectroscopy with attenuated total reflectance. *Journal of the American Oil Chemists Society* 71 (4):371–377. doi: 10.1007/Bf02540516.

Saito, M., and Takashi, K. 1976. Studies on the relationship between the nutritive value and the structure of polymerized oils. X. Structures and toxicity of heat-polymerized oils. 1. *Journal of Japan Oil Chemists Society* 25 (2):79–86. doi: 10.5650/jos1956.25.79.

Sebedio, J. L., Chardigny, J. M., Juaneda, P., Giraud, M. C., Nour, M., Christie, W. W., and Dobson, G. A. 1996a. Nutritional impact and selective incorporation of cyclic fatty acid monomers in rats during reproduction. In *Oils, Fats, Lipids 1995: Proceedings of the 21st World Congress of the International Society for Fat Research (ISF)*—The Hague, October 1995, edited by W. A. M. Castenmiller, 307–310. Bridgewater: P. J. Barnes & Associates.

Sebedio, J. L., Chardigny, J. M., and Malpuech-Brugere, C. 2007. Physiological effects of trans and cyclic fatty acids. In *Deep Frying: Chemistry, Nutrition and Practical Applications*, edited by M. D. Erickson, 205–228. Urbana, IL: AOCS Press.

Sebedio, J. L., Dobarganes, M. C., Marquez, G., Wester, I., Christie, W. W., Dobson, G., Zwobada, F., Chardigny, J. M., Mairot, T., and Lahtinen, R. 1996b. Industrial production of crisps and prefried french fries using sunflower oils. *Grasas Y Aceites* 47 (1–2):5–13.

Sebedio, J. L., and Grandgirard, A. 1989. Cyclic fatty acids: Natural sources, formation during heat treatment, synthesis and biological properties. *Progressin Lipid Research* 28 (4):303–36.

Sebedio, J. L., and Juaneda P. 2007. Isomeric and cyclic fatty acids as a result of frying. In *Deep Frying: Chemistry, Nutrition and Practical Applications*, edited by M. D. Erickson, 57–86. Urbana, IL: AOCS Press.

Sebedio, J. L., Kaitaranta, J., Grandgirard, A., and Malkki, Y. 1991. Quality assessment of industrial prefried French fries. *Journal of the American Oil Chemists Society* 68 (5):299–302. doi: 10.1007/Bf02657680.

Sebedio, J. L., Lequere, J. L., Semon, E., Morin, O., Prevost, J., and Grandgirard, A. 1987a. Heat-treatment of vegetable-oils: 2. Gc-Ms and Gc-FTIRspectra of some isolated cyclic fatty-acid monomers. *Journal of the American Oil Chemists Society* 64 (9):1324–1333. doi: 10.1007/Bf02540791.

Sebedio, J. L., Prevost, J., and Grandgirard, A. 1987b. Heat-treatment of vegetable-oils: 1. Isolation of the cyclic fatty-acid monomers from heated sunflower and linseed oils. *Journal of the American Oil Chemists Society* 64 (7):1026–1032. doi: 10.1007/Bf02542443.

Sherazi, S. T. H., Younis Talpur, M., Mahesar, S. A., Kandhro, A. A., and Arain, S. 2009. Main fatty acid classes in vegetable oils by SB-ATR-Fourier transform infrared (FTIR) spectroscopy. *Talanta* 80 (2):600–606. doi: 10.1016/j.talanta.2009.07.030.

Sinelli, N., Casale, M., Di Egidio, V., Oliveri, P., Bassi, D., Tura, D., and Casiraghi, E. 2010. Varietal discrimination of extra virgin olive oils by near and mid infrared spectroscopy. *Food Research International* 43 (8):2126–2131. doi: 10.1016/j.foodres.2010.07.019.

Tapp, H. S., Defernez, M., and Kemsley, E. K. 2003. FTIR spectroscopy and multivariate analysis can distinguish the geographic origin of extra virgin olive oils. *Journal of Agricultural and Food Chemistry* 51 (21):6110–6115. doi: 10.1021/jf030232s.

van de Voort, F. R., Sedman, J., and Russin, T. 2001. Lipid analysis by vibrational spectroscopy. *European Journal of Lipid Science and Technology* 103 (12):815–826. doi: 10.1002/1438-9312(200112)103:12<815::AID-EJLT1111815>3.0.CO;2-P.

Vatèla, J. M., Sébédio, J. L., and Le Quéré, J. L. 1988. Cyclic fatty acid monomers: Synthesis and characterization of methyl ω-(2-alkylcyclopentyl) alkenoates and alkanoates. *Chemistry and Physics of Lipids* 48 (1–2):119–128. doi: 10.1016/0009-3084(88)90139-9.

Velasco, J., Marmesat, S., and Dobarganes, M. C. 2008. Chemistry of frying. In *Advances in Deep-Fat Frying of Foods*, edited by S. G. Sumnu and S. Sahin, 33–56. CRC Press: FL (Taylor and Francis).

Voort, F. R., Ismail, A. A., Sedman, J., and Emo, G. 1994. Monitoring the oxidation of edible oils by Fourier transform infrared spectroscopy. *Journal of the American Oil Chemists Society* 71 (3):243–253. doi: 10.1007/BF02638049.

Wehling, R. L. 2010. Infrared spectroscopy. In *Food Analysis*, edited by S.S. Nielsen, 407–420. New York: Springer.

White, P. J. 1991. Methods for measuring changes in deep-fat frying. *Food Technology* 45 (2):75–80.

Wolff, R. L. 1992. trans-Polyunsaturated fatty acids in French edible rapeseed and soybean oils. *Journal of the American Oil Chemists Society* 69 (2):106–110. doi: 10.1007/BF02540558.

Yang, H., Irudayaraj, J., and Paradkar, M. M. 2005. Discriminant analysis of edible oils and fats by FTIR, FT-NIR and FT-Raman spectroscopy. *Food Chemistry* 93 (1):25–32. doi: 10.1016/j.foodchem.2004.08.039.

Yildiz, G., Wehling, R. L., and Cuppett, SL. 2001. Method for determining oxidation of vegetable oils by near-infrared spectroscopy. *Journal of the American Oil Chemists Society* 78 (5):495–502. doi: 10.1007/s11746-001-0292-1.

Yu, Q. T., Liu, B. N., Zhang, J. Y., and Huang, Z. H. 1989. Chemical modification in mass-spectrometry. 8. Location of double-bonds in fatty-acids of fish oil and rat testis lipids—Gas chromatography-mass spectrometry of the oxazoline derivatives. *Lipids* 24 (1):79–83. doi: 10.1007/Bf02535269.

Zhang, J. Y., Wang, H. Y., Yu, Q. T., Yu, X. J., Liu, B. N., and Huang, Z. H. 1989. The structures of cyclopentenyl fatty acids in the seed oils of flacourtiaceae species by GC-MS of their 4,4-dimethyloxazoline derivatives. *Journal of the American Oil Chemists Society* 66 (2):242–246. doi: 10.1007/BF02546068.

Zhang, J. Y., Yu, Q. T., Liu, B. N., and Huang, Z. H. 1988. Chemical modification in mass spectrometry IV—2-alkenyl-4,4-dimethyloxazolines as derivatives for the double bond location of long-chain olefinic acids. *Biological Mass Spectrometry* 15 (1):33–44. doi: 10.1002/bms.1200150106.

10 Chemistry and Safety of Mycotoxins in Food

Junping Wang, Gang Xie, and Shuo Wang

CONTENTS

Mycotoxins are secondary metabolites and represent a diverse group of chemical substances produced by fungi, which come from the Greek word "Mykes" and Latin "Toxicum." More than 400 mycotoxins have been identified. Cereals and the other crops are susceptible to fungal attacks and may be contaminated by mycotoxins, either in the field or during transportation, storage, and processing. Cereal is usually contaminated by mycotoxins including aflatoxins (AFs), ochratoxin, zearalenone, fumonisins, deoxynivalenol (DON), T-2 toxin, sterigmatocystin, cyclopiazonic acid, ergot alkaloids, and patulin and tenuazonic acid.

From 430BC to the eighteenth century, ergot fungus-contaminated rye and ergot poisoning caused a large number of disability and death in Europe. From 1950 to 1975, scientists have isolated many fungus toxins. In 1960, in Britain, poisoning of 100,000 birds was caused by a fluorescent substance in the feed containing Turkey peanut powder and the substance was produced by *Aspergillus flavus*. This toxin was named aflatoxin. Numerous studies have shown that the intake of foods with mycotoxins and its derivatives easily leads to diseases in human beings or animals, such as cancer, liver damage, kidney dysfunction, Kaschin-beck disease, Keshan disease, abortion and fertility problems, and so on.

The major mycotoxins in food are regulated in more than 100 countries. Aflatoxin B1 (AFB1), deoxynivalenol (vomitoxin, DON), zearalenone (ZEN), and ochratoxin A (OTA) in food are also stipulated in China.

According to the report of UN Food and Agriculture Organization (FAO), about 25% of food crops in the world are contaminated with mycotoxins each year.[1] Investigation of wheat products contaminated by fusarium toxins in the United Kingdom (2001–2006) indicated that the percentage of samples exceeding the newly introduced legal mycotoxin limits varied between 0.4% and 11.3% over the five-year period.[2] In China, grains are often contaminated with mycotoxins. In 2009, in China, DON and ZEN contamination rate was between 53.42% and 100% in wheat and corn, respectively. In 2010, DON and ZEN contamination rate was between 69.3% and 96.8% for wheat flour and corn, respectively. The ratio for corn samples with the above-limit ZEN contamination in 2009 and 2010 was 15.56% and 10.70%, respectively.[3]

Mycotoxin contamination in food products is an inevitable problem. Every year mycotoxin contamination causes large loss of foods, as well as results in human illnesses. As a consequence, research on mycotoxin and its derivatives has become very important. The research may include metabolic mechanism of mycotoxins, environmental conditions for their formation, detection technology and instruments, early warning, mechanism of toxication, risk assessment, regulation, degradation and disposal technology, and so on. In recent years, new approaches were used in researching grain and oil mycotoxins, including molecular biology, biological engineering, molecular toxicology, statistics, food engineering, chemoimmunology, instrumental analysis, microbial metabolomics, and so on. Mycotoxin research is helpful in controlling mycotoxin and reducing their hazardous effects. In this chapter, the chemical structure, harmful effects of the selected mycotoxins, including AFs, ochratoxin, zearalenone, fumonisins, DON, and T-2 toxin, are reviewed and discussed, along with the analytical methods for their detection. The chemical and biochemical mechanisms for mycotoxin formation are also discussed using grains as examples.

10.1 COMMON MYCOTOXINS IN FOODS

Mycotoxins are low-molecular weight toxic secondary metabolites that are produced by fungi, including many important pathogenic and food-spoilage species of *Aspergillus*, *Fusarium*, and *Penicillium*. These toxins have been recognized as causal factors associated with sickness and death among animals and humans. The chemical toxicity and associated diseases, collectively termed mycotoxicoses, indirectly result from the ability of fungi to infect crop species, thereby contaminating the foods ingested by both livestock and humans. Pathogenesis by the fungi usually occurs either before or after harvest, under preferred growth conditions, such as high moisture and optimal temperature. Seven categories of mycotoxins that include AFs (AFB1), fumonisins (fumonisin B1 [FB1]), ochratoxins (OTA), zearalenone, DON, T-2 toxin, and citreoviridin affect food safety significantly. Mycotoxins produced by diverse fungi have different chemical structures and different pathways metabolized by human and animal, which result in different toxicities (Figure 10.1).[4]

10.1.1 AFLATOXIN

AFs are derived from *A. flavus*, parasitic *Aspergillus*, and *Aspergillus nomius*. Under appropriate temperature and humidity conditions, AFs may frequently contaminate improperly stored nuts (especially peanuts), grains, meals, and other foods. Discovered after an outbreak of the Turkey X disease in England in 1960, the naturally occurring AFs were classified as carcinogenic to humans by the World Health Organization's (WHO) International Agency for Research on Cancer (IARC) (Group 1).

10.1.1.1 Physical and Chemical Characteristics

The common structure feature of the four major AFs is a dihydrodifurano or tetrahydrodifurano group fused to a substituted coumarin group (Figure 10.2). AFs with molecular weights 312–346 and melting points 200–300°C are soluble in

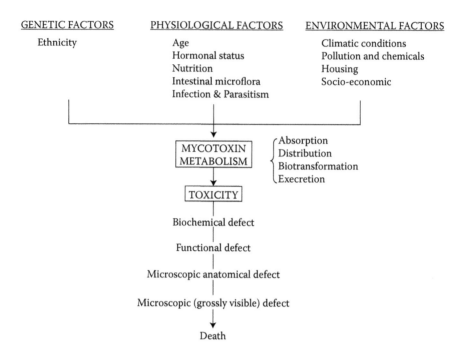

GENETIC FACTORS   PHYSIOLOGICAL FACTORS   ENVIRONMENTAL FACTORS

Ethnicity   Age   Climatic conditions
                         Hormonal status   Pollution and chemicals
                         Nutrition   Housing
                         Intestinal microflora   Socio-economic
                         Infection & Parasitism

MYCOTOXIN METABOLISM
⎧ Absorption
⎨ Distribution
⎩ Biotransformation
  Execretion

TOXICITY

Biochemical defect

Functional defect

Microscopic anatomical defect

Microscopic (grossly visible) defect

Death

FIGURE 10.1 A simplified representation of selected general relationships in a mycotoxicosis.

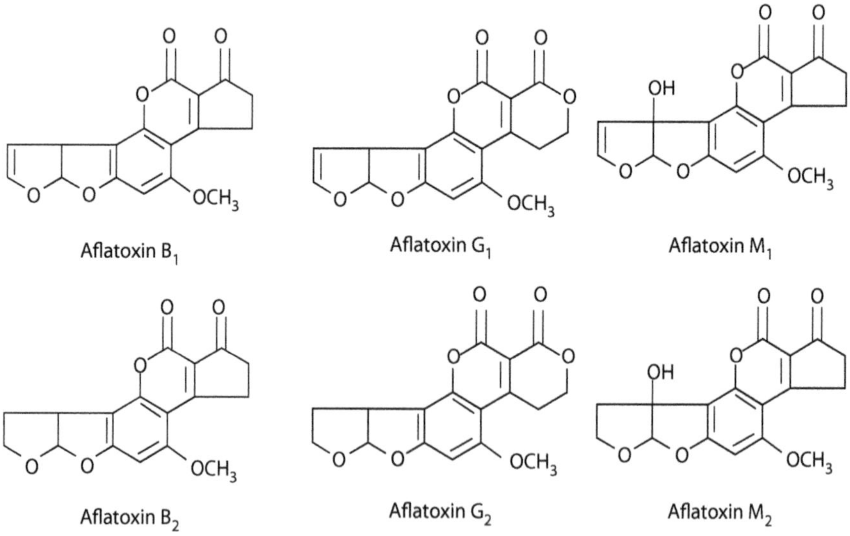

Aflatoxin B₁

Aflatoxin G₁

Aflatoxin M₁

Aflatoxin B₂

Aflatoxin G₂

Aflatoxin M₂

FIGURE 10.2 Chemical structure of aflatoxin B (AFB1 and AFB2), aflatoxin G (AFG1 and AFG2), and aflatoxin M (AFM1 and AFM2).

chloroform, methanol, dimethyl sulfoxide, ethanol, acetone, and other polar organic solvents. AF is poorly soluble in water, hexane, and petroleum ether. Their solubility in water is about 10–20 mg/L. AF is, in general, thermally stable, with a decomposition temperature as high as 280°C. The naturally occurring AFs are identified in physicochemical assays as intensely blue (AFs B1 and B2) or blue-green (AFs G1 and G2) fluorescent compounds under long-wave ultraviolet light.

10.1.1.2 Toxicity and Mechanism

AFs are toxigenic, carcinogenic, mutagenic, and teratogenic in many animal species. The LD_{50} values (the dosage level that causes 50% of a group to die) of AF for animals is between 0.5 and 10 mg per kilogram of body weight. AFB1 is usually the most abundant naturally occurring member of the family, and used for most studies on the pharmacological activity of AF. AFB1 is also the most known hepatocarcinogenic agent, although the liver by no means is the only organ susceptible to AF carcinogenesis. AF is listed as a probable human carcinogen by the IARC of WHO. The AFs were evaluated repeatedly by Joint FAO/WHO Expert Committee on Food Additives (JECFA), and the results showed that AFs might lead to human liver cancer. The intake of dietary AFs should be reduced to the lowest possible level to reduce their potential harm to humans. Each year, about 250,000 people in some of Asian and African countries die of liver cancer because of the excessive intake of AF (1.4 mg/d).[5]

10.1.2 Ochratoxin A

OTA is produced by *Aspergillus* and *Penicillium*. The major species for food OTA production includes *Aspergillus ochraceus, Aspergillus carbonarius, Aspergillus melleus, Aspergillus sclerotiorum, Aspergillus sulphureus,* and *Pichia verrucossum. Aspergillus niger* and *Pichia purpurascens* may also be important OTA producers. Ochratoxins can cause human disease and cancer. OTA (Figure 10.3), *N*-[(3R)-(5-chloro-3, 4-dihydro-8-hydroxy-3-methyl-1-oxo-7-isocoumarin) carbonyl]-L-phenylalanine, has been included in the level 2b suspected carcinogens by the WHO IARC. They are natural contaminations possibly in corn, barley, wheat, oats, sorghum, beans, peas, green beans, peanuts, bread, olive, beer, feed, meat, cheese, milk powder, hay, raisins, and nuts. In China, there are few foodstuff samples containing OTA in amounts greater than the maximum limits.[6]

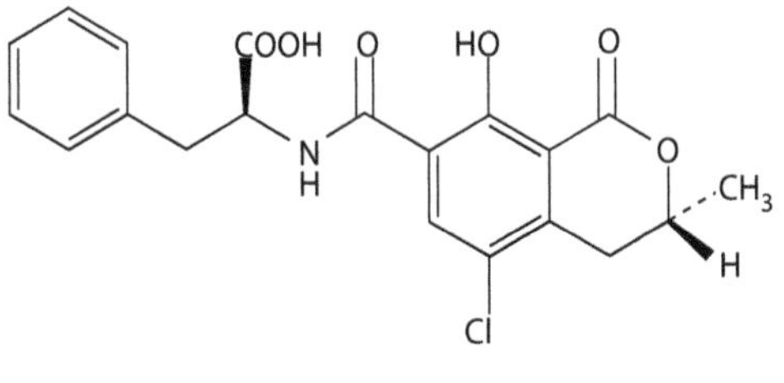

FIGURE 10.3   Chemical structure of ochratoxin A.

10.1.2.1 Physical and Chemical Characteristics

Ochratoxin A is a colorless crystalline compound and can crystallize in xylene, while its sodium salt is soluble in water. As an acid, it can dissolve in chloroform, methanol, and acetonitrile, and slightly soluble in water and diluted bicarbonate solution. Frozen in the alcohol, away from light, Ochratoxin A can be kept for at least 1 year. With heat resistance, Ochratoxin A is generally not damaged with the conventional heating processing for food preparation. Under UV light blue, at different pH or polar solvents they have different absorptions, with peak absorptions at 213 and 332 nm in ethanol. Its fluorescent maximum emission wavelength is 428 nm. Ochratoxin A may undergo acid hydrolysis, and generate acid phenylalanine, and optical active lactone. Methylation may occur in methanol-HCl.

10.1.2.2 Toxicity and Mechanism

OTA contains chlorine atoms, and is highly toxic. It mainly causes liver and kidney poisoned, and can cause mutations, teratogenicity, and inhibition of protein synthesis. Through the rumen and gut microbiota, Ochratoxin A is converted into Ochratoxin α (nontoxic to humans and animals), and is discharged in urine and feces.

OTA may generate reactive oxygen species, which may induce lipid peroxidation, change antioxidant enzyme activity, and consequently induce DNA damage and have toxic effects on the human body. OTA can also induce brain oxidative stress and DNA damage, and the serious drain of brain striatum dopamine.

OTA can restrain the respiration of rat liver cells, cause the consumption of ATP, and alter the morphology of mitochondria. Possible mechanisms were proposed that OTA may competitively inhibit the mitochondrial membrane carrier proteins, and suppress phosphate transportation, resulting in mitochondria or OTA inhibition of the succinic acid salt-related electronic activities and affect the electron transportation chain.

OTA can be a critical factor to upregulate or downregulate the expression of proteins (including enzymes) involved in a number of metabolic pathways, which affects the signal transductions in cells and alters the key gene expressions, showing its genetic toxicity. It may also induce cellular apoptosis, and a nonsequence of DNA synthesis (unscheduled DNA short) and damage.[7] OTA might lead to the urethral epithelial tumors, known as the Balkan endemic nephropathy. It may interact with other substances in food and biological systems, which may determine its ultimate toxicity. The related research is relatively less, and further research is needed.

10.1.3 Deoxynivalenol

Deoxynivalenol (vomitoxin, DON) is a type B trichothecene, an epoxy-sesquiterpenoid. It has a 12, 13 epoxy group, 3 OH functional groups, and an α,β-unsaturated ketone group. The chemical name is 12, 13-epoxy-3, 4, 15-trihydroxytrichothec-9en-8one (Figure 10.4). Structural analogues may include fructus trichosanthis scythe alcohol, 3- and 5-acetyl oxygen fructus trichosanthis scythe bacteria alcohol, melons wilting alcohol, acetic acid, scirpentriol, T-2 tetraol, and so on. In the grain

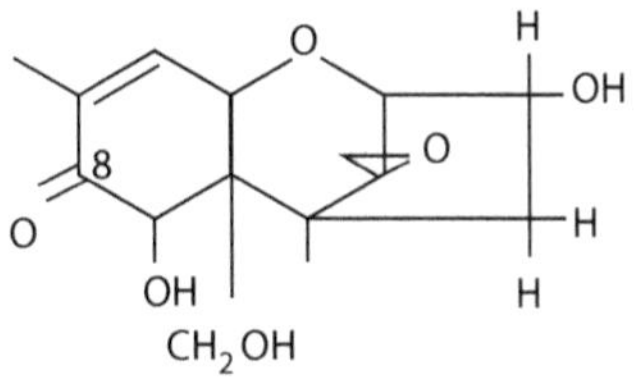

FIGURE 10.4   Chemical structure of deoxynivalenol.

pollutions, 3- and 15-acetyl deoxidization fructus trichosanthis bacteria alcohol unit can happen at the same time.[8] This mycotoxin was often found in grains such as wheat, barley, oats, rye, and maize, but rarely present in rice, sorghum, and triticale. A direct relationship between the incidence of fusarium head blight and contamination of wheat with DON has been established. An increased amount of moisture toward harvest time has been associated with lower amount of vomitoxin in wheat grain due to leaching of toxins. Furthermore, DON contents are significantly affected by the susceptibility of cultivars toward *Fusarium* species, preceding crop culture, tillage practices, and fungicide usage. DON can cause nausea, vomiting, gastrointestinal discomfort, diarrhea, and headache in humans. In 1993, IARC listed DON in Class 3, meaning "cannot be classified as having a carcinogenic effect in humans" of the material.[9]

10.1.3.1   Physical and Chemical Characteristics

DON is a colorless needle crystal with a melting point of 151–152°C, and soluble in water, ethanol, methanol, acetonitrile, and other polar solvents. It is generally stable to heat, and has to be heated to 110°C or above to be destroyed. Heating at 121°C under high pressure for 25 min only destroyed a portion, but heating at 210°C for 30–40 min can destroy all. Also noted was that the dry acid treatment does not reduce its toxicity, but alkaline or high pressure processing can damage part of the toxin. General cooking and food processing cannot reduce its toxicity, but washing grain with distilled water three times may reduce the DON toxin content by 65–69%. When grain was washed with 1 mol/L sodium carbonate solution, DON toxin can be reduced by 72–74%. Soaking in 0.1 mol/L sodium carbonate solution for 24–72 h, grain DON toxin levels can be reduced by 42–100%.

10.1.3.2   Toxicity and Mechanism

DON can cause animal poisoning, and cause nausea, vomiting, gastrointestinal discomfort, diarrhea, and headache in humans. In 1993, IARC listed DON in Group 3.[9] Low-dose exposure to DON may affect human immune function and the selected phosphoprotein may potentially serve as biomarkers for DON exposure. The digestive tract is a target for mycotoxin DON, a major cereal grain contaminant of public health concern in Europe and North America. Pig, the most sensitive species to DON toxicity, can be regarded as the most relevant animal model for investigating the intestinal effects of DON.

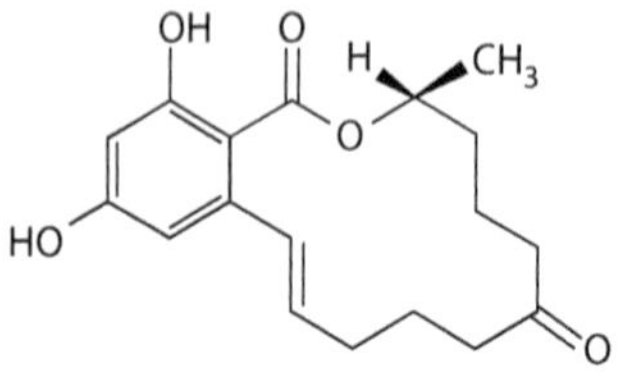

FIGURE 10.5   Chemical structure of zearalenone.

10.1.4   ZEARALENONE

Zearalenone (Figure 10.5) is a mycotoxin produced by *Fusarium graminearum*, *Fusarium culmorum, Fusarium oxysporum, Fusarium roseum, Fusarium moniliforme, Fusarium avenaceum, Fusarium equiseti,* and *Fusarium nivale* and other Fusarium on corn, wheat, barley, oats, and sorghum as substrates. It is a nonsteroidal compound that exhibits estrogen-like activity in certain farm animals such as cattle, sheep, and pigs. Zearalenone is a phenolic resorcyclic acid lactone with potent estrogenic properties, produced primarily by Fusarium. Zearalenone is a phytoestrogenic compound known as 6-(10-hydroxy-6-oxo-*trans*-1-undecenyl)-b-resorcylic acid L-lactone. Alcohol metabolites of ZEN (i.e., a-zearalenol and b-zearalenol) are also estrogenic.[10]

10.1.4.1   Physical and Chemical Characteristics

Zearalenone is a white crystalline. It exhibits blue-green fluorescence when excited by a long wavelength UV light (360 nm) and a more intense green fluorescence when excited with a short wavelength UV light (260 nm). In methanol, its UV absorption primarily occurs at 236 ($e = 29,700$), 274 ($e = 13,909$), and 316 nm ($e = 6020$). The maximum fluorescence in ethanol occurs with an irradiation at 314 nm and with an emission wavelength at 450 nm. The infrared spectrum has a maximum absorption at 970 nm. Solubility in water is about 0.002 g/100 mL. It is slightly soluble in hexane and progressively more so in benzene, acetonitrile, methylene chloride, methanol, ethanol, and acetone. It is also soluble in aqueous alkali.

10.1.4.2   Toxicity and Mechanism

Zearalenone has strong reproductive toxicity and teratogenic effects in animals, and can cause infertility and miscarriage. 1 ng/kg ZEN can induce feminization in animals, and a high concentration (50–100 ng/kg) of ZEN has showed an adverse impact for pregnancy, ovulation, fetal development, and viability of newborn animals. ZEN has immunotoxicity and tumorigenesis, and might lead to liver lesions and suppress lung function in rats.[11] Zearalenone can reduce womb and influence uterine tissue morphology in females, leading to reproductive diseases. A recent study has found that it may alter estrogen regulation and increase the incidence of female breast cancer. IARC classified it as Group 3 carcinogens. ZEN has estrogen-like poisoning effect, and can bind to estrogen receptors and affect cell nucleus, and stimulate breast cancer cell growth. The mean dietary intakes for ZEN have been estimated at 20 ng/kg b.w./day for consumers in Canada, Denmark, and Norway and at 30 ng/kg b.w./day for those in the United States. The JECFA established a provisional maximum tolerable daily intake (PMTDI) for ZEN at 0.5 µg/kg of body weight.[12] The induction of phenotypic alterations by zearalenone administered in utero and in the neonatal period at doses as low as 0.2 µg/kg suggests that zearalenone could contribute to the induction of breast endocrine disorders.[13] As a-zearalenol shows the strongest estrogenic activity, the preferential production and basal transfer of this metabolite suggests that intestinal cells may contribute to the manifestation of the adverse effects for zearalenone.

10.1.5 FUMONISIN

Fumonisin is produced by *F. moniliforme, Fusarium proliferatum, Fusarium verticillioides, F. oxysporum,* and so on. Fumonisin can cause equine leuco-encephalomalacia (ELEM), nerve poisoning, porcine pulmonary edema, and human esophageal and liver cancers, and so on, and is classified as Group 2b carcinogen by the IARC in 1993.[9] Fumonisin is a frequent natural contaminant of many foodstuffs such as corn and its products, rice, sorghum, noodles, and beer. So far, 28 kinds of fusarium analogues have been identified. They are divided into four groups (Figure 10.6), namely A, B, C, and P groups. Fumonisin is the most abundant one among the wild-type strains, with FB1 accounting for about 70% of the total. FB1 is the most toxic fusarium, with molecular formula $C_{34}H_{59}NO_{15}$ (and molecular mass 721).[14]

10.1.5.1 Physical and Chemical Characteristics

Fumonisin is a white hygroscopic powder. Since there are four free carboxyl and amino groups in the molecule, fumonisin is hydrophilic and soluble in water. Fumonisin is also soluble in acetonitrile–water and methanol, but insoluble in chloroform or hexane. FB1 and FB2 are stable during storage at −18°C and 25°C, respectively, but not stable above this temperature range. In acetonitrile–water solution (1:1, v/v) at 25°C, fumonisin can be stored for 6 months. It is not stable in methanol, and might degrade to single or double methyl ester.

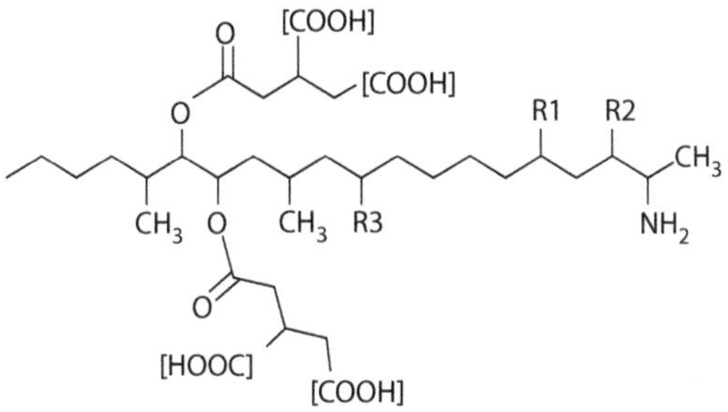

FIGURE 10.6 Structural formula of fumonisin B1–B4. FB1: R1 = OH; R2 = OH; R3 = OH; FB2: R1 = OH; R2 = OH; R3 = H; and FB3: R1 = OH; R2 = OH; R3 = H; FB4: R1 = H; R2 = OH; R3 = H.

10.1.5.2 Toxicity and Mechanism

Fumonisin has cell toxicity, and is also a carcinogenic substance. By inhibiting the sphingosine *N*-acyltransferase activity, FB1 breaks sphingolipid metabolism. FB1 may also inhibit cellular protein phosphatase and arginosuccinate synthetase, sphingomyelin, and metabolism of protein and urea. The carcinogenic mechanism is an unscheduled DNA synthesis that might be initiated by the built-up sphingoid base. Sphingolipids, which consist of hundreds of molecules, play a key role in cell adhesion, differentiation, growth, and death. Fumonisin can damage lipid metabolism. In India's acute abdominalgia, diarrheal diseases, FB1 was the cause of poisoning. In addition, the esophagus cancer in China might be associated with Fumonisin intake according to the International Cancer Research Organizations.[15]

10.1.6 T-2 TOXIN

T-2 toxin ($C_{24}H_{34}O_9$, with molecular weight 466.51, Figure 10.7) is a type B trichothecene mycotoxin. The T-2 toxin is produced primarily by *F. graminearum*, *Fusarium tricinctum*, *F. oxysporum*, *Fusarium poae*, and so on, and has immunotoxicity and hematotoxicity to humans and animals, which can result in nausea, vomiting, abdominal pain, abdominal distention, diarrhea, bloody stools, lightheadedness, chills, and other illness. In 1993, IARC classified T-2 toxin as Group 3 toxic substance. T-2 toxin is identified in wheat, corn, oats, barley, rice, soybeans, and other grains, and in their processed products. The chemical structures of trichothecenes are derived from a 12, 13-epoxytrichothec-9-ene ring system. Individual groups of these mycotoxins differ in various functional groups.[16] To date, 148 trichothecenes are known, 83 of which are nonmacrocyclic and 65 are macrocyclic compounds.[17]

10.1.6.1 Physical and Chemical Properties

T-2 toxin is a white needle crystal, with a melting point 150–151°C, soluble in chloroform, acetone, ethyl acetate, and other polar solvents, soluble in water, and does not show fluorescence under ultraviolet light. General food cooking methods do not destroy their structures, and they are stable for 6–7 years at room temperature or heated to 200°C for 1–2 h. Sodium hypochlorite can reduce the toxicity of T-2 toxin, the oxygen ring and double bond are considered to be bioactive units.

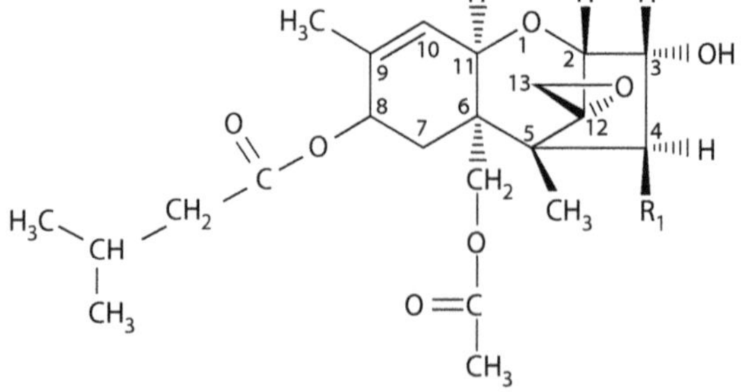

FIGURE 10.7   Structural formula of type A-trichothecenes T-2.

10.1.6.2 Toxicity and Mechanism

T-2 toxin has been known for a number of years to cause toxicosis in farm animals. In addition, they are associated with several human diseases. The correlation between the presence of certain trichothecene mycotoxins and adverse physiological responses indicates the need for specific methods to detect these mycotoxins in various commodities. Chromatographic methods or, alternatively, enzyme-linked immunosorbent assays (ELISAs) are preferred for their detection.

In 1993, T-2 toxin was classified as Group 3 toxic substance by IARC. JECFA 56th conference for the first time evaluated experimental data for T-2 toxin. The committee concluded that the safety of food contaminated with T-2 toxin could be evaluated from the lowest observed effect level (LOEL) of 0.029 mg/kg per body weight per day for changes in leukocyte and erythrocyte counts in the 3-week feeding study in pigs. The committee used this LOEL and a safety factor of 500 to derive a PMTDI of 60 ng/kg per body weight for T-2 toxin.[9]

Multiple mycotoxin contamination often occurs in food. The combined effects of mycotoxins are of major concern due to the widespread occurrence of human exposure. The low concentrations of multiple mycotoxins might cause strong inhibitory effects on the immune functions when they are concurrent in food.[18]

10.2 FACTORS INFLUENCING CONTENT IN FOODS AND FOOD INGREDIENTS

Grain and oil seeds can be infected by fungal during plant growth, harvest, storage, transportation, and food processing, following the insect or rodent damage of seed integrity or other biological and mechanical damage. If water activity is above 0.65 and the temperature is appropriate, the fungus may thrive and produce mycotoxin contamination. The mycotoxin may not be a single, but can be multi-mycotoxins. In grains, mycotoxins may also occur in conjugated forms, either soluble (masked mycotoxins) or incorporated into/associated with/attached to macromolecules (bound mycotoxins). These conjugated mycotoxins can be released due to the metabolism in living plants, fungi, and mammals or during food processing (Figure 10.8).[19]

In addition to the suitable moisture and temperature, other factors including, but not being limited to, oxygen, drought, waterlogging, illumination, carbon sources, nitrogen sources, pH value, and fungi interspecific competition may be important

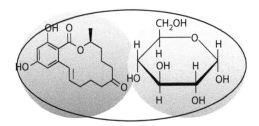

FIGURE 10.8   Zearalenone-4-glucose.

for mycotoxin production by fungi. Some mycotoxins have a clear-cut role both in generating a pathogenetic process, that is, fumonisins and some trichothecenes, and in competing with other organisms, that is, patulin. In other cases, such as AFs, more than one role can be hypothesized.

10.2.1 CLIMATE CHANGE

Climate change may affect the contents of mycotoxins in food. The 2007 Intergovernmental Panel on Climate Change Report is reinterpreted herein to account for what may occur with mycotoxins. Warmer weather, heat waves, greater precipitation, and drought may affect mycotoxin content of grain. The effects of moisture on mycotoxins in crops are more ambiguous than those for temperature according to the in vivo data. In vitro data on fungal growth and mycotoxin production may not relate directly to the situation in the field or post-harvest conditions, but are useful for baseline assumptions. The effects of climate in various regions of the world, that is, Africa, Europe, Asia, Latin America, and North America, are considered higher risk for mycotoxin contamination. Crops introduced to exploit altered climate may be subject to fewer mycotoxin producing fungi (the "Parasites Lost" phenomenon). UV radiation may cause fungi to mutate on crops and produce different mycotoxins. There is relevant information about climate change on AFs, DON, and OTA, but data on patulin are not available.

10.2.2 STORAGE CONDITION

Poor storage may result in mycotoxin contamination. Storage of agricultural products in less-developed economies is often less satisfactory, such as uninsulated metal silos with moisture migration, buildings with leaky roofs, or earthen floors, or in outdoor wooden bins. The inability of *F. graminearum* and related species to grow below $0.9a_w$ means that the fusarium toxin levels may not rise during storage. Very high moisture occurs due to water ingress, result in high water activities of grain, which may significantly increase fusarium toxin levels. Storage in well-constructed silos to prevent moisture migration could limit mycotoxin production. The influence of high carbon dioxide and low oxygen concentration on growth of the foodborne fungal species, including *Mucor plumbeus, F. oxysporum, Byssochlamys fulva, Byssochlamys nivea, Penicillium commune, Penicillium roqueforti, A. flavus, Eurotium chevalieri,* and *Xeromyces bisporus,* was investigated. The production of AF, patulin, and roquefortine C was greatly reduced under the atmospheres.

10.2.3 PROCESSING TECHNOLOGY

Food processing, by ingredient split-flow or biological degradation, may reduce mycotoxin content; however, when processing conditions is not well controlled, mycotoxin levels are also likely to be increased. In some parts of China, gluten processing can reduce the content of DON, and the gluten, alcohol, waste water, and other by-products may contain DON at a level less than the maximum allowed content value. Milling of maize grains and separation of germ and bran substantially increases the acceptance rate of maize flour. Thermal processing below 150°C has little effect on fusarium toxin concentrations, but extrusion, used extensively in the production of breakfast cereals and snack foods, substantially reduces fusarium toxin levels, especially in the presence of glucose. In Central America, the process of nixtamalization removes almost all fumonisins as well as AFs, resulting in tortillas and other maize-based foods being possibly free of these mycotoxins.

10.3 POSSIBLE APPROACH TO REDUCE MYCOTOXIN CONTENT IN CEREALS

Cereals are a very important part of human and animal diet. Many approaches have been developed to reduce the mycotoxin levels in cereals, including development of resistant cultivars, physical detoxification, chemical and biodegradation methods, as well as adding nutrients, and so on. Physical methods mainly include heat treatment, microwave and ultraviolet, rinse, solvent extraction, ultrafiltration/infiltration, degerming processing, and the adsorption of additives. Examples of biological methods are microbial biotransformation function and enzyme degradation and destruction of mycotoxin structure, to reduce the toxicity of mycotoxins. Nutrients may be used to suppress the adsorption of mycotoxins or to increase the body repairing ability against mycotoxins to reduce their toxic effects.

10.3.1 Aflatoxins

AF detoxification may, generally, be achieved with physical detoxification, chemical and biodegradation methods, as well as use of nutrients, and so on. For instance, attapulgite can adsorb AFB1. The adsorbent is added in diets for adsorption of AF in animal gut, breaking absorption of AF, and reducing mycotoxin harm to animals. As another example, bentonites have been used in AF-contaminated feeds and diets to reduce their bioavailability. Under dry conditions, major adsorbing mechanism between AFB1 and smectite is ion–dipole interactions, and association between exchangeable cations and carbonyl groups. Under humid conditions, H-bonding with the exchangeable-cation hydration-shell water plays a role. Attapulgite adsorption can solve the problem of feed mass production.

The seed extract of Ajowan (*Trachyspermum ammi* (L.) Sprague ex Turrill) showed the maximum degradation of AFG1 up to 65%. The dialyzed *T. ammi* extract was more effective than the crude extract, capable of degrading >90% of the toxin. Significant level of degradation of other AFs, AFB1 (61%), AFB2 (54%), and AFG2 (46%), by the dialyzed *T. ammi* extract was also observed. Time course study of AFG1 detoxification by dialyzed *T. ammi* extract showed that more than 78% degradation occurred within 6 h and 91% degradation occurred 24 h after incubation.[20] McKenzie and others researched ozone degradation effect of the mycotoxins.[21] The 2% of ozone could rapidly degrade AFB1 or AFG1, while 20% of ozone could degrade AFB2 and AFG2. *Equisetum arvense* and *Stevia rebaudiana* extracts were tested for their control of mycotoxigenic fungi in maize. The results showed that both *E. arvense* and *S. rebaudiana* extracts could be developed as an alternative treatment to control aflatoxigenic mycobiota in moist maize.[22] Gamma radiation may become an alternative for reducing mycotoxins in the feed for broiler chickens, which cause marked economic losses for rural producers. The maximum toxin reductions, at 18% moisture content and 30 kGy, were 50.6%, 39.2%, 47.7%, and 42.9% for AFB1, AFB2, AFG1, and AFG2, respectively.[23]

Biodegradation of AF refers to the damage of AF molecular toxicity groups by microbial secondary metabolites or enzyme decomposition, and producing nontoxic degradation products at the same time. Currently, many researchers have found that thousands of microorganisms including bacteria, fungi, and yeast can reduce AF content. Hernandez-Mendoza reported that eight strains of *Lactobacillus casei* can bind AFB1 in aqueous solution.[24] The strain with the highest AFB1 binding was *L. casei L30*, which bound 49.2% of the available AF (4.6 µg/mL). The binding was to a limited degree (0.6–9.2% release) reversible; the *L. casei* 7R1–AFB1 complex exhibited the greatest stability. *L. casei* L30, an isolate from humans, was the strain that was least sensitive to the inhibitory effects of bile salts.

Vitamin deficiency can aggravate AF toxicity, whereas vitamin supplementation may reduce the risk of AF poisoning. It was found that in the compound feed double the amount of vitamin, especially vitamins A, D, E, and K, can ease the poisoning effect of AF. Adding niacin and nicotinic acid amine in feed can enhance the activity of glutathione transferase, and increase its detoxification effect. Dietary consumption of apiaceous vegetables inhibits CYP1A2 activity in humans, and it has been demonstrated that some compounds in those vegetables act as potent inhibitors of human CYP1A2 and cause reduced hCYP1A2-mediated mutagenicity of AFB. SFN has been shown to protect animals from AFB-induced tumors, to reduce AFB in humans in vivo, and to reduce efficiently AFB adduct formation in human hepatocytes. Chlorophyllin has been shown to significantly reduce genotoxic AFB biomarkers in humans, and it therefore holds promise as a practical means of reducing the incidence of AFB-induced liver cancer (Figure 10.9).[25]

10.3.2 Ochratoxins

OTA is stable, and has a strong toxicity. The present detoxication approaches generally include inactivation, damaging the structure, and converting the toxic form into nontoxic material.

Espejo and others found that activated carbon at 0.24 g/L dose was the most effective treatment to reduce OTA, up 70% in sweet wines.[26] Amino methane, hydrogen peroxide, sodium hydroxide, and calcium hydroxide ammonium salt can effectively reduce the OTA in the matrix. Ozone can also reduce the OTA. Gamma irradiation is an important means of decontamination of food commodities, especially spices. The maximum toxin reductions found at 18% moisture content and 30 kGy was 55.2% for OTA.[23]

Carboxypeptidase and some metal enzymes can degrade OTA. *Lactobacillus acidophilus VM 20, Phaffia rhodozyma,* and *Bacillus licheniformis* can degrade OTA in vitro, with a degrade rate being as high as over 95%. Microbial flora such as cattle and sheep rumen, mice colon and cecum, and human intestinal microbes can degrade OTA to certain degree. Some materials such as melatonin (*N*-acetyl-5-methoxyl

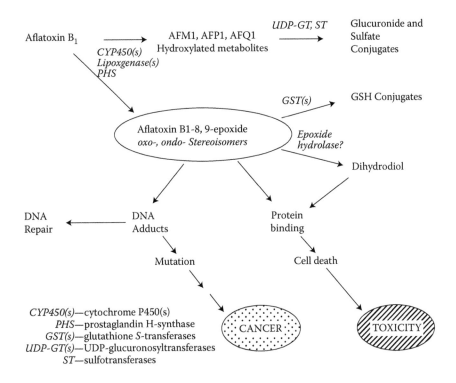

FIGURE 10.9 Aflatoxin B1 for DNA damage and repair.

primary amine), catechin derivatives, vitamin E, and so on can reduce the toxic effects of OTA.

Cleaning or screening can reduce OTA concentration (2–3%) of cereal grains. OTA reduction was similar for the two contamination levels: 95.1% and 97.2% with the rotating cylinder and 81.3% and 79.2% with the fluidized bed at the maximal roasting time. The OTA degradation kinetics differed between the two processes. The complete degradation of OTA within the limit of this study (230°C) was not observed, but the rotating cylinder roasting was the most efficient technical process for the OTA reduction in a commercial roasted dark coffee (88%).[27]

Yeast can effectively reduce the OTA produced by inhibiting *Aspergillus ochraceus* growth on the malt extract AGAR culture medium; the three kinds of yeasts were proved to be able to inhibit OTA produced. Cecchini and others studied the OTA concentration change during yeast fermentation of white grape juice and red grape juice, and found that the OTA content decreased after the alcoholic fermentation. Besides, barbecue, frying, and hypochlorous acid salting were found to reduce OTA content.[28]

10.3.3 DEOXYNIVALENOL

Although DON is heat resistant, DON content could be reduced significantly by thermal treatment. The ultraviolet rays can also degrade DON in the feed. Secondary and higher intensity ultraviolet irradiation gradually reduced the DON content in a time-dependent manner. Greater intensity of ultraviolet light resulted in a more rapid reduction in DON content. DON may be completely removed after 5 s of microwave-induced plasma processing.

Treating moldy corn samples with alkali for 1 and 18 h, the DON concentration of corn was decreased by 9% and 85%, respectively. Treating with 0.1 mol/L NaOH for 1 h, DON broke down into three kinds of toxic isomers. The chemical mechanism may be alkaline damage of the 12, 13-DON epoxy group that plays an important role in toxicity. NaClO can turn DON to a single nontoxic product at room temperature. Chlorine at 1% concentration can reduce the DON content in corn. Treatment of wheat with 2% ascorbic acid for 24 h reduced DON content by 50%. Under the condition of 100°C and 22% water, 1% Na_2S_2O saturated steam treatment for 15 min, DON content in the mildew wheat can be reduced from 7.6 to 0.28 mg/kg. $NaHSO_3$ and $Na_2S_2O_5$ can convert DON into DON-sulfonate, reducing the toxicity of DON. Glycosylation can also reduce DON toxicity.[29]

A microbe in carp, at 15°C for 96 h, can convert DON into lower toxicity degradation products, and another microbe, at 4–25°C and pH 4.5–10.5, has efficient transformation ability.[30] An agrobacterium belonging to the E3 + 39 in the soil can convert DON into three kinds of products; one of them is 3-ketone-DON, with a toxicity about 1/10 of that for DON.[31] The negative impact of DON to cows can be reduced by treating the animal feed with optimal coli BBSH797. At present, optimal coli BBSH797 is the only commercial strain.[32] LS100 bacteria can convert DON into DOM-1. The toxicity of DOM-1 is 1/55 of that for DON. An *Aspergillus*, isolated from soil, can utilize DON as the only carbon source in inorganic salt medium, which transformed 94.4% of DON into a lower toxic component.[33] The available options of managing FHB include the use of fungicides, cultural practices, resistant cultivars, and biological agents. Chitosan and fluorescent pseudomonad can significantly reduce the DON content in wheat and barley both in greenhouse and field, and the degradation rates were ≥74% and ≥79%, respectively.[34]

Wet grinding can remove the DON in the wheat flour since DON can be dissolved in the liquid, and deshalling of wheat grains may reduce the DON content in the flour too. It was found that the first removal (removed 10% grain) might reduce DON content by 45%, and after the second removal (removed 35% grain group), DON content was only 30%, indicating that grinding process can effectively remove DON content from the contaminated grains.[35]

10.3.4 ZEARALENONE

Some of the adsorbents can adsorb zearalenone (ZEN). These adsorbents could be added in animal feed, to adsorb ZEN in animal intestines. For instant, glucan, extracted from beer yeast cell wall, can effectively combine with ZEN in the aqueous solution of different pH in vitro. Many researchers have also found that activated carbon and modified montmorillonite, in particular, montmorillonite, kaolinite, and cholestyramine also have good binding capacity with ZEN in vitro. However, some important nutrients of feed can be adsorbed and lost as well, such as amino acids and vitamins.

It should be paid attention that 10% (w/w) of ozone may completely degrade ZEN in water in 15 s. High concentrations of H_2O_2 solution can degrade ZEN of grain feed, and H_2O_2 concentration and action time had a greater influence on ZEN degradation efficiency. A good detoxification effect needs adequate H_2O_2 concentration and treatment time.[36] Ozone and H_2O_2 have detoxin effects, and suitable for grain and feed with high ZEN concentration, but can cause great damage to nutrition value and can produce undesirable chemical compounds in the meantime.

Alcohol dehydrogenase of yeast can transform ZEN into zearalenol. Some bacillus and mold can also reduce zearalenone to zearalenol. A lactone enzyme from *Gliocladium roseum* may catalyze the hydrolysis of the ester bond in ZEN and convert it into a lower estrogen activity metabolite. In addition, *Trichosporon mycotoxinivorans* can degrade ZEN into carbon dioxide and other metabolites with weak ultraviolet absorption.[37]

10.3.5 FUMONISINS

Screening impurity and unsound kernel can reduce fumonisins of the grain. Density separation technology and washing can remove fumonisins. UV, ultra-high pressure, microwave, and NaCl treatments were tested for their potential to remove AFB1 and FB1 in purple corn. UV and ultra-high-pressure treatments had no effect on the two toxins; microwave treatment could remove more than 90% of toxins in 12 min, while 30% NaCl soaking could remove 99% of toxins in 10 min. Therefore, NaCl soaking is a simple and economic method for detoxication.[38]

Fumonisins B1 degradation rate in solution is more than 90% at 10 kGy. Fumonisins B1 can be degraded by radiation in a dose-dependent manner.[39] Reducing sugar treatment at high temperature, alkaline hydrolysis, and hydrogen peroxide or saleratus can degrade fumonisins. It was also reported that ammonia treatment could degrade 30–45% of the natural pollution fumonisins B1 of corn.[40] At high pressure (60 lb/inch²) and low temperature (20°C), ammonia can effectively reduce fumonisin contents, with a degradation rate of 79%.[41] In addition, enzymes, such as the ones produced by saccharomycetes, may degrade fumonisins effectively by oxidative deamination. Interestingly, puffing and wet-milling can degrade fumonisins, and the degrading effect is influenced by temperature and extrusion screw speed.

10.4 ANALYTICAL METHODS

Lots of analytical methods can be used to detect the mycotoxin in grain and oil. With the progress of science and technology, mycotoxin detection methods are advanced as well. The commonly used methods may include, but are not limited to, thin layer chromatography (TLC), ELISA, high-performance liquid chromatography (HPLC), gas chromatographic (GC) method, liquid chromatography/mass spectroscopy (LC/MS), GC/MS, capillary electrophoresis, fluorescence spectrophotometry, ultraviolet spectroscopy, lateral horizontal flow chromatography immune colloidal gold technique, biosensor (biochip) test, x-ray method, near-infrared reflectance (NIR) technique, and so on. Depending on the application, the analytical methods can be grouped into the rapid screening methods and the laboratory accurate methods for regulation and quality assurance. As grain and oil foods may contain multiple mycotoxins, their simultaneous determination technology using LC/MS has been developed. LC/MS is an accurate, reliable, and highly-sensitive method, and requires less pretreatment and no derivatization step. It has been widely used in mycotoxin detections. Capillary electrophoresis, developed in the 1980s, combines the advantages of HPLC and conventional electrophoresis technique, and is rapid and possible for high-throughput mycotoxin measurements. Lateral horizontal flow chromatography immune colloidal gold technique (LF-GICA) can determine mycotoxins in about 10 min.

10.4.1 Rapid Determination of Mycotoxin

Rapid analysis of mycotoxin is mainly used in food trade (Figure 10.10),[42] such as ELISA, LF-GICA, fluorescence spectrophotometry, ultraviolet spectroscopy, biosensor (biochip) test, x-ray method, NIR technique, and so on.

10.4.1.1 Lateral Horizontal Flow Chromatography Immune Colloidal Gold Technique

Lateral horizontal flow chromatography immune colloidal gold technique (LF-GICA) is a solid-phase membrane immune analysis method that combines immune colloidal gold technique and chromatography technology. LF-GICA was developed in the 1980s. LF-GICA is a rapid, highly sensitive and selective, stable and easy operating assay, which does not need any special equipment. For food and feed, the total analysis time is not more than 10 min.

Mycotoxins in the samples first react with the monoclonal antibody on the surface of the colloidal gold particles. If the content of mycotoxins is more than the up-limit value, antibody on the colloidal gold site may be saturated, leading to an off scale during colloid gold particle chromatography and the gold particles move out of the line area and reach the detecting area in order to react with anti-MIgG (quality online) and produce a single red stripe. If the samples do not contain mycotoxin or the levels are below the threshold, antibody on colloidal gold reacts with the compounds in the line area, presents a red stripe, and the quality control line is also a red stripe. Due to the improvement of the materials used, LF-GICA is continuously improved for its accuracy and adaptability. This strip could have a detection limit of 2.0 ng/mL for AFB1 in food and feed sample extracts. Additionally, the entire analysis is completed within 10 min. Because of the cross reaction between the antibody and AFB1, colloidal gold still needs to be improved for accuracy.

10.4.1.2 Biosensor (Biochip)

With bio-recognition elements in combination with various transductions, scientists have developed a new means of analysis with high selectivity. For instance, a novel AFB1 bioassay was created by introducing a lipomyces kononenkoae α-amylase gene into a strain of *S. cerevisiae* capable of expressing the human cytochrome P450 3A4 (CYP3A4), and the cognate human CYP450 reductase. The amylase-expressing strain could detect AFB1 at 2 ng/mL concentration, and was more sensitive than the

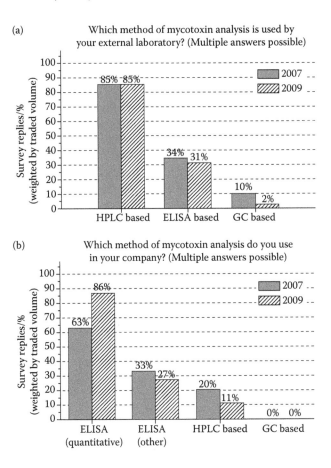

FIGURE 10.10 Method of mycotoxin analysis. (From Ansari AA, Kaushik A, Solanki PR, Malhotra BD. Nanostructured zinc oxide platform for mycotoxin detection. *Bioelectrochemistry* 77, 2010: 75–81. With permission from Elsevier.)

dextranase-expressing strain. AF G1 could be detected at 2 μg/mL, and the trichothecene mycotoxin T-2 toxin was detectable at 100 ng/mL concentration.[43]

Another example, nanostructured zinc oxide (nano-ZnO) film was deposited onto indium–tin–oxide (ITO) glass plate for co-immobilization of rabbit-immunoglobulin antibodies (r-IgGs) and bovine serum albumin (BSA) for OTA detection. Electrochemical impedimetric response of BSA/r-IgGs/nano-ZnO/ITO immunoelectrode obtained as a function of OTA concentration has a linearity in the range of 0.006–0.01 nM/dm^3, a detection limit of 0.006 nM/dm^3, a response time of 25 s, and a sensitivity of 189 Ω/nM/dm^{3}cm^{-2} with a regression coefficient of 0.997.[44]

10.4.1.3 Near-Infrared Reflectance

Application of NIR spectroscopy for rapid mycotoxin detection in grain and oil has been developed. NIR detection has the characteristic of rapid, nondestructive, and suitable for mycotoxin screening for grain trade.

Fernández-Ibañez assessed the utility of NIR spectroscopy for rapid detection of AFB1.[46] The resulting spectroscopic models demonstrated that NIR technology is an excellent alternative for rapid AFB1 detection in cereals. The best predictive model to detect AFB1 in maize was obtained using standard normal variate and detrending (SNVD) as scatter correction ($r^2 = 0.80$ and 0.82; SECV = 0.211 and 0.200 for grating and FT-NIRS instruments, respectively). In the case of barley, the best predictive model was developed using SNVD on the dispersive NIR instrument ($r^2 = 0.85$ and SECV = 0.176) using spectral data as log $1/R$ for FT-NIR ($r^2 = 0.84$ and SECV = 0.183).[47] In the rapid competitive direct ELISA, AFB1 at 0.15 ng/mL caused 50% inhibition (IC_{50}) of binding AFB1 horseradish peroxidase to the antibodies. Effective onsite detection of AFB1 has also been developed based on a rapid and sensitive antibody gold nanoparticle. In addition, NIR absorptions at 1709 and 1743 cm^{-1} could be used for a rapid assessment of DON content.[46]

10.4.1.4 Microsphere Array Technology and Indirect Competition Theory

Immune magnetic beads is a new type of material based on immunology and magnetic carrier technology, and is a kind of magnetic microspheres that was over the surface of specific ligands (antibodies and active chemical groups).

The method for quantitative analysis of AFB1 was established using the microsphere array technology platform. The yield of the coupled microballoon sphere was between 70.08% and 80.80%, and the best dosage of the coupling antigen was 100 µg. The median fluorescence intensity (MFI) detected by the same dosage of the coupling antigen and the identical concentration of the monoclonal anti-AFB1 increased along with the extending incubating time and became stable finally. When the incubating time was 24 and 48 h, the MFI was the biggest with 2.0 ng/µL concentration of the monoclonal anti-AFB1. Referring to the international current standard of IC_{50}, the standard curve showed that the detection sensitivity was 2.33 ng/mL (0.1165 ng), the detection linear equation was $y = 0.0006x + 0.0014$, the cross-reactivity rate of the monoclonal anti-AFB1 with AFB2, AFG1, and ZEN was 9.34%, 2.88%, and <0.10%, respectively.[47]

Carboxyl-modified microspheres, which enwrapped fluorescent dye, were conjugated with the artificial antigen AFB1-BSA. AFB1 was used as a positive control to compete with the AFB1-BSA antigen on the surface of the microspheres for the anti-AFB1 McAb. A fluorescein isothiocyanate-labeled IgG reporter antibody was added to react specifically with the anti-AFB1 McAb on the microspheres. The detection limit of AFB1 reached 0.03 ng/mL, with a good linearity ranging from 0.05 to 1.0 ng/mL. The cross-reactivity rates were less than 1.0% with other toxins such as maflatoxins B2, aflatoxins G1, aflatoxins G2, aflatoxins M1, and aflatoxins M2. The recovery of AFB1 from artificially contaminated corn samples was from 89% to 92%, with CVs from 6.8% to 9.0%.[48]

10.4.1.5 Fluoroimmunoassay

Immunofluorescence technique (fluoroimmunoassay) is a new detection technology based on biochemistry, microscopy, and immunology. It combines the accuracy of the microscopy and the specificity and sensitivity of the immunological detection.

A novel fluoroimmunoassay for AFB1 determination was developed based on the high performance of glutathione-stabilized CdTe/CdS quantum dots. When AFB1 concentration is in the range of 0.68–40 pmol/L, their luminescent intensity showed a good linear relationship with AFB1 concentration, with a correlation coefficient of 0.9914 (R^2). The detection limit was 0.3 pmol/L. The method has shown high sensitivity and stability, and has been successfully applied to the determination of trace AFB1 in rice vinegar samples.[49]

An indirect competitive detection method of DON in cereals was established by Zhou and others using a polyclonal antibody with nano-homogeneous time-resolved fluorescence immunoassay technology. The detection concentration ranged from 0.007 to 100 ng/mL, and the detection limit was 0.007 ng/mL. The intra- and inter-assay CVs of the assay were both below 5%. The recoveries of DON spiked in corn, wheat, and beer were 84–110%, 67–100%, and 74–100%, respectively.[50]

Zhang and others developed an indirect competitive time-resolved fluoroimmunoassay (TRFIA) for the quantitative determination of FB1 in cereals. The sensitivity was 0.05 ng/mL, and the assay range was 1.0–1000 ng/mL with intra- and inter-assay CVs less than 10%. The average recoveries for FB1 were 104.3% for inter-assay and 100.7% for intra-assay in corn samples. In FB1-TRFIA, there was a slight interaction between FB1 and FB2, and there was no interaction between FB1 and other mycotoxins, such as DON, ZEN, and T2. The results indicated that the assay was of high specificity.[51]

10.4.2 Quantitative Determination of Mycotoxins

HPLC and ultra-high performance liquid chromatography (UPLC) with the characteristics of high resolution and sensitivity have been widely used in the mycotoxin analysis.

10.4.2.1 Aflatoxin

HPLC and UPLC are common detection methods for AF. Xie and others established a rapid and environment-friendly method for determining AFs (B_1, B_2, G_1, and G_2) in grains without any derivatization. The detection limits for AFB_1, AFB_2, AFG_1, and AFG_2 were 0.15, 0.05, 0.4, and 0.06 µg/kg, respectively. The linear detection ranges for AFB_1, AFB_2, AFG_1, and AFG_2 were 0.4–60.0, 0.2–15.0, 1.5–60, and 0.2–15.00 µg/kg with correlation coefficients (R^2) of 0.9999, 0.9999, 0.9998, and 0.9992, respectively. Recovery in grains (corn, rice, wheat) spiked with AFs was in the range of 77.4–104.2%, with the relative standard deviation of 1.8–8.9%.[52]

10.4.2.2 Ochratoxin A

A method for the clean-up of a mycotoxin, that is, OTA, from cereal extracts was developed employing a new molecularly imprinted polymer (MIP) as selective sorbent for solid-phase extraction. Recovery was close to 100%. The high selectivity from the MIP was demonstrated by comparing to the selectivity of an immunoaffinity cartridge for a wheat sample clean-up.[53]

10.4.2.3 Deoxynivalenol

A rapid and environment-friendly analytical method was established for DON determination in grain samples. The linearity of the method was validated for a concentration range between 0.5 and 10 ng. The limit of detection was 0.15 ng, and the limit of quantification was 0.5 ng. The determined average recovery rates fluctuated between 84% and 98% (grain) with relative standard deviations of between 2.86% and 5.97%.[54]

10.4.2.4 Zearalenone

A high-field asymmetric waveform ion mobility spectrometry (FAIMS)-based method for the determination of the mycotoxin ZEN and its metabolites α-zearalenol (α-ZOL), β-zearalenol (β-ZOL), and β-zearalanol (β-ZAL) in a corn meal (maize) matrix has been described.[55] The detection limits achieved using the FAIMS device coupled with electrospray ionization and MS detection were 0.4 ng/mL for ZEN and 3 ng/mL for α-ZOL + β-ZOL and β-ZAL.

Food legislation requires methods for food quality and safety control. Reliable and simple analytical methods are in high demand to make enforcement of the regulations possible. The Association of Official Analytical Chemists (AOAC International) and the European Standardization Committee (CEN) and the European Equivalent of ISO have a number of standardized methods for the analysis of mycotoxins. Analytical methodology is prerequisite for effective food law enforcement. The fact that mycotoxins can have serious adverse effects on humans and animals has led many countries to establish regulations on mycotoxins in food and feed in the last decades to safeguard the health of humans and guarantee the economical interests of producers and traders. More than 100 countries have developed specific limits for mycotoxins in foodstuffs and feedstuffs. For food safety and security and public health, mycotoxins, particularly the carcinogenic mycotoxins, should be excluded from food as much as possible. Additional research for mycotoxins is still needed. These researches may include, but are not limited to hazard identification and characterization, exposure assessment, sampling procedures, methods of analysis, rapid warning system, trade contacts, food supply chain, and so on.

REFERENCES

1. Bhat R, Rai RV, Karim AA. Mycotoxins in food and feed: Present status and future concerns. *Compr Rev Food Sci F* 9, 2010: 57–81.
2. Edwards SG. Investigation of fusarium mycotoxins in UK wheat production. HGCA Project Report 2007, No. 413. London: HGCA, 2007.
3. Wang W, Shao B, Zhu JH, Yu HX, Li FQ. Dietary exposure assessment of some important Fusarium toxins in cereal-based products in China. *J Hygiene Res* 39, 2010: 709–714.
4. Bryden, WL. Aflatoxin and animal production: An Australian perspective. *Food Technol Aust* 34, 1982: 216–223.
5. Wild CP, Hudson GJ, Sabbioni G, Chapot B, Hall AJ, Wogan GN, Whittle H, Montesano R, Groopman JD. Dietary intake of aflatoxins and the level of albumin bond aflatoxin in peripheral blood in The Gambia, West Africa. *Cancer Epidemiol Biomarkers Prev* 1, 1992: 229–234.
6. Benford D, Boyle C, Dekant W, Fuchs E, Gaylor DW, Hard G, McGregory DB, et al. http://www.inchem.org/documents/jecfa/jecmono/v47je04.htm
7. Xiao H, Srinivasa M, Marquar RR, Li S, Vodela JK, Frohlich AA, Kemppainen BW. Toxicity of ochratoxin A, its opened form, and several of its analogs: Structure–activity relationships. *Toxicol Appl Pharmacol* 137, 1996: 182–192.

8. Gautam P, Dill-Macky R. Type I host resistance and trichothecene accumulation in fusarium-infected wheat heads. *Am J Agric Anim Sci* 6, 2011: 231–241.
9. Joint FAO/WHO Expert Committee. Evaluation of certain mycotoxins in food: Fifty-sixth report of the Joint FAO/WHO Expert Committee on Food Additives. Geneva: World Health Organization, 2002.
10. Tobias JZ, Frank HN, Ewa SP, Knapp S, Oppermann U, Maier ME. Discovery of a potent and selective inhibitor for human carbonyl reductase 1 from propionate scanning applied to the macrolide zearalenone. *Bioorgan Med Chem* 17, 2009: 530–536.
11. Minervini F, Dell-Aquila ME, Maritato F, Minoia P, Visconti A. Toxic effects of the mycotoxin zearalenone and its derivatives on in vitro maturation of bovine oocytes and 17 beta-estradiol levels in mural granulosa cell cultures. *Toxicol In Vitro* 15, 2001: 489–495.
12. Abdellah Z, Jose MS, Juan CM, Jordi M. Review on the toxicity, occurrence, metabolism, detoxification, regulations and intake of zearalenone: An oestrogenic mycotoxin. *Food Chem Toxicol* 45, 2007: 1–18.
13. Patrick B, Claire B, Jitka D, Sabine B, Nadège M, Magali M, Jean-Francois M, Mohamed BC, Christian LJ. Fetal and neonatal exposure to the mycotoxin zearalenone induces phenotypic alterations in adult rat mammary gland. *Food Chem Toxicol* 48, 2010: 2818–2826.
14. Bezuidenhout SC, Gelderblom WCA, Gorst-Allman CP, Marthinus Horak R, Walter FOM, Spiteller G, Vleggaar R. Structure elucidation of the fumonisins, myeotoxins from *Fusarium moniliforme. J Chem Soc Chem Commun* 11, 1988: 743–745.
15. Wang E, Ross FP, Wilson TM, Riley RT, Merrill AH Jr. Inhibition of sphingolipid biosynthesis by fumonisins. Implications for diseases associated with *Fusarium moniliforme. J Biol Chem* 266 (22), 1992: 14486–14490.
16. Yoshio U. Trichothecenes: Chemical, biological and toxicological aspects. Japan: Kodansha, 1983.
17. Kotal F, Holadová K, Hajšlová J, Poustka J, Radová Z. Determination of trichothecenes in cereals. *J Chromatogr A* 830, 1999: 219–225.
18. Tammer B, Lehmann I, Nieber K, Altenburger R. Combined effects of mycotoxin mixtures on human T cell function. *Toxicol Lett* 170, 2007: 124–133.
19. Berthiller F, Schuhmacher R, Adam G, Krska R. Formation, determination and significance of masked and other conjugated mycotoxins. *Anal Bioanal Chem* 395, 2009: 1243–1252.
20. Rethinasamy V, Selvaraj V, Ayyathurai V, Paranidharan V, Samiyappan R, Iwamoto T, Friebe B, Muthukrishnan S. Detoxification of aflatoxins by seed extracts of the medicinal plant, *Trachyspermum ammi* (L.) Sprague ex Turrill—Structural analysis and biological toxicity of degradation product of aflatoxin G1. *Food Control* 21, 2010: 719–725.
21. McKenzie KS, Sarr AB, Mayura K, Bailey RH, Miller DR, Rogers TD, Norred WP. Oxidative degradation and detoxification of mycotoxins using a novel source of ozone. *Food Chem Toxicol* 35, 1997: 807–820.
22. Garcia D, Ramos AJ, Sanchis V, Marín S. Effect of Equisetum arvense and *Stevia rebaudiana* extracts on growth and mycotoxin production by *Aspergillus flavus* and *Fusarium verticillioides* in maize seeds as affected by water activity. *Int J Food Microbiol* 153, 2012: 21–27.
23. Jalili M, Jinap S, Noranizan MA. Aflatoxins and ochratoxin a reduction in black and white pepper by gamma radiation. *Radiat Phys Chem* 81, 2012: 1786–1788.
24. Slizewska K, Nowak A, Libudzisz Z, Blasiak J. Probiotic preparation reduces the faecal water genotoxicity in chickens fed with aflatoxin B1 contaminated fodder. *Res Vet Sci* 89, 2010: 391–395.
25. Gross-Steinmeyer K, Eaton DL. Dietary modulation of the biotransformation and genotoxicity of aflatoxin B1. *Toxicology* 299, 2012: 69–79.

26. Espejo FJ, Armada S. Effect of activated carbon on ochratoxin A reduction in "Pedro Ximenez" sweet wine made from off vine dried grapes. *Eur Food Res Technol* 229, 2009: 255–262.

27. Castellanos-Onorio O, Gonzalez-Rios O, Guyot B, Fontana TA, Guiraud JP, Schorr-Galindo S, Durand N, Suárez-Quiroz M. Effect of two different roasting techniques on the Ochratoxin A(OTA) reduction in coffee beans (*Coffea arabica*). *Food Control* 22, 2011: 1184–1188.

28. Cecchini F, Morassut M, Moruno EG, Stefano RD. Influence of yeast strain on ochratoxin A content during fermentation of white and red must. *Food Microbiol* 23, 2006: 411–417.

29. Young JC, Subryan LM, Potts D, McLaren ME, Gobran FH. Reduction in levels of deoxynivalenol in contaminated wheat by chemical and physical treatment. *J Agric Food Chem* 34, 1986: 461–465.

30. Shu G, He JW, Young JC, Zhu HH, Li XZ, Ji C, Zhou T. Transformations of trichothecene mycotoxins by microorganisms from fish digestate. *Aquaculture* 290, 2009: 90–95.

31. Shima J, Takase S, Takahashi Y, Iwai Y, Fujimoto H, Yamazaki M, Ochi K. Novel detoxification of the trichothecene mycotoxin deoxynivalenol by a soil bacterium isolated by enrichment culture. *Appl Environ Microbiol* 63, 1997: 3825–3830.

32. Hochsteiner W, Schuh M, Luger K, Baumgartner W. Influence of mycotoxins contaminated feed blood parameters and milk production. *Berliner und Münchener Tierärztliche Wochenschrift* 113, 2000: 14–21.

33. He CH, Fan YH, Liu GF, Zhang HB. Isolation and identification of a strain of *Aspergillus tubingensis* with deoxynivalenol biotransformation capability. *Int J Mol Sci* 9, 2008: 2366–2375.

34. Khan MR, Doohan FM. Comparison of the efficacy of chitosan with that of a fluorescent pseudomonad for the control of Fusarium head blight disease of cereals and associated mycotoxin contamination of grain. *Biol Control* 48, 2009: 48–54.

35. Ríos G, Pinson-Gadais L, Abecassis J, Zakhia-Rozis N, Lullien-Pellerin V. Assessment of dehulling efficiency to reduce deoxynivalenol and Fusarium level in durum wheat grains. *J Cereal Sci* 49, 2009: 387–392.

36. McKenzie KS, Sarr AB, Mayura K, Bailey RH, Miller DR, Rogers TD, Norred WP, Voss KA, Plattner RD, Kubena LF, Phillips TD. Oxidative degradation and detoxification of mycotoxins using a novel source of ozone. *Food Chem Toxicol* 35, 1997: 807–820.

37. Molnar O, Schatzmayr G, Fuchs E, Prillinger H. *Trichosporon mycotoxinivorans* sp. nov., a new yeast species useful in biological detoxification of various mycotoxins. *Syst Appl Microbiol* 27, 2004: 661–671.

38. Xiao LX, Wang F, Yu HT, Zhao XY, Hu XS. Detoxication of aflatoxin B1 and fumonisin B1 in purple porn. *Food Sci* 32, 2011: 114–117.

39. Yang J. Research on irradiation-induced degradation of mycotoxins in agricultural products. Thesis. Beijing: Chinese Academy of Agricultural Sciences, 2009.

40. Norred WP, Voss KA, Bacon CW, Riley RT. Effectiveness of ammonia treatment in detoxification of fumonisin-contaminated corn. *Food Chem Toxicol* 29, 1991: 815–823.

41. Park DL, Rua SM, Mirocha CJ, EI-Sayed Abd-Alla AM, Weng CY. Mutagenic potential of fumonisin contaminated corn following ammonia decontamination procedure. *Mycopathologia* 117, 1992: 105–108.

42. Siegel D, Babuscio T. Mycotoxin management in the European cereal trading sector. *Food Control* 22, 2011: 1145–1153.

43. Li XM, Millson S, Coker R, Evans I. A sensitive bioassay for the mycotoxin aflatoxin B1, which also responds to the mycotoxins aflatoxin G1 and T-2 toxin, using engineered baker's yeast. *J Microbiol Meth* 77, 2009: 285–291.

44. Ansari AA, Kaushik A, Solanki PR, Malhotra BD. Nanostructured zinc oxide platform for mycotoxin detection. *Bioelectrochemistry* 77, 2010: 75–81.

45. Fernández-Ibañez V, Soldado A, Martínez-Fernández A, Roza-Delgado B. Application of near infrared spectroscopy for rapid detection of aflatoxin B1 in maize and barley as analytical quality assessment. *Food Chem* 113, 2009: 629–634.

46. Abramović B, Jajić I, Abramović B, Ćosić J, Jurić V. Detection of deoxynivalenol in wheat by Fourier transform infrared spectroscopy. *Acta Chim Slovenica* 54, 2007: 859–867.

47. Song HJ, Liu SY, Ma HR, Jiang D, Cao YY. The quantitative analysis of AFB1 by using the microsphere array technology and indirect competition theory. *Chin Agric Sci Bull* 27, 2011: 144–150.

48. Li YN, Wu HY, Zheng YY, Guo YH. Determination of aflatoxin B1 based on a flow cytometric microsphere immunoassay. *J Fuzhou Univ* (Natural Science Edition) 40 (2012): 126–131.

49. Zhou XY, Yang SP, Li ZJ, Sun XL, Liu JK. Study on the preparation of high-performance CdTe/CdS core/shell quantum dots using glutathione as stabilizer and its application to detect aflatoxin B1 in rice vinegar by fluoroimmunoassay. *Chin J Anal Lab* 31, 2012: 1–6.

50. Zhou B, Gao L, Jin J, Chen Y, Huang B, Zhao WG. Quantitative determination of deoxynivalenol using light initiated chemiluminescence assay. *Sci Technol Food Ind* 31, 2010: 338–342.

51. Zhang J, Zhu L, Chen Y, Hang B, Zhao WG. Ultra-sensitive determination of fumonisin B1 by time-resolved fluoroimmunoassay. *J Environ Health* 26, 2009: 1112–1115.

52. Xie G, Wang SX, Zhang Y. Rapid analysis of aflatoxins (B1,B2,G1,G2) in grain by immuno-affinity clear-up column and ultra performance liquid chromatography without derivation. *Chin J Anal Chem* 41, 2013: 223–228.

53. Ali WH, Derrien D, Alix F, Pérollier C, Lépine O, Bayoudh S, Chapuis-Hugon F, Pichon V. Solid-phase extraction using molecularly imprinted polymers for selective extraction of a mycotoxin in cereals. *J Chromatogr A* 1217, 2010: 6668–6673.

54. Xie G, Wang SX, Zhang Y. Rapid analysis of deoxynivalenol in grain by ultra performance liquid chromatography with immuno-affinity column clear-up. *J Instrum Anal* 30, 2011: 1362–1366.

55. McCooeye M, Kolakowski B, Boison J, Mester Z. Evaluation of high-field asymmetric waveform ion mobility spectrometry mass spectrometry for the analysis of the mycotoxin zearalenone. *Anal Chim Acta* 627, 2008: 112–116.

11 Chemicals and Safety of Chemical Contaminants in Seafood

Pierina Visciano

CONTENTS

11.1 INTRODUCTION

The term "seafood" generally covers a heterogeneous group of aquatic organisms both from the marine environment and also freshwater, including mollusks, crustaceans, and all types of finfish. It represents a major source of the very long-chain polyunsaturated n-3 fatty acids, but also of other potentially protective components, such as vitamin D, vitamin B_{12}, calcium, selenium, iodine, choline and taurine, as well as a well-balanced amino acid composition (Lund, 2013). Several studies (Calder and Yaqoob, 2009; Costa, 2007; Daviglus et al., 2002; Olsen, 2004; Patterson, 2002) have suggested links between seafood consumption and reduction in risk of many chronic diseases including, but are not being limited to, coronary heart disease, high blood pressure, stroke, age-related maculopathy, some cancers, rheumatoid arthritis, and other inflammatory diseases, as well as preterm delivery, obesity, and deficit of childhood cognitive development. It is widely recommended that human diet should include at least two portions of fish a week, and maternal fish consumption during pregnancy is beneficial for fetal neurodevelopment (Daniels et al., 2004).

However, despite the beneficial effects that its consumption provides, seafood can harbor a number of chemical hazards, such as heavy metals and persistent organic pollutants (polycyclic aromatic hydrocarbons, insecticides like aldrin, dieldrin, DDT, polychlorinated biphenyls, dioxins, and furans), which are known to cause adverse effects in humans both during development and in adulthood (Marques et al., 2011).

Currently, marine pollution is a global environmental concern. Various human activities on land, water, and air contribute to the contamination of seawater, sediments, and organisms with potentially toxic substances. Contaminants can be natural substances or artificially produced compounds. On being discharged into the sea, they stay in the water in dissolved form or they are removed from the water column through sedimentation to the bottom sediments (Funes et al., 2006). They are consumed by aquatic organisms due to bio-accumulation and bio-magnification in the food chain as original compounds or their metabolites, causing concern for the potential concentrations in animals at the top of the food chain (Papagiannis et al., 2004).

Fish can contribute significantly to the dietary exposure to some contaminants, such as methylmercury, persistent organochlorine compounds, brominated flame retardants, and organotin compounds (EFSA, 2005). The most important ones of these are methylmercury and the dioxin-like compounds, for which high-level consumers of certain fish may exceed the provisional tolerable weekly intake even without taking into account other sources of dietary exposure. Species, season, diet, location, lifestage, and age have a major impact on both nutrient and contaminant profiles and levels of fish. Contaminants in fish derive predominantly from their diet, and the levels of bio-accumulation of contaminants are higher in fish that are higher in the food chain (EFSA, 2005).

The focus of this chapter is on the main chemical contaminants in seafood, posing a particular attention to their sources, the toxicity for humans, and possible tools for risk assessment and reduction.

11.2   HEAVY METALS

The role of heavy metals in marine ecosystems has been intensively investigated. Metallic elements from natural and anthropogenic sources are environmentally ubiquitous, readily dissolved, and transported by water. It is known that water chemistry influences the bio-accumulation of trace elements into primary producers, and as this is the basis of the food chain, it ultimately determines the resultant levels in higher trophic organisms (Mason et al., 2000). Several studies have shown that the accumulation of heavy metals in tissues is primarily dependent on the needs, sex, size, and molt of marine animals and also on the water concentrations of metals and the exposure period, while other environmental factors such as salinity, pH, hardness, and temperature also play significant roles in metal accumulation (Bryan and Langston, 1992; Canli and Atli, 2003; Kalay et al., 1999; Zhang and Wong, 2007). Nonessential metals (e.g., lead, cadmium, and mercury) are held to be the most dangerous, since continuous exposure of marine organisms to their low concentrations may result in bio-accumulation, and subsequent transfer to human beings through seafood (Bilandžić et al., 2011).

Mercury is a metal that is released into the environment from both natural and anthropogenic sources, such as burning household and industrial wastes and fossil fuels such as coal. Once released, it undergoes a series of complex transformations and cycles between atmosphere and ocean, ocean and land. In the aquatic ecosystem, mercury tends to be absorbed by particles and deposited into sediments. There, under certain conditions, bacteria convert metallic or elemental forms of mercury into methylmercury (Ullrich et al., 2001). The dissolved organic matter in aquatic environments is known to bind trace metals strongly, affecting their speciation, solubility, mobility, and toxicity. Methylmercury can be adsorbed in particles or from the water by small creatures such as shrimp and other invertebrates, which are then consumed by predators including fish. Among fish, benthic and predatory pelagic species accumulate this form of mercury. It must be noted that this accumulation occurs primarily through food chain transfer (bio-magnification) and not through direct uptake from water or sediments (Perugini et al., 2009).

Mercury occurs in three chemical forms: elemental or metallic mercury, inorganic mercury, and organic mercury. Methylmercury is by far the most common form in the food chain. The other organic mercury compounds, such as phenylmercury, thiomersal, and merbromin (also known as Mercurochrome), have been used as fungicides and in pharmaceutical products (EFSA, 2008). The largest source of mercury exposure for most people in developed countries is by inhalation of mercury vapor due to the continuous release of elemental mercury from dental amalgam. Exposure to methylmercury mostly occurs via diet. Methylmercury collects and concentrates especially in the aquatic food chain, making populations with a high intake of fish and seafood particularly vulnerable (Richardson et al., 2011).

In human blood, methylmercury is accumulated to a large extent (>90%) in the erythrocytes, where it is bound to the cysteinyl residues of hemoglobin. It distributes in all tissues and is able to cross the blood–brain and placental barriers. The mechanisms underlying the high sensitivity of the developing brain to methylmercury exposure can be attributed to the disturbance of the highly regulated processes during brain development, including the very fast and strongly coordinated cell proliferation, differentiation, and migration (EFSA, 2012a). In addition, in neural stem cells exposed to nanomolar concentrations of methylmercury, long-term inherited effects associated with a decrease in global DNA methylation have been recently reported (Bose et al., 2012). In numerous *in vitro* and *in vivo* studies, disruption of cellular redox homeostasis by an increased level of reactive oxygen and nitrogen species, leading to cumulative oxidative stress, has been shown to play a key role in methylmercury- and mercuric mercury-induced toxicity. The underlying mechanism that is involved seems to be related to alterations in mitochondrial functions (Garrecht and Austin, 2011), resulting in increased cellular superoxide anion and subsequently hydrogen peroxide and hydroxylradical levels, and a disturbance of the cellular oxidative defense capacity, as shown by decreased glutathione levels and impaired superoxide dismutase, glutathione reductase, and glutathione peroxidase activities. The observed symptoms in humans exposed in utero include mental retardation, cerebral palsy, deafness, blindness, and dysarthria. Chronic, low-dose prenatal methylmercury exposure from maternal consumption of fish has been associated with more subtle endpoints of neurotoxicity in children. These effects include poor performance on neurobehavioral tests, fine motor function, language, visual–spatial abilities, and verbal memory (NRC, 2000).

Cadmium occurs naturally in the environment as a result of volcanic emissions and weathering of rocks. In addition, anthropogenic sources (wastewater, waste incineration, and diffuse pollution of agricultural soils caused by the use of fertilizers) have increased the background levels of cadmium in soil, water, and living organisms (EFSA, 2004).

The molecular mechanisms of cadmium toxicity are not well understood yet. Cadmium is known to enhance lipid peroxidation by increasing the production of free radicals, which leads to tissue damage and cellular death, in several organs, mainly lung and brain (Mendez-Armenta et al., 2003). The ability of cadmium to generate free radicals also leads to the oxidation of nucleic acids and alteration of DNA repair mechanisms, alterations of membrane structure/function, and inhibition of energy metabolism (Castro-Gonzalez and Mendez-Armenta, 2008). The sensitive targets of cadmium toxicity are the kidney and bone following oral exposure, and kidney and lung following inhalation exposure (ATSDR, 2012). The renal toxicity of cadmium includes tubular proteinuria (increased excretion of low-molecular weight proteins), decreased resorption of other solutes (increased excretion of intracellular tubular enzymes such as N-acetyl-β-glucosaminidase, amino acids, glucose, calcium, inorganic phosphate), evidence of increased glomerular permeability (increased excretion of albumin), increased kidney stone formation, and decreased glomerular filtration rate (ATSDR, 2012; Noonan et al., 2002). The respiratory response to cadmium is similar to the response seen with other agents that produce oxidative damage. At the cellular level, catalase, superoxide dismutase, non-protein sulfhydryl, glucose-6-phosphate dehydrogenase, and glutathione peroxidase are decreased in response to cadmium lung insults (ATSDR, 2012).

Lead is one of the most ubiquitous and useful metals known to humans, and it is detectable in practically all phases of the inert environment and in all biological systems. Environmental levels of lead have increased more than 1000-fold over the past three centuries as a result of human activity. Lead is a naturally occurring element and it is usually found combined with two or more other elements to form lead compounds (ATSDR, 2007). It reaches the aquatic system due to superficial soil erosion and atmospheric deposition. The main mechanism that regulates the concentration of this metal is the absorption into the sediments and/or suspended particles.

In humans, the central nervous system is the major target organ for lead toxicity. In adults, lead-associated neurotoxicity was found to affect central information processing, especially for visual–spatial organization and short-term verbal memory, to cause psychiatric symptoms and to impair manual dexterity (EFSA, 2010). There is considerable evidence demonstrating that the developing brain is more vulnerable to the neurotoxicity of lead than the mature brain. During brain development, lead interferes with the trimming and pruning of synapses, migration of neurons, and neuron/glia interactions. Alterations of any of these processes may result in failure to establish appropriate connections between structures and eventually in permanently altered functions (ATSDR, 2007). In children, an elevated blood lead level is inversely associated with a reduced Intelligence Quotient score and reduced cognitive functions up to at least seven years of age (EFSA, 2010). Lead nephrotoxicity is characterized by proximal tubular nephropathy, glomerular sclerosis, and interstitial fibrosis and related functional deficits, including enzymuria, low- and high-molecular weight proteinuria, impaired transport of organic anions and glucose, and depressed glomerular filtration rate (Diamond, 2005). The adverse hematological effects of lead are mainly the result of its perturbation of the heme biosynthesis pathway. A potential consequence of the inhibition of heme synthesis is a decreased formation of mixed function oxidases in the liver resulting in impaired metabolism of endogenous compounds, as well as impaired detoxification of xenobiotics (ATSDR, 2007). Mitochondrial cytochrome oxidase is another heme-requiring protein that could be affected by heme synthesis inhibition. In addition, tryptophan pyrrolase, a hepatic heme-requiring enzyme system, is inhibited via the reduction in the free hepatic heme pool. This could ultimately lead to increased levels of the neurotransmitter serotonin in the brain and increased aberrant neurotransmission in serotonergic pathways. Inhibition of heme synthesis also results in increased levels of δ-aminolevulinic acid, which has a structure similar to that of the inhibitory neurotransmitter gamma-aminobutyric acid (GABA), and, therefore, interferes with GABA neurotransmission (ATSDR, 2007).

Arsenic is the 20th most abundant element in the Earth's crust and is primarily associated with igneous and sedimentary rocks where it occurs mostly in inorganic forms. It is used in the manufacture of glassware, metal alloys, microelectronics, agricultural pesticides, and wood preservatives. It is released through mineral processing and fossil fuel combustion. Arsenic is also mobilized naturally through volcanic, geothermal, and microbiological processes, and by weathering of crustal rocks (Peshut et al., 2008). Arsenic exists in four oxidation states: arsenate (As^V), arsenite (As^{III}), arsenic (As^0), and arsine (As^{-III}). The toxicity of arsenic to organisms depends on its concentration and speciation, and inorganic arsenic species are generally more toxic than organoarsenic species (Rahman et al., 2012). However, recent toxicological studies indicate that some organic arsenic species, including monomethylarsonic acid and dimethylarsinic acid, are more toxic than previously estimated (Kenyon and Hughes, 2001; Petrick et al., 2001; Wanibuchi et al., 2004).

In marine animals, the predominant form of arsenic is arsenobetaine, a trimethylated pentavalent compound. Other organic arsenicals found in seafood include simple methylated compounds, particularly methylarsonate, dimethylarsinate, and trimethyl arsine oxide, and more complex organic compounds such as arsenocholine and arsenosugars. About 90% of the arsenic in human diet comes from seafood, of which only a small proportion occurs in inorganic forms; the majority consists of complex organic compounds that generally have been regarded as nontoxic. However, recent studies have documented the formation of carcinogenic metabolites in experimental rodents (Borak and Hosgood, 2007).

The chronic ingestion of arsenic (mainly through water or food) can cause a series of severe health disorders, from skin lesions to neuropathology, reproductive effects, and various kinds of cancers (Kapaj et al., 2006). The toxicity of trivalent arsenic is related to its high affinity for the sulfhydryl groups of biomolecules such as glutathione and lipoic acid and the cysteinyl residues of many enzymes (Aposhian and Aposhian, 2006). The formation of As(III)–sulfur bonds results in various harmful effects by inhibiting the activities of enzymes, such as glutathione reductase, glutathione peroxidases, thioredoxin reductase, and thioredoxin peroxidase (Chang et al., 2003; Lin et al., 2001; Schuliga et al., 2002).

Current European regulations focus on regulating microbiological agents, phycotoxins, and some chemical contaminants. Since 2006, these regulations have been compiled under the name of "Hygiene Package." Among heavy metals, the European legislation (Regulation 1881/2006/EC; Regulation 629/2008/EC and Regulation 420/2011/EU) set maximum levels only for three metals (lead, cadmium, and mercury), while the USFDA has included a further three metals (chromium, arsenic, and nickel) in the list (USFDA, 1993a,b,c). The maximum levels for lead and cadmium and for mercury in seafood (Regulation 835/2011/EU) are reported in Tables 11.1 and 11.2, respectively.

TABLE 11.1
Maximum Levels for Lead and Cadmium in Seafood According to European Legislation

Foodstuffs	mg/kg Wet Weight	Official Journal of the European Union
Lead		
Muscle meat of fish	0.30	Regulation 1881/2006/EC
Crustaceans: muscle meat from appendages and abdomen. In case of crabs and crab-like crustaceans (*Brachyura* and *Anomura*) muscle meat from appendages	0.50	Regulation 420/2011/EU
Bivalve mollusks	1.5	Regulation 1881/2006/EC
Cephalopods (without viscera)	1.0	Regulation 1881/2006/EC
Cadmium		
Muscle meat of fish excluding species listed below	0.050	Regulation 1881/2006/EC
Bonito (*Sardasarda*)	0.10	Regulation 629/2008/EC
Common two-banded seabream (*Diplodus vulgaris*)		
Eel (*Anguilla anguilla*)		
Grey mullet (*Mugillabrosus labrosus*)		
Horse mackerel or scad (*Trachurus* species)		
Louvaror luvar (*Luvarus imperialis*)		
Mackerel (*Scomber* species)		
Sardine (*Sardina pilchardus*)		
Sardinops (*Sardinops* species)		
Tuna (*Thunnus* species, *Euthynnus* species, *Katsuwonus pelamis*)		
Wedge sole (*Dicologogloss acuneata*)		
Bullet tuna (*Auxis* species)	0.20	Regulation 629/2008/EC
Anchovy (*Engraulis* species)	0.30	Regulation 629/2008/EC
Swordfish (*Xiphias gladius*)		
Crustaceans: muscle meat from appendages and abdomen. In case of crabs and crab-like crustaceans (*Brachyura* and *Anomura*) muscle meat from appendages	0.50	Regulation 420/2011/EU
Bivalve mollusks	1.0	Regulation 1881/2006/EC
Cephalopods (without viscera)	1.0	Regulation 1881/2006/EC

TABLE 11.2
Maximum Levels for Mercury in Seafood According to European Legislation

Foodstuffs	mg/kg Wet Weight	Official Journal of the European Union
Mercury		
Fishery products and muscle meat of fish excluding species listed below. The maximum level for crustaceans applies to muscle meat from appendages and abdomen. In case of crabs and crab-like crustaceans (*Brachyura* and *Anomura*), it applies to muscle meat from appendages	0.50	Regulation 420/2011/EU
Anglerfish (*Lophius* species)	1.0	Regulation 629/2008/EC
Atlantic catfish (*Anarhichas lupus*)		
Bonito (*Sardasarda*)		
Eel (*Anguilla* species)		
Emperor, orange roughy, rosy soldier fish (*Hoplostethus* species)		
Grenadier (*Coryphaenoides rupestris*)		
Halibut (*Hippoglossus hippoglossus*)		
Kingklip (*Genypterus capensis*)		
Marlin (*Makaira* species)		
Megrim (*Lepidorhombus* species)		
Mullet (*Mullus* species)		
Pink cusk eel (*Genypterus blacodes*)		
Pike (*Esoxlucius*)		
Plain bonito (*Orcynopsis unicolor*)		
Poor cod (*Tricopterus minutes*)		
Portuguese dogfish (*Centroscymnus coelolepis*)		
Rays (*Raja* species)		
Redfish (*Sebastesmarinus, S. mentella, S. viviparus*)		
Sail fish (*Istiophorus platypterus*)		
Scabbard fish (*Lepidopus caudatus, Aphanopus carbo*)		
Seabream, pandora (*Pagellus* species)		
Shark (all species)		
Snake mackerel or butterfish (*Lepidocybium flavobrunneum, Ruvettus pretiosus, Gempylus serpens*)		
Sturgeon (*Acipenser* species)		
Swordfish (*Xiphias gladius*)		
Tuna (*Thunnus* species, *Euthynnus* species, *Katsuwonus pelamis*)		
Food supplements	0.10	Regulation 629/2008/EC

11.3 PERSISTENT ORGANIC POLLUTANTS

11.3.1 POLYCYCLIC AROMATIC HYDROCARBONS

Polycyclic aromatic hydrocarbons (PAHs) constitute a large class of organic compounds, with a similar structure, containing two or more fused aromatic rings made up of carbon and hydrogen atoms. They may be formed and released during incomplete combustion or pyrolysis of organic matter, industrial processes, and other human activities. Soils, surface water, and sediments may be contaminated by atmospheric fallout or deposition from sewage and oil and gasoline spills. Their sources, natural and mostly anthropogenic, in the environment are numerous and include the following: stubble burning and spreading of contaminated sewage sludge on agricultural fields; exhausts from mobile sources (motor vehicles and aircrafts); industrial plants (e.g., aluminum foundries, incinerators); wood preservation, use of tar-coated wood; domestic heating with open fireplaces; burning of coal for thermal and electric energy; burning of automobile tires or of creosote treated wood; tobacco smoke; oil pollution of surface waters and soils; forest fires; and volcanic eruptions (Perugini and Visciano, 2009). They are ubiquitous contaminants in the marine environment derived from uncontrolled petroleum spills, marine transport, discharges from ships, and urban runoff, all incomplete combustion at high temperature or pyrolytic processes involving fossil fuels and atmospheric deposition (Perugini et al., 2007).

Over 100 PAHs have been identified. They occur as complex mixtures throughout the environment in air, water, and soil. PAHs are hydrophobic compounds and they become more hydrophobic as molecular weight increases (Juhasz and Naidu, 2000). Due to their low water solubility, PAHs show a high affinity for the organic fraction, and in water they are adsorbed on particulate matter, which can be deposited as sediments (Geffard et al., 2003). As lipophilic compounds they can easily cross lipid membranes and have the potential to bio-accumulate in aquatic organisms (Billiard et al., 2002). Their fate in marine organisms is considered to be species-dependent. Generally, the metabolic capacity in edible aquatic species appears to be best developed in fish, intermediate in crustaceans, and least in mollusks (James, 1989; Stegeman and Lech, 1991). Filtering organisms, such as mussels or oysters, accumulate PAHs with elimination rates much lesser than those observed in vertebrates (Livingstone, 1994).

Exposure of humans to single PAHs does not occur because PAHs are always encountered as complex mixtures. The fact that exposure to PAHs is always due to a mixture, which is not always of constant composition, makes the assessment of health consequences difficult (Carpenter et al., 2002). PAHs containing up to four fused benzene rings are known as light PAHs and those containing more than four benzene rings are called heavy PAHs. Heavy PAHs are more stable and more toxic than light ones. Some PAHs have been demonstrated to be carcinogenic and mutagenic, while those PAHs that have not been found to be carcinogenic may act as synergists (Wenzl et al., 2006). Benzo(a)pyrene (BaP) is usually highlighted, although it comprises less than 5% of the total amount of PAHs present in the atmosphere, since it is thought to be the most toxic PAH. The International Agency for Research on Cancer (IARC) considers BaP to be a known animal carcinogen and a probable human carcinogen (Group 2A). The carcinogenicity of BaP has been attributed to the reaction of BaP metabolites, primarily the diol epoxides, with DNA (Visciano and Perugini, 2009). PAHs exhibit their biological effects through metabolic activation by cytochrome P450-related enzymes (CYP) to electrophilic species, capable of reacting with the nucleophilic sites of DNA to form adducts. The formation and persistence of carcinogen–DNA adducts are believed to be critical events for the initiation of neoplasia in target cells. The risk of mutations arising from DNA adducts does not depend only on the adduct level but also on their mutagenic potential as well as on the DNA sequence of neighboring adducts (Binková and Šrám, 2004).

Aside from carcinogenicity, PAHs toxicity is linked to other effects such as endocrine disruption (estrogenic and antiestrogenic activity), immunosuppression, cardiovascular dysfunction, atherosclerosis, neurotoxicity, and adverse health outcomes in children born to mothers who are particularly exposed to these compounds (Visciano and Perugini, 2009).

Because of the widespread distribution of PAHs in the environment, most foods contain measurable levels of PAHs, generally in the parts per billion or microgram per kilogram range. Although foods such as smoked or barbecued meat and fish may contain relatively high levels, unless the diet consists of very frequent consumption of such foods, it is cereals and vegetables, and their fats and oils that make the major contributions to human dietary exposure (Perugini and Visciano, 2009). Food contaminated with PAHs largely arises from production practices, although environmental contamination is also an issue. This holds true, especially for vegetables and grains that do not absorb significant PAHs from the soil; nevertheless, there could be other sources of contamination, such as particles from the air (especially when whole-grain products, for instance bread and breakfast cereals, are produced). Moreover, grains and raw products for oil production may be contaminated with PAHs through artificial drying and heating during processing, if precautionary measures are not taken, such as indirect drying and good temperature control (Wenzl et al., 2006). The maximum levels for PAHs in seafood are reported in Table 11.3.

11.3.2 Organochlorine Chemicals

The three main categories of organochlorine compounds present as contaminants in the environment are insecticides and agrochemicals, as exemplified by the hydrocarbon insecticide DDT, electricity supply transformer liquid coolants, for example, the polychlorinated biphenyls (PCBs) and the contaminants polychlorinated dibenzodioxins (PCDDs) and polychlorinated dibenzofurans (PCDFs). 2,3,7,8-Tetrachlorodibenzo-*p*-dioxin (TCDD), first detected in commercial formulations of the defoliant 2,4,5-trichlorophenoxyacetic acid, is the "bench-mark" compound used for relating other PCDDs, PCDFs, and PCBs. The other sources of organochlorine compounds in the environment include the herbicides 2,4-dichloro- and 2,4,5-trichloro-phenoxyacetic acids, the chlorophenoxyacetic acids, and the fungicide hexachlorobenzene (HCB). The chlorophenoxyacetic acids and HCB are no longer permitted for use but continue to be detected in marine life (Smith and Gangolli, 2002).

Organochlorine chemicals adsorbed on particulates and on underwater sediments provide a ready source for the intake of these compounds by crustaceans and bottom feeders and, in the food chain, they can reach high concentrations in top-level predators (Porte and Albaigés, 1993). The distribution and uptake of organochlorine compounds from aquatic environments by marine organisms is dynamic, complex, and subject to seasonal variations and local conditions (McDowell Capuzzo et al., 1989; Vassilopoulou and Georgakopoulos-Gregoriades, 1993). Thus, fish and other fishery

TABLE 11.3

Maximum Levels for POPs in Seafood According to European Legislation

Foodstuffs	Maximum Levels		Official Journal of the European Union
Benzo(a)pyrene, benz(a)anthracene, benzo(b)fluoranthene and chrysene	Benzo(a)pyrene (µg/kg wet weight)	Sum of benzo(a)pyrene, benz(a)anthracene, benzo(b)fluoranthene and chrysene (µg/kg wet weight)	
Muscle meat of smoked fish and smoked fishery products excluding fishery products listed below.	5.0 until 31.8.2014	30.0 as from 1.9.2012 until 31.8.2014	Regulation 835/2011/EU
The maximum level for smoked crustaceans applies to muscle meat from appendages and abdomen. In case of smoked crabs and crab-like crustaceans (*Brachyura* and *Anomura*), it applies to muscle meat from appendages	2.0 as from 1.9.2014	12.0 as from 1.9.2014	
Smoked sprats and canned smoked sprats (*Sprattus sprattus*); bivalve molluscs (fresh, chilled or frozen); heat-treated meat and heat-treated meat products[a] sold to the final consumer	5.0	30.0	Regulation 835/2011/EU
Bivalve mollusks (smoked)	6.0	35.0	Regulation 835/2011/EU
Dioxins and PCBs	Sum of dioxins (WHO-PCDD/F-TEQ)	Sum of dioxins and dioxin-like PCBs (WHO-PCDD/F-PCB-TEQ)	
Muscle meat of fish and fishery products and products thereof, excluding eel. The maximum level for crustaceans applies to muscle meat from appendages and abdomen. In case of crabs and crab-like crustaceans (*Brachyura* and *Anomura*), it applies to muscle meat from appendages	4.0 pg/g wet weight	8.0 pg/g wet weight	Regulation 420/2011/EU
Muscle meat of eel (*Anguilla anguilla*) and products thereof	4.0 pg/g wet weight	12.0 pg/g wet weight	Regulation 1881/2006/EC
Marine oils (fish body oil, fish liver oil, and oils of other marine organisms intended for human consumption)	4.0 pg/g fat	10.0 pg/g fat	Regulation 1881/2006/EC

[a] Meat and meat products that have undergone a heat treatment potentially resulting in formation of PAH, that is, only grilling and barbecuing.

products caught from coastal regions and estuaries close to regions of industrialization, for instance, California, the north-western Atlantic, Japan, and the Baltic and Mediterranean Seas, can be particularly at risk of contamination, especially those that are nonmigratory (Strandberg et al., 1998). Seafood in the human diet contributes a significant proportion of the total intake of organochlorine compounds, particularly fish with higher fat levels, due to their ability to bio-accumulate in fatty tissues and bio-magnify (Ritter et al., 1995).

11.3.2.1 Chlorinated Insecticides

Dichlorodiphenyltrichloroethane (DDT) has been widely used as a pesticide to control insects in agriculture and insects that carry diseases such as malaria. Dichlorodiphenyldichloroethylene (DDE) and dichlorodiphenyl-dichloroethane (DDD) or TDE are chemicals similar to DDT that contaminate commercial DDT preparations. DDE and DDD enter the environment as contaminant of the commercial DDT formulation or as breakdown product of DDT. The main target organs are the nervous system and liver. It also affects hormonal tissues, reproduction, fetal development, and the immune system. DDT including p,p'-DDE and DDD cause tumors mainly in the liver of experimental animals and are mostly negative in genotoxicity studies. DDT is classified by IARC as a possible carcinogenic (group 2B) substance to humans (EFSA, 2006a).

DDT, aldrin, chlordane, dieldrin, endrin, heptachlor, HCB, mirex, and toxaphene are organochlorine pesticides that, together with polychlorinated dibenzo-p-dioxins, furans, and biphenyls, have been the subject of the Stockholm Convention on Persistent Organic Pollutants (POPs), which was held in May 2001; this treaty calls for an immediate ban on the production, import, export, and use of most of these POPs as well as disposal guidelines (Barr and Needham, 2002). Aldrin and dieldrin (a metabolite of aldrin as well as a marketed pesticide) are both fat-soluble persistent and bio-accumulating organochlorine insecticides. The conversion of aldrin to dieldrin occurs much more rapidly than the subsequent biotransformation and elimination of dieldrin, resulting in the accumulation of dieldrin in lipid-rich tissues. The dominant toxic effects are in the nervous system and liver, the latter mainly following chronic exposure. Aldrin and dieldrin are approximately equally toxic. They are not genotoxic or teratogenic. Due to high level of bio-accumulation in the aquatic food chain, fish-derived products, particularly fish oil, were identified to contain the highest levels of dieldrin (EFSA, 2006b).

Lindane may produce kidney and liver changes in some experimental animals and, more importantly, there have been several cases of aplastic anemia with a clear association between their recent or, less frequently, remote exposure. Occupational overexposure to aldrin produces excitation of the nervous system (Moreno Frías et al., 2001).

11.3.2.2 Polychlorinated Biphenyls, Dibenzodioxins, and Dibenzofurans

The first category of organochlorine compounds of technological importance was the polychlorinated biphenyls (PCBs), synthesized in the late nineteenth century. The PCBs are a group of 209 related compounds (known as congeners), which differ only in the number of chlorine atoms attached to the parent biphenyl molecule.

PCBs and related chemicals follow the dissolving of these substances present in the atmosphere from burning waste, in soils, and landfills by rainfall, and by the uncontrolled discharge of contaminated industrial and sewage effluents (Smith and Gangolli, 2002). The polychlorinated dibenzo-para-dioxins (PCDDs) and PCDFs are two series of almost planar tricyclic aromatic compounds with very similar chemical properties. The number of chlorine atoms can vary between 1 and 8. The number of positional congeners is quite large; in all there are 75 PCDDs and 135 PCDFs (WHO, 2000). PCDDs and PCDFs are complex mixtures present as adventitious contaminants formed in poorly controlled manufacturing conditions in the production of various chlorinated compounds and also in combustion and incineration processes (IPCS, 1989).

Several PCBs exhibit toxicological effects similar to dioxins (mediated by activation of the aryl hydrocarbon receptor, AhR), and are thus called dioxin-like PCBs. All these compounds are standardized against the most potent dioxin, TCDD, and potential toxicities are calculated using toxic equivalency factors. TCDD is classified as a group 1 carcinogen, that is, carcinogenic to human (Steenland et al., 2004), and cancer risk is the greatest toxicological concern for all dioxins and dioxin-like PCBs. Additional adverse effects of these compounds include immunotoxicity (Birnbaum and Tuomisto, 2000), reproductive toxicity, endocrine disruption, and developmental neurotoxicity (Hays and Aylward, 2003). Dioxin exposure may also be associated with an increased risk of diabetes (Fujiyoshi et al., 2006).

In view of the complexities in chemical compositions, particularly of PCBs, PCDDs, and PCDFs mixtures in the environment and in the food chain, the concept of toxic equivalency factor (TEF) was developed by the World Health Organization (WHO) jointly with the European Centre for Environment and Health. The term TEF indicates approximately one-half to one *order of magnitude* estimate of the toxic potency of a compound relative to TCDD (Smith and Gangolli, 2002). The TEF values of compounds are based on the following criteria: the compound must show structural relationship to PCDDs and PCDFs; the compound must bind to the AhR; the compound must elicit AhR-mediated biochemical and toxic responses; the compound must be persistent and accumulate in the food chain (WHO, 2000). On the general assumption that the metabolic disposition, tissue distribution, body burden, and toxicity of congeners of PCBBs, PCDFs, and PCBs are estimated on an additive basis, the quantitative levels of these dioxin-like congeners can be expressed as toxic equivalents (TEQs). The TEQ of a congener is obtained as a product of the concentration of the congener multiplied by the TEF value, and the total TEQ value of a dioxin-like mixture present in a food matrix is obtained by the summation of each of the individual TEQ values of the congeners (Smith and Gangolli, 2002).

Many environmental chemicals have no measurable estrogenic/antiestrogenic activity in simple *in vitro* systems yet produce significant activity *in vivo*. This is the case with numerous compounds that are activated by any one of the P450 cytochromes (P450 or CYP). P450s are the stalwarts of our detoxification system and, as such, are involved in the metabolism of most xenobiotics as well as the metabolism of steroid hormones. Xenobiotics and steroid hormones also induce P450s. Environmental mixtures may contain chemicals that induce P450s, are metabolized by P450s, or both. Induction of CYP1A1 and/or CYP1B1 occurs when xenobiotics such as TCDD or PAHs bind the AhR, and may result in antiestrogenic activity through increased metabolism and depletion of endogenous estrogens (Spink et al., 1990). TCDD is also the environmental toxicant whose adverse effects on male reproduction have been the most studied (Carpenter et al., 2002).

A monitoring study on dietary exposure to dioxins and PCBs showed that the major contributor was either the food category fish and seafood products or meat and meat products in the groups of adolescent, adult, elderly, and very elderly (EFSA, 2012b). It was followed by milk and dairy products and animal and vegetable oils and fats. Indeed, for some groups of infants, toddlers, and other children, milk and dairy products and/or foods for infants and young children were the major contributors to total exposure (EFSA, 2012b). The maximum levels for these compounds in seafood are reported in Table 11.3.

11.3.3 Polybrominated Diphenyl Ethers

Polybrominated diphenyl ethers (PBDEs) have been used as flame retardants in many manufactured items and have received great attention due to their ubiquitous environmental distribution and bio-accumulation potential (Hites, 2004). They are structurally similar to PCBs, chlorinated compounds with similar physical and chemical properties. It may be interesting to note that both groups of compounds are transportable in the same way and to the same extent through air and water. They are persistent in the environment, build up in living organisms, and last for a long time in animal bodies. PBDEs are potent thyroid disruptors. They alter the thyroid hormone homeostasis, and they have been suggested to disrupt brain development resulting in permanent neurologic damage in mice (Shinsuke, 2004). There are at least three possible mechanisms by which PBDEs can adversely affect brain development: thyroid hormone disruption, disruption of second messenger communications, and alteration of neurotransmitter systems (McDonald, 2002).

The dominant food category for human dietary intake is represented by fish and other seafood, followed by meat and meat products, fats and oils, milk and dairy products, and eggs and egg products. Supplements, such as fish oil, for example, cod liver oil, are another source of PBDE exposure (EFSA, 2011a).

Recently, focus has shifted to structural analogues of PBDEs, such as hydroxylated (OH) and methoxylated (MeO) PBDEs. OH-PBDEs are structurally similar to the thyroid hormone thyroxin (T4). The toxicity of OH-PBDEs has been investigated and was considered more potent than that of PBDEs (Sun et al., 2013). The effects of OH-PBDEs on organisms include oxidative phosphorylation disruption, neurotoxicity, and thyroid disruptions (Canton et al., 2008; Harju et al., 2007; Meerts et al., 2001; Mercado-Feliciano and Bigsby, 2008).

11.4 RISK ASSESSMENT

Generally, the evaluation of the toxicological risk by contaminants is carried out through comparison of measured or presumed levels of a certain class of substances in the environment with the levels imposed by law or following guidelines proposed by different international organizations. This approach is obviously a generic one, as it tends to protect the world population globally and it cannot in any way account for the different eating habits and the different consumption rates of local populations (Binelli and Provini, 2004).

Moreover, the way food chemical contaminants are currently controlled is questionable. So far, the limits set by authorities for the presence of chemical contaminants in food products and risk assessment analysis are mostly evaluated in raw products, however most food products are cooked before consumption. How seafood is cooked can influence the amount of contaminants ingested by consumers (Salama et al., 1998; Trotter and Corneliussen, 1989; Zabik et al., 1992, 1996). This can arise by removal of parts likely to contain contaminants before cooking (e.g., the hepatopancreas of crabs or skin of fish), by loss into water when simmering, and by loss of oils and juice from fish when grilling. The studies performed so far, to compare the availability of chemical contaminants in raw and cooked/ processed seafood products, highlighted strong variations according to the cooking procedure and species. Such variations may be due to a decrease in weight of foodstuff resulting from loss of water, volatiles, and to a lesser extent of other gross sample constituents (lipids, carbohydrates, and proteins) or the volatilization or solubilization of contaminants, as heat from cooking (except frying) melts some fat in seafood, thus allowing contaminated fat to drip away and to decrease the content of contaminants in the edible part (Bandarra et al., 2009; Devesa et al., 2001; Rey-Salgueiro et al., 2009).

The degree of accumulation or removal of contaminants from seafood products strongly varies according to species; for example, higher mercury removal in marine fish species than in freshwater species (Devesa et al., 2001; Limaverde-Filho et al., 1999), cooking process (Ersoy et al., 2006), cooking conditions such as time, temperature, and medium of cooking (Perello et al., 2008). During thermal processing, the application of heat hastens protein degradation and loss of weight and water, and therefore chemical contaminants might also be affected by the heat applied (Burger et al., 2003; Cabañero et al., 2004). The effect of seafood cooking in the bioaccessibility of chemical contaminants is also different according to the chemical contaminant analyzed, being higher for arsenic in cooked seafood and lower for other toxicants (Marques et al., 2011). Mercury concentrations showed a significant increase in different portions of Norway lobster after the boiling process, with the greatest values in the following decreasing order: white meat > brown meat > exoskeleton (Perugini et al., 2013). Due to local consumer habits, some groups of European coastal populations also eat the so-called brown meat of large crustaceans, namely the soft meat from the body of the crab that adheres to the inside of the hard upper shell, hepatopancreas and in some cases, gonads and roe, but no maximum limit has been set for brown meat of crustaceans, although it is known that these parts can contain high levels of toxic trace elements (Noël et al., 2011).

Maternal supplementation with omega-3 fatty acids during pregnancy and lactation was reported to increase children's intelligence quotient at four years of age (Helland et al., 2003). Supplementation of the maternal diet with fish oil has been shown to increase docosahexaenoic acid levels in breast milk and to slightly increase early language development (Lauritzen et al., 2005). In the process of fish oil production, organochlorine compounds, together with other lipid-soluble substances present in the fish, are unavoidably extracted and concentrated in the final products. Fish oil dietary supplements and medicinal products containing fish oils intended for administration to infants and young children have been found to contain significant amounts of PCBs and related organochlorine residues (Jimenez et al., 1996; Stringer et al., 1996).

The health impact of chemical hazards in food is estimated by comparing dietary exposure to toxicological levels of concern. Exposure assessments combine data on concentrations of a chemical substance present in food with the quantity of those foods consumed. To improve chemical concentration data, statistically based sampling frames could be developed in collaboration with Member States for their monitoring programs and a coordinated approach of using data from Total Diet Studies (TDSs) could be promoted. A TDS consists of selecting, collecting, and analyzing commonly consumed foods purchased at retail level, processing the food as for consumption, pooling the prepared food items into representative food groups, homogenizing the pooled samples, and analyzing them for harmful and beneficial chemical substances. TDSs are designed to cover the average diet or the most commonly consumed foods, based on data from dietary surveys, in a country or by a specific population group (EFSA, 2011b). The TDS has been used by many countries as a food risk assessment because it is the most reliable way to estimate dietary intake and exposure to pollutants in different population groups. The TDS also assesses the impact of common household cooking methods on the decomposition of the least stable chemical compounds and the formation of new chemical compounds. Data on the concentration of contaminants, minerals, and nutrients combined with dietary consumption data help assess the levels of exposure to chemicals through dietary intake (Millour et al., 2010). In 2011, EFSA, FAO, and WHO jointly published a guidance document providing best practices on methods and protocols for a future collaborative and harmonized TDS worldwide with particular focus on Europe (EFSA/FAO/WHO, 2011).

With respect to the different population groups for which special consideration is needed, infants and young children are considered the most exposed. This is due to the fact that infants and young children present the highest food consumption levels per kilogram body weight (Lowik, 1996; WHO, 2009). In relation to vulnerability, infants, young children, pregnant, and lactating women are also considered important population groups within most areas related to food safety studies (WHO, 2009). Data on the consumption patterns of the elderly, especially of those older than 75 years, are of particular interest when biological agents are considered, whereas, in most of the cases, they are not a priority when chemical substances are under consideration. Their vulnerability relates to a diminished efficiency of the immune system with increasing age that makes them more at risk of infection and more severely affected by communicable diseases. In order to identify consumption patterns that might be associated with a higher risk of exposure, subjects with special dietary habits due to their personal choice (e.g., vegetarians) or special dietary requirements due to health problems (e.g., diabetics and celiacs) can also be considered. Information should also be collected in order to assess exposure in subjects belonging to different ethnic groups and different socioeconomic strata, since these two aspects might be correlated with particular consumption patterns (EFSA, 2011b).

11.5 CONCLUSION

The diffusion of toxic contaminants is expected to rise in the future due to the effect of climate changes (Marques et al., 2010) and there is little that can be done to reduce levels, other than to continue the policies for restriction of use and lowering of release into the environment. More pragmatic, realistic, and harmonized risk assessment and management analysis of food chemical contaminants are required to ensure a higher level of food safety for consumers. In this way, it is necessary to develop reliable tools that are able to quantify toxicity, bioaccessibility, and uptake of chemical contaminants in seafood. The feeding habits are also important to potentiate the risks, particularly when consumers do not have a diversified diet and regularly eat food with high levels of chemical contaminants. The fact that the limits set by authorities are not exceeded should not, by itself, be considered as sufficient evidence of food safety, because there are still a lot of uncertainties linked to the toxicological assessment of contaminants, and even worst, synergistic or antagonistic effects of different contaminants, nutrients, microbial flora, occasionally present in the gastrointestinal tract, which have not been taken into consideration. Until new scientific researches and strategies for assessment and control of food chemical contaminants are completely achieved, the precautionary approach is still the only way to limit the consumers' exposure. Moreover, a better information and knowledge of the problem could also be able to protect vulnerable groups of people.

REFERENCES

Aposhian, H.V., Aposhian, M.M. 2006. Arsenic toxicology: Five questions. *Chem. Res. Toxicol.* 19, 1–15.

ATSDR, 2007. *Toxicological Profile for Lead*, U.S. Department of Health and Humans Services, Public Health Service, Centres for Diseases Control, Atlanta, GA.

ATSDR, 2012. *Toxicological Profile for Cadmium*, U.S. Department of Health and Human Services, Public Health Service, Centres for Diseases Control, Atlanta, GA.

Bandarra, N.M., Batista, I., Nunes, M.L. 2009. Chemical composition and nutritional value of raw and cooked black scabbardfish (*Aphanopus carbo*). *Sci. Mar.* 73S2, 105–113.

Barr, D.B., Needham, L.L. 2002. Analytical methods for biological monitoring of exposure to pesticides: A review. *J. Chromatogr.* B 778, 5–29.

Bilandžić, N., Đokić, M., Sedak, M. 2011. Metal content determination in four fish species from the Adriatic Sea. *Food Chem.* 124, 1005–1010.

Billiard, S.M., Hahn, M.E., Franks, D.G., Peterson, R.E., Bols, N.C., Hodson, P.V. 2002. Binding of polycyclic aromatic hydrocarbons (PAHs) to teleost aryl hydrocarbon receptors (AHRs). *Comp. Biochem. Physiol.* Part B 133, 55–68.

Binelli, A., Provini, A. 2004. Risk for human health of some POPs due to fish from Lake *Iseo*. *Ecotox. Environ. Safe.* 58, 139–145.

Binková, B., Šrám, R.J. 2004. The genotoxic effect of carcinogenic PAHs, their artificial and environmental mixtures (EOM) on human diploid fibroblasts. *Mut. Res.* 547, 109–121.

Birnbaum, L.S., Tuomisto, J. 2000. Non-carcinogenic effects of TCDD in animals. *Food Addit. Contam.* 17, 275–288.

Borak, J., Hosgood, H.D. 2007. Seafood arsenic: Implications for human risk assessment. *Reg. Toxicol. Pharmacol.* 47, 204–212.

Bose, R., Onishchenko, N., Edoff, K., Janson Lang, A.M., Ceccatelli, S. 2012. Inherited effects of low-dose exposure to methylmercury in neural stem cells. *Toxicol. Sci.* 130, 383–390.

Bryan, G., Langston, W.J. 1992. Bioavailability, accumulation and effects of heavy metals in sediments with special reference to United Kingdom estuaries: A review. *Environ. Poll.* 76, 89–131.

Burger, J., Dixon, C., Boring, C.S., Gochfeld, M. 2003. Effect of deep frying fish on risk from mercury. *J. Toxicol. Environ. Health Part* A 66, 817–828.

Cabañero, A.I., Madrid, Y., Cámara, C. 2004. Selenium and mercury bioaccessibility in fish samples: An *in vitro* digestion method. *Anal. Chim. Acta* 526, 51–61.

Calder, P.C., Yaqoob, P. 2009. Omega-3 polyunsaturated fatty acids and human health outcomes. *BioFactors* 35(3), 266–272.

Canli, M., Atli, G. 2003. The relationships between heavy metal (Cd, Cr, Cu, Fe, Pb, Zn) levels and the size of six Mediterranean fish species. *Environ. Poll.* 121, 129–136.

Canton, R.F., Scholten, D.E.A., Marsh, G., De Jong, P.C., Van den Berg, M. 2008. Inhibition of human placental aromatase activity by hydroxylated polybrominated diphenyl ethers (OH-PBDEs). *Toxicol. Appl. Pharm.* 227, 68–75.

Carpenter, D.O., Arcaro, K., Spink, D.C. 2002. Understanding the human health effects of chemical mixtures. *Environ. Health Perspect.* 110 (Suppl. 1), 25–42.

Castro-Gonzalez, M.I., Mendez-Armentab, M. 2008. Heavy metals: Implications associated to fish consumption. *Environ. Toxicol. Pharmacol.* 26, 263–271.

Chang, K.N., Lee, T.C., Tam, M.F., Chen, Y.C., Lee, L.W., Lee, S.Y., Lin, P.J., Huang, R.N. 2003. Identification of galectin I and thioredoxin peroxidase II as two arsenic-binding proteins in Chinese hamster ovary cells. *Biochem. J.* 371, 495–503.

Costa, L.G. 2007. Contaminants in fish: Risk–benefits considerations. *Arh. Hig. RadaToksikol.* 58, 367–374.

Daniels, J.L., Longnecker, M.P., Rowland, A.S., Golding, J. 2004. The ALSPAC Study Team—University of Bristol Institute of Child Health. Fish intake during pregnancy and early cognitive development of offspring. *Epidemiology* 15(4), 394–402.

Daviglus, M., Sheeshka, J., Murkin E. 2002. Health benefits from eating fish. *Comments Toxicol.* 8, 345–374.

Devesa, V., Macho, M.L., Jalón, M., Urieta, I., Muñoz, O., Suñer, M.A., López, F., Vélez, D., Montoro, R. 2001. Arsenic in cooked seafood products: Study on the effect of cooking on total and inorganic arsenic concentrations. *J. Agr. Food Chem.* 49, 4132–4140.

Diamond, G.L. 2005. Risk assessment of nephrotoxic metals. In: Tarloff J, Lash L, eds. *The Toxicology of the Kidney*. London: CRC Press, pp. 1099–1132.

EFSA, 2004. Opinion of the Scientific Panel on contaminants in the food chain on a request from the Commission related to cadmium as undesirable substance in animal feed. *EFSA J.* 72, 1–24.

EFSA, 2005. Opinion of the Scientific Panel on contaminants in the food chain on a request from the European Parliament related to the safety assessment of wild and farmed fish. *EFSA J.* 236, 1–118.

EFSA, 2006a. Opinion of the Scientific Panel on contaminants in the food chain on a request from the Commission related to DDT as an undesirable substance in animal feed. *EFSA J.* 433, 1–69.

EFSA, 2006b. Opinion of the Scientific Panel on contaminants in the food chain on a request from the Commission related to aldrin and dieldrin as undesirable substance in animal feed. *EFSA J.* 285, 1–43.

EFSA, 2008. Mercury as undesirable substance in animal feed. Scientific Opinion of the Panel on contaminants in the food chain. *EFSA J.* 654, 1–74.

EFSA, 2010. Scientific opinion on lead in food. *EFSA J.* 8(4):1570.

EFSA, 2011a. Scientific Opinion on polybrominateddiphenylethers (PBDEs) in food. *EFSA J.* 9(5):2156.

EFSA, 2011b. Overview of the procedures currently used at EFSA for the assessment of dietary exposure to different chemical substances. *EFSA J.* 9(12):2490.

EFSA, 2012a. Scientific opinion on the risk for public health related to the presence of mercury and methylmercury in food. *EFSA J.* 10(12):2985.

EFSA, 2012b. Scientific report of EFSA. Update of the monitoring of levels of dioxins and PCBs in food and feed. *EFSA J.* 10(7):2832.

EFSA/FAO/WHO, 2011. Towards a harmonized total diet study approach: A guidance document. *EFSA J.* 9(11):2450.

Ersoy, B., Yanar, Y., Küçükgülmez, A., Çelik, M. 2006. Effects of four cooking methods on the heavy metal concentrations of sea bass fillets (*Dicentrarchuslabrax* Linne, 1785). *Food Chem.* 99, 748–751.

Fujiyoshi, P.T., Michalek, J.E., Matsumura, F. 2006. Molecular epidemiologic evidence for diabetogenic effects of dioxin exposure in U.S. Air Force veterans of the Vietnam war. *Environ. Health Perspect.* 114, 1677–1683.

Funes, V., Alhama, J., Navas. J.I., Lopez-Barea, J., Peinado, J. 2006. Ecotoxicological effects of metal pollution in two mollusk species from the Spanish South Atlantic littoral. *Environ. Poll.* 139, 214–223.

Garrecht, M., Austin, D.W. 2011. The plausibility of a role for mercury in the etiology of autism: A cellular perspective. *Toxicol. Environ. Chem.* 93, 1251–1273.

Geffard, O., Geffard, A., His, E., Budzinski, H. 2003. Assessment of the bioavailability and toxicity of sediment-associated polycyclic aromatic hydrocarbons and heavy metals applied to *Crassostreagigas* embryos and larvae. *Mar. Pollut. Bull.* 46, 481–490.

Harju, M., Hamers, T., Kamstra, J.H., Sonneveld, E., Boon, J.P., Tysklind, M., Andersson, P.L. 2007. Quantitative structure–activity relationship modeling on *in vitro* endocrine effects and metabolic stability involving 26 selected brominated flame retardants. *Environ. Toxicol. Chem.* 26, 816–826.

Hays, S.M., Aylward, L.L. 2003. Dioxin risks in perspectives: Past, present, and future. *Reg. Toxicol. Pharmacol.* 37, 202–217.

Helland, I.B., Smith, L., Saarem, K., Saugstad, O.D., Drevon, C.A. 2003. Maternal supplementation with very-long-chain n-3 fatty acids during pregnancy and lactation augments children's IQ at 4 years of age. *Pediatrics* 111, 39–44.

Hites, R.A. 2004. Polybrominated diphenyl ethers in the environment and in people: A meta-analysis of concentrations. *Environ. Sci. Technol.* 38, 945–956.

IPCS (International Programme on Chemical Safety) 1989. *Environmental Health Criteria 88. Polychlorinated Dibenzo-Para-Dioxins and Dibenzofurans*. Geneva: WHO.

James, M. O. 1989. Biotransformation and deposition of PAH in aquatic invertebrates. In: Varanasi U, ed. *Metabolism of Polycyclic Aromatic Hydrocarbons in the Aquatic Environment*. Boca Raton, FL: CRC Press, pp. 69–92.

Jimenez, B., Wright, C., Kelly, M., Startin, J.R. 1996. Levels of PCDDs, PCDFs and non-ortho PCBs in dietary supplement fish oil obtained in Spain. *Chemosphere* 32, 461–467.

Juhasz, A.L., Naidu, R. 2000. Bioremediation of high molecular weight polycyclic aromatic hydrocarbons: A review of the microbial degradation of benzo(a)pyrene. *Int. Biodeter. Biodegrad.* 45, 57–88.

Kalay, M., Ay, O., Canli, M. 1999. Heavy metal concentrations in fish tissues from the Northeast Mediterranean Sea. *Bull. Environ. Contam. Toxicol.* 63, 673–681.

Kapaj, S., Peterson, H., Liber, K., Bhattacharya, P. 2006. Human health effects from chronic arsenic poisoning—A review. *J. Environ. Sci. Health* 41A, 2399–2428.

Kenyon, E.M., Hughes, M.F. 2001. A concise review of the toxicity and carcinogenicity of dimethylarsinic acid. *Toxicology* 160, 227–236.

Lauritzen, L., Jorgensen, M.H., Olsen, S.F., Straarup, E.M., Michaelsen, K.F. 2005. Maternal fish oil supplementation in lactation: Effect on developmental outcome in breastfed infants. *Reprod. Nutr. Dev.* 45, 535–547.

Limaverde-Filho, A.M., Campos, R.C., Goes, V., Pinto, R.A.G. 1999. Avaliação da perda de mercúrio total empeixes antes e apósosprocessos de fritura e cocção. *Ciencia y Tecnologia Alimentaria* 19, 19–22.

Lin, S., Del Razo, L.M., Styblo, M., Wang, C., Cullen, W.R., Thomas, D.J. 2001. Arsenicals inhibit thioredoxin reductase in cultured rat hepatocytes. *Chem. Res. Toxicol.* 14, 305–311.

Livingstone, D. R. 1994. Recent developments in marine invertebrate organic xenobiotic metabolism. *Toxicol. Ecotoxicol. News* 1, 88–95.

Lowik, M.R.H. 1996. Possible use of food consumption surveys to estimate exposure to additives. *Food Addit. Contam.* 13(4), 427–442.

Lund, E.K. 2013. Health benefits of seafood. Is it just the fatty acids? *Food Chem.* 140, 413–420.

Marques, A., Lourenço, H.M., Nunes, M.L., Roseiro, C., Santos, C., Barranco, A., Rainieri, S., Langerholc, T., Cencic, A. 2011. New tools to assess toxicity, bioaccessibility and uptake of chemical contaminants in meat and seafood. *Food Res. Int.* 44, 510–522.

Marques, A., Nunes, M.L. Moore, S.K., Strom, M.S. 2010. Climate change and seafood safety: Human health implications. *Food Res. Int.* 43, 1766–1779.

Mason, R.P., Laporte, J.M., Andres, S. 2000. Factors controlling the bioaccumulation of mercury, methylmercury, arsenic, selenium, and cadmium by freshwater invertebrates and fish. *Arch. Environ. Contam. Toxicol.* 38, 283–297.

McDonald, T.A. 2002. A perspective on the potential health risks of PBDEs. *Chemosphere* 46, 745–755.

McDowell Capuzzo, J., Farrington, J.W., Rantamki, P., Hovey Clifford, C., Lacaster, B.A., Leavitt, D.F., Jia, X. 1989. The relationship between lipid composition and seasonal difference in the distribution of PCBs in *Mytilus edulis* L. *Mar. Environ. Res.* 28, 259–264.

Meerts, I.A.T.M., Letcher, R.J., Hoving, S., Marsh, G., Bergman, A., Lemmen, J.G., van der Burg, B., Brouwer, A. 2001. *In vitro* estrogenicity of polybrominated diphenyl ethers, hydroxylated PBDEs, and polybrominated bisphenolA compounds. *Environ. Health Perspect.* 109, 399–407.

Mendez-Armenta, M., Villeda-Hernandez, J., Barroso-Moguel, R., Nava-Ruiz, C., Jimenez-Capdeville, M.E., Rios, C. 2003. Brain regional lipid peroxidation and metallothionein levels of developing rats exposed to cadmium and dexamethasone. *Toxicol. Lett.* 144, 151–157.

Mercado-Feliciano, M., Bigsby, R.M. 2008. Hydroxylated metabolites of the polybrominated diphenyl ether mixture DE-71 are weak estrogen receptor-alpha ligands. *Environ. Health Perspect.* 116, 1315–1321.

Millour, S., Noël, L., Chekri, R., Vastel, C., Kadar, A., Guérin, T. 2010. Internal quality controls applied in inductively coupled plasma mass spectrometry multi-elemental analysis in the second French Total Diet Study. *Accred. Qual. Assur.* 15, 503–513.

Moreno Frías, M., GarridoFrenich, A., Martínez Vidal, J.L., Mateu Sánchez, M., Olea, F., Olea, N. 2001. Analyses of lindane, vinclozolin, aldrin, *p,p'*-DDE, *o,p'*-DDT and *p,p'*-DDT in human serum using gas chromatography with electron capture detection and tandem mass spectrometry. *J. Chromatogr. B* 760, 1–15.

Noël, L., Chafey, C., Testu, C., Pinte, J., Velge, P., Guérin, T. 2011. Contamination levels of lead, cadmium and mercury in imported and domestic lobsters and large crab species consumed in France: Differences between white and brown meat. *J. Food Compos. Anal.* 24, 368–375.

Noonan, C.W., Sarasua, S.M., Campagna, D., Kathman, S.J., Lybarger, J.A., Mueller, P.W. 2002. Effects of exposure to low levels of environmental cadmium on renal biomarkers. *Environ. Health Perspect.* 110, 151–155.

NRC (National Research Council) 2000. *Committee on the Toxicological Effects of Methylmercury: Toxicological Effects of Methylmercury*. Washington, DC: National Academy Press.

Olsen, S.F. 2004. Is supplementation with marine omega-3 fatty acids during pregnancy a useful tool in the prevention of preterm birth? *Clin. Obstet. Gynecol.* 47, 768–774.

Papagiannis, I., Kagalou, I., Leonardos, J., Petridis, D., Kalfakakou, V. 2004. Copper and zinc in four freshwater fish species from Lake Pamvotis (Greece). *Environ. Int.* 30, 357–362.

Patterson, J. 2002. Introduction-comparative dietary risk: Balance the risks and benefits of fish consumption. *Comments Toxicol.* 8, 337–344.

Perello, G., Martí-Cid, R., Llobet, J.M., Domingo, J.L. 2008. Effects of various cooking processes on the concentrations of arsenic, cadmium, mercury and lead in foods. *J. Agr. Food Chem.* 56, 11262–11269.

Perugini, M., Visciano, P. 2009. Occurrence of polycyclic aromatic hydrocarbons in foods and consumer safety. In: *Consumer Product Safety Issues.* Hauppauge, NY: NOVA Science Publishers Inc., pp. 109–123.

Perugini, M., Visciano, P., Manera, M., Abete, M.C., Gavinelli, S., Amorena, M. 2013. Contamination of different portions of raw and boiled specimens of Norway lobster by mercury and selenium. *Environ. Sci. Pollut. Res.* 20, 8255–8262.

Perugini, M., Visciano, P., Manera, M., Turno, G., Lucisano, A., Amorena, M. 2007. Polycyclic aromatic hydrocarbons in marine organisms from the Gulf of Naples, Tyrrhenian Sea. *J. Agric. Food Chem.* 55, 2049–2054.

Perugini, M., Visciano, P., Manera, M., Zaccaroni, A., Olivieri, V., Amorena, M. 2009. Levels of total mercury in marine organisms from Adriatic Sea, Italy. *Bull. Environ. Contam. Toxicol.* 83, 244–248.

Peshut, P.J., Morrison, R.J., Brooks, B.A. 2008. Arsenic speciation in marine fish and shellfish from American Samoa. *Chemosphere* 71, 484–492.

Petrick, J.S., Jagadish, B., Mash, E.A., Aposhian, H.V. 2001. Monomethylarsonous acid (MMAIII) and arsenite: LD50 in hamsters and *in vitro* inhibition of pyruvate dehydrogenase. *Chem. Res. Toxicol.* 14, 651–656.

Porte, C., Albaigés, J. 1993. Bioaccumulation patterns of hydrocarbons and polychlorinated biphenyls in bivalves, crustaceans, and fishes. *Arch. Environ. Con. Toxicol.* 26, 273–281.

Rahman, M.A., Hasewaga, H., Lim, R.P. 2012. Bioaccumulation, biotransformation and trophic transfer of arsenic in the aquatic food chain. *Environ. Res.* 116, 118–135.

Regulation 1881/2006/EC. Commission Regulation (EC) No. 1881/2006 of 19 December 2006 setting maximum levels for certain contaminants in foodstuffs. *Off. J. Eur. Union* L 364, 5–24.

Regulation 420/2011/EU Commission Regulation (EU) No. 420/2011 of 29 April 2011 amending Regulation (EC) No. 1881/2006 setting maximum levels for certain contaminants in foodstuffs. *Off. J. Eur. Union* L 111, 3–6.

Regulation 629/2008/EC. Commission Regulation (EC) No. 629/2008 of 2 July 2008 amending Regulation (EC) No. 1881/2006 setting maximum levels for certain contaminants in foodstuffs. *Off. J. Eur. Union* L 173, 6–9.

Regulation 835/2011/EU. Commission Regulation (EU) No. 835/2011 of 19 August 2011 amending Regulation (EC) No. 1881/2006 setting maximum levels for polycyclic aromatic hydrocarbons in foodstuffs. *Off. J. Eur. Union* L 215, 4–8.

Rey-Salgueiro, L., Martínez-Carballo, E., García-Falcón, M.S., Simal-Gándara, J. 2009. Survey of polycyclic aromatic hydrocarbons in canned bivalves and investigation of their potential sources. *Food Res. Int.* 42(8), 983–988.

Richardson, G.M., Wilson, R., Allard, D., Purtill, C., Douma, S., Graviere, J. 2011. Mercury exposure and risks from dental amalgam in the US population, post-2000. *Sci. Total Environ.* 409, 4257–4268.

Ritter, L., Solomon, K.R., Forget, J., Stemeroff, M., O'Leary, C. 1995. A review of selected persistent organic pollutants. *The International Programme on Chemical Safety (IPCS). Inter-Organization Programme for the Sound Management of Chemicals (IOMC)* 145pp.

Salama, A.A., Mohamed, M.A.M., Duval, B., Potter, T.L., Levin, R.E. 1998. Polychlorinated biphenyl concentration in raw and cooked North Atlantic bluefish (*Pomatomus saltatrix*) fillets. *J. Agr. Food Chem.* 46, 1359–1362.

Schuliga, M., Chouchane, S., Snow, E.T. 2002. Upregulation of glutathione-related genes and enzyme activities in cultured human cells by sublethal concentration of inorganic arsenic. *Toxicol. Sci.* 70, 183–192.

Shinsuke, T. 2004. PBDEs, an emerging group of persistent pollutants. *Mar. Poll. Bull.* 49, 369–370.

Smith, A.G., Gangolli, S.D. 2002. Organochlorine chemicals in seafood: Occurrence and health concern. *Food Chem. Toxicol.* 40, 767–779.

Spink, D.C., Lincoln, D.W., Dickerman, H.W., Gierthy, J.F. 1990. 2,3,7,8-Tetrachlorodibenzo-p-dioxin causes an extensive alteration of 17β-estradiol metabolism in MCF-7 breast tumor cells. *Proc. Natl. Acad. Sci. USA* 87, 6917–6921.

Steenland, K., Bertazzi, P., Baccarelli, A., Kogevinas, M. 2004. Dioxin revisited: Developments since the 1997 IARC classification of dioxin as a human carcinogen. *Environ. Health Perspect.* 112, 1265–1268.

Stegeman, J.J., Lech, J.J. 1991. Cytochrome P-450 monooxygenase systems in aquatic species: Carcinogens metabolism and biomarkers for carcinogen and pollutant exposure. *Environ. Health Perspect.* 90, 101–109.

Strandberg, B., Strandberg, L., Van Bavel, B., Bergqvist, P.A., Broman, D., Falanysz, J., Naf, C., Papakosta, O., Rolff, C., Rappe, C. 1998. Concentrations and spatial variations of cyclodienes and other organochlorines in herring and perch from the Baltic Sea. *Sci. Total Environ.* 215, 69–83.

Stringer, R., Jacobs, L.M.N., Johnston, P.A., Wyatt, C.L., Santillo, D. 1996. Organochlorine residues in fish oil dietary supplements. *Organohalogen Comp.* 28, 551–556.

Sun, J., Liu, J., Liu, Y., Jiang, G. 2013. Levels and distribution of methoxylated and hydroxylatedpolybrominated diphenyl ethers in plant and soil samples surrounding a seafood processing factory and a seafood market. *Environ. Poll.* 176, 100–105.

Trotter, W.J., Corneliussen, P.E. 1989. Levels of polychlorinated biphenyls and pesticides in bluefish before and after cooking. *J. Assoc. Off. Anal. Chem.* 72, 501–503.

Ullrich, S.M., Tanton, T.W., Abdrashitova, S.A. 2001. Mercury in the aquatic environment: A review of factors affecting methylation. *Crit. Rev. Environ. Sci. Technol.* 31, 241–293.

USFDA (US Food and Drug Administration) 1993a. *Guidance Document for Chromium in Shellfish.* Washington, DC: DHHS/PHS/FDA/CFSAN/Office of Seafood.

USFDA (US Food and Drug Administration) 1993b. *Guidance Document for Arsenic in Shellfish.* Washington, DC: DHHS/PHS/FDA/CFSAN/Office of Seafood.

USFDA (US Food and Drug Administration) 1993c. *Guidance Document for Nickel in Shellfish.* Washington, DC: DHHS/PHS/FDA/CFSAN/Office of Seafood.

Vassilopoulou, V., Georgakopoulos-Gregoriades, E. 1993. Factors influencing the uptake of PCBs and DDTs in red mullet (*Mullus barbatus*) from Pagassitikos Gulf, Central Greece. *Mar. Poll. Bull.* 26, 285–287.

Visciano, P., Perugini, M. 2009. Polycyclic aromatic hydrocarbons toxicity in animals and humans. In: *Polycyclic aromatic hydrocarbons: Pollution, Health Effects and Chemistry.* Hauppauge, NY: NOVA Science Publishers Inc., pp. 51–66.

Wanibuchi, H., Salim, E.I., Kinoshita, A., Shen, J., Wei, M., Morimura, K., Yoshida, K., Kuroda, K., Endo, G., Fukushima, S. 2004. Understanding arsenic carcinogenicity by the use of animal models. *Toxicol. Appl. Pharmacol.* 198, 366–376.

Wenzl, T., Simon, R., Kleiner, J., Anklam, E. 2006. Analytical methods for polycyclic aromatic hydrocarbons (PAHs) in food and the environment needed for new food legislation in the European Union. *TRAC* 25 (7), 716–725.

WHO 2000. Chapter 5.11 Polychlorinated dibenzodioxins and dibenzofurans. In: *Air Quality Guidelines*, Second Edition. Copenhagen, Denmark: WHO Regional Office for Europe, pp. 1–21.

WHO 2009. Dietary exposure assessment of chemicals in food (Chapter 6). *Principles and Methods for the Risk Assessment of Chemicals in Food. Environmental Health Criteria 240.* FAO/WHO. International Programme on Chemical Safety (IPCS). Geneva: WHO.

Zabik, M.E., Booren, Al., Zabik, M.J., Welch, R., Humphrey, H. 1996. Pesticide residues, PCBs and PAHs in baked, charbroiled, salt boiled and smoked Great Lakes lake trout. *Food Chem.* 55, 231–239.

Zabik, M.E., Harte, J.B., Zabik, M.J., Dickmann, G. 1992. Effect of preparation and cooking on contaminants distribution in crustaceans: PCBs in blue crab. *J. Agr. Food Chem.* 40, 1197–1203.

Zhang, I., Wong, M.H. 2007. Environmental mercury contamination in China: Sources and impacts. *Environ. Int.* 33, 108–121.

12 Chemical Contaminants in Meat Products

Yao Tang, Xihong Li, Bing Zhang, and Rong Tsao

CONTENTS

12.1   INTRODUCTION

As the global population continues to grow and consumers rely more heavily on processed and semi-processed food products, the demand for safe and healthy foods has become greater than ever before. Meat and meat products are not only good and important sources of many macronutrients, such as protein and fats, but also micronutrients such as vitamins and minerals including iron and zinc. Processed meat has several advantages over raw meat as they are convenient to prepare and serve; they have longer shelf-life and can be made into different forms of food with diverse tastes.

Meat products are exposed to environmental pollutants, pesticides, antibiotics, and process-induced toxins (PITs) from the field to fork, at almost every step of the food value chain. Chemical contaminants in meat products are of great public concern and a serious food chemical safety issue. Every year, over 320,000 human cases on chemical contaminants in food have been reported in the European Union (EU), while millions of illnesses in the United States can be attributed to contaminated foods (EFSA and ECDC, 2013).

Meat products can be contaminated at any point along the production process (Figure 12.1). Controls and effective measures (i.e., control of raw meat, reduction in the additive and package material hazards, and optimized processing conditions) can be taken to minimize these problems in cooked, cured, dry-cured, and packaged

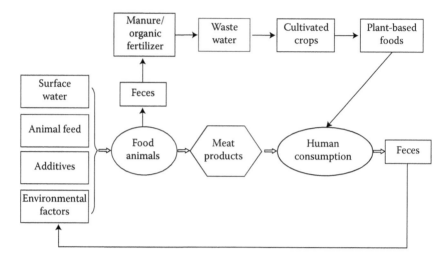

FIGURE 12.1 The main transmission routes of chemical contaminants from animals to human being.

meat products. Failure mode and effect analysis system and other regulations developed and implemented by producers can also help assess or prioritize vulnerable meat production chain steps to reduce or eliminate the vulnerability caused by the various contaminants. Related laws have been established to restrict and prohibit the use of hazardous additives, and to penalize the violators. This chapter is intended to provide readers with information on the major contaminants in raw meat and processed meat products and how strategies can be developed to reduce and eliminate contaminants thereof.

12.2 ANTIBIOTICS

Synthetic and natural antibiotics have been used in animal agriculture to improve growth rates and production as early as the World War II era; right after the large-scale production of penicillin was possible (Jukes, 1975). Antibiotics were used during meat processing to inhibit or retard growth of microorganisms for prolonging shelf-life of meat products (McEwen, 2006). The permit for using antibiotics was approved by the U.S. Food and Drug Administration (FDA) in 1951 (Jones and Ricke, 2003). Subsequently, the use of antibiotics, such as tetracycline, penicillin, streptomycin, and bacitracin, in animal feed became popularized and quickly spread to other countries around the world (Jukes, 1975). By 2011, 13,542,030 kg antibiotics were produced in the United States annually for uses in farm animals (FDA, 2011). Their usage has become essential in reducing the production cost. Antibiotics at low levels are beneficial to animal and poultry production (Jones and Ricke, 2003). Antibiotics are also good growth promoters with improved feeding efficiency, and are effective against bacterial infections in horses, sheep, goats, and pigs (McEwen, 2006).

One of the largest concerns for antibiotics use in animal production is developing resistant bacteria, not only directly in animals, but indirectly in humans consuming meat and meat products with residual antibiotics. Other toxicological consequences were also found by antibiotics administration; for example, rats fed with basal diet containing >500 ppm chlortetracycline showed acute gastrointestinal disturbance and bilateral testicular atrophy, occasionally with degenerative changes in the seminiferous tubules in male weanling rats (Dessau and Sullivan, 1961). The residues of antibiotics in meat products can induce gastrointestinal symptoms, infectious diseases, skin rashes, and thrush. Specific effects include nephrotoxicity associated with aminoglycosides and staining of the teeth attributable to tetracycline and other adverse effects to human (Dancer, 2004). Additionally, the animal waste discharged from the farm or processing plants may contain significant amount of residual antibiotics and resistant bacteria, which would pollute the ground water, soil, plants and other animals, and eventually humans (Solhaug et al., 2004).

Some antibiotics are quite stable in liquid media (such as milk), and certain others such as neomycin are even tolerant to heat treatment. Antibiotics can pass through the body and accumulate in the blood, muscle tissues, and body fluids; thus becoming difficult to be eliminated (Dancer, 2004). Antibiotics such as aminoglycosides and aminocyclitols can interfere with protein synthesis in bacterial cell membrane by binding irreversibly to ribosome. Apramycin, gentamicin, kanamycin, streptomycin, dihydrostreptomycin, gentamicin, neomycin, and spectinomycin are most commonly detected in meat products. These antibiotics are found to accumulate mostly in human kidney after meat consumption (Solhaug et al., 2004).

Chloramphenicols, including chloramphenicol, thiamphenicol, and florfenicol, are broad-spectrum antibiotics against both Gram-positive and Gram-negative bacteria as well as other groups of microorganisms by interacting with ribosome to inhibit organisms from synthesizing proteins. Residues of chloramphenicol have been found in plasma, muscle, fat, liver, kidney, and skin of human subjects who consumed meat products (Kowalski et al., 2008). The lactam antibiotics, occurring in beta, gamma, and delta forms, are also one of the most widely used antimicrobial drugs, which inhibit the bacterial *trans*-peptidase enzymes that are essential for the construction of peptidoglycan of the cell wall. Penicillins have also been found in agricultural animals, and have been found to be teratogenic in rats, causing limb malformations in offspring (Demain and Sanchez, 2009).

12.3   GROWTH HORMONES

Growth hormones (GH) are a family of somatolactogenic hormones that include prolactin and placental lactogen. In swine, poultry, and livestock, GH can enhance animal production effectively, such as daily weight gain, feed efficiency by increasing muscle growth, milk production and all while reducing net adipose tissue deposition and hepatic amino acid degradation. Additionally, GH can increase animals' ability to deliver more glucose and amino acids to the muscles and increase insulin-like growth factor I synthesis (Meyer, 2001). Zeranol, testosterone, diethylstilbestrol, 3,17β-dihydroxyestra-1,3,5(10)-triene(oestradiol-17-β), progesterone, trenbolone acetate, and melengestrol acetate are the primary hormones used as growth promoters in farm animals.

Zeranol, resorcyclic acid lactone, and zearalenone are mycotoxins with estrogenic effect produced by *Fusarium* species on several grains, and have been widely used as growth promoters to improve animal performance in ruminants (Dusietal, 2009). In animals, zeranol is usually metabolized into two stable metabolites with less biological activities; β-zearalanol (Taleranol, predominantly) and zearalanone (minor) (Songsermsakul et al., 2013). The bovine liver, muscle, and uterine tissues are the places where biotransformation of zeranol occurs. Zeranol has been found to have only insignificant adverse effects on humans, with a low acute toxicity and carcinogenicity (Dusi et al., 2009). Considering the overall exposure, it is a weak estrogen naturally present in our food supply.

Testosterone and its acetate, propionate or isobutyrate esters are steroid hormones used for growth-promoting purposes to increase bone and muscle mass since the 1950s. After entering the metabolic pathways, the synthetic testosterone is undistinguished from endogenously produced molecule, and similarly binds to the androgen receptors. Testosterone may cause traumatic damage to the central nervous system, inflammatory neurodegenerative diseases, cervical–uterine tumors in female rats, and prostate cancer in male rats, albeit without a dose–response relationship (Tehranipour and Moghimi, 2010).

Diethylstilbestrol (DES) was a synthetic estrogen additive to cattle feed. Due to its teratogenic, mutagenic, and carcinogenic properties, the use of DES has been banned since 1973 in the United States and 1981 in the EU (Chen et al., 2011). In fact, the European Commission also banned the use of any anabolic steroids in agricultural animals. The use of GH also changes the sensory quality of meat products. GH can lead to reduced protein breakdown in postmortem by producing tougher with a high level of connective tissue and collagen crosslinks. In addition, the overuse of growth promoters may compromise the quality grades of beef carcasses and increase the incidence of dark cutting carcasses.

12.4   OTHER ADDITIVES TO FEED AND RAW MEAT

Vitamin E (tocopherol) and vitamin C (ascorbic acid) are commonly used to function in postmortem muscle (meat) as important antioxidant additives in animals. While these antioxidant vitamins provide many health benefits to humans, alarm has been raised in recent years concerning supplementing mixed synthetic vitamins in certain population groups. The Women's Angiographic Vitamin and Estrogen Study reported that vitamin E plus vitamin C may have played a role in increasing the mortality rate (400 IU of vitamin E twice daily plus 10 IU of vitamin C twice daily) and an increased cardiovascular mortality rate in postmenopausal women with coronary disease compared with the placebo women (Waters et al., 2002). Although consumers will not normally receive the high dosage used in such studies from meat products, considering that some are taking additional vitamin supplements, consumers need to be alerted. Butylated hydroxytoluene (BHT) and butylated hydroxyanisole (BHA) are synthetic antioxidants often added to meat products during processing. Besides the carcinogenicity and toxicity at high doses in rodents and monkeys, BHA and BHT may also act as prooxidants at high concentrations. For most animals, the LD_{50} value of these synthetic antioxidants has been reported to be ca. 2000 mg/kg (Bouayed and Bohn, 2010).

Many flavoring agents are commonly added into the animal feed for improving feed intake. Monosodium glutamate (MSG), a flavor enhancer, along with sodium phosphate, sodium citrate, and sodium lactate has been used in meat products. Several studies have proven that MSG is digested and utilized in the small intestine (Rezaei et al., 2013). Acute toxicity is rare; however, published reports have indicated that MSG can cause many disorders. That is to say, MSG may play roles in a number of pathological conditions including addiction, stroke, epilepsy, brain trauma, neuropathic pain, schizophrenia, anxiety, depression, and degenerative disorders such as Parkinson's disease, Alzheimer's disease, Huntington's disease, and amyotrophic lateral sclerosis (Eweka et al., 2013).

Nitrates and nitrites can restore and treat meat products by lowering the water activity and inhibit the growth of *Clostridium botulinum* and other spoilage microorganisms (EFSA, 2008). The use of these salts in cured meat has been implicated in chemical toxicities and formation of carcinogens after digestion and absorption. Oxidation of nitric to nitrite and then from nitrite to nitrate contributes to the pool of NO_x (a generic term for nitrogen oxides NO and NO_2) compounds. These compounds serve as signaling molecules systemically or as local substrates for nitric oxide production. Meanwhile, nitrates can be reduced to nitrites by enzymes in the mouth and gastrointestinal tract (Hord et al., 2009). The nitrate intake from cured meat products has been estimated to be on an average of 35–44 mg/person per day for a 60-kg human (EFSA, 2008). Nitrite is rapidly distributed in the plasma and binds to erythrocytes after absorption, as the bioavailability of dietary nitrate and nitrite is close to 100%. Therefore, the potential hazardous influence of nitrate and nitrite on human health should be closely monitored. Potassium sorbate (ca. 2000–2600 ppm) is used in meat products to reduce carcinogenic *N*-nitrosamine formation, maintain color, flavor, and heat stability while lowering nitrite concentration in meat products (Keeton, 2011).

Synthetic dyes are widely used to improve meat coloring properties, uniformity, stability, and cost saving. Long-term exposure to these colorants, including tart azine, new red, amaranth, Ponceau 4R, sunset yellow FCF, allura red AC, and brilliant blue FCF, has been shown to induce human health problems such as indigestion, anemia, and allergic reactions as asthma and urticaria, DNA damage, pathological lesions in the brain, kidney, spleen, liver, tumors, and cancer (Moutinho et al., 2007).

12.5 TOXINS FORMED DURING MEAT PROCESSING AND STORAGE

Meat is known to be an excellent source of protein, vitamins, and minerals with superior quantities and bioavailability as compared to plant foods. Conversely, consumption of meat, particularly red meat, has been slowly declining in the past decade, mainly due to its high fat content and other controversial health effects. Concerns over consuming high quantities of meat have been raised in recent studies. In particular, the presence of heterocyclic amines (HCAs) in cooked meat has drawn concern within the health community, as these compounds are known carcinogens in long-term animal experiments with rodents and monkeys (Cheng et al., 2006).

HCAs are chemicals found at trace levels in cooked beef as well as in many other cooked meat products, namely turkey, chicken, pork, and fish. High-temperature cooking methods such as barbequing, pan frying, and grilling increase the levels of HCAs (Cheng et al., 2006). There are two structurally distinct groups of HCAs. Polar HCAs, such as 2-amino-3-methylimidazo[4,5-f] quinoline (IQ), 2-amino-3,4-dimethylimidazo[4,5-f] quinoline (MelQ), and 2-amino-3,8-dimethylimidazo[4,5-f]quin-oxaline (MelQx), are classified as probable human carcinogens by the International Agency for Research on Cancer (IARC) (Dong et al., 2011). Nonpolar HCAs, such as 2-amino-1-methyl-6-phenylimidazo[4,5-b]pyridine (PhIP), 3-amino-1, 4-dimethyl-5H-pyrido [4,3-b] indole (Trp-P-1), and 3-amino-1-methyl-5H-pyrido[4,3-b] indole (Trp-P-2), have been shown to induce liver tumors in mice and rats for urinary bladder cancer (Wei et al., 2002). The formation of the HCAs was dependent on time, temperature, and precursors and followed first-order kinetics in real meat model (Dong et al., 2011).

The content of HCAs exposure from the diet is not well established, as no comprehensive survey has been done to evaluate their prevalence in the marketplace, let alone home-cooking techniques. Levels of HCAs in actual food samples can be highly variable, even when consistency is taken into consideration during study design. For example, the concentrations of HCAs in well-done steaks can vary by more than an order of magnitude even when the steaks are obtained from the same restaurants on different days. One study found that the levels of the HCA compound PhIP in restaurant chicken Caesar salads varied from 0.27 to 11.6 ppb, a difference of over 40-fold (Sullivan et al., 2008). Therefore, precise monitoring of cooking conditions, reproducible sampling methods, and replicate analyses are essential to minimize the formation of HCAs in cooked muscle meat. An average HCAs intake is 26 ng/kg body weight per day for the U.S. population, whereas in European countries, the estimated intake levels were lower, about 2.5 ng/kg body weight per day to 5 ng/kg body weight per day (Fu et al., 2011). The addition of antioxidants to barbequed meat has shown promise as a means to reduce formation of HCAs. A number of studies have exposed that foods rich in flavonoids and other dietary antioxidants reduced the formation of the heterocyclic amines PhIP and MeQI in cooked ground meat patties (Friedman et al., 2009). Vitamin E was found to be one of the strongest antioxidant inhibitor of HCAs formation. Addition of 1% vitamin E to the ground beef reduced HCAs from 48.0% to 79.2% compared to controls (Balogh et al., 2000). The exact mechanism for HCAs formation is not well understood; however, it is suggested that there is a strong oxidative component, such as extracts of finger root, rosemary, and turmeric could be used to decrease the HCA formation (Puangsombat et al., 2011).

Another important group of process-induced toxins are polycyclic aromatic hydrocarbons (PAHs). Typically, PAHs can be produced from burning wood, fossil fuels, and other carbon-containing materials, and are environmental pollutants. Food processing such as smoking, barbequing, and grilling of meat can also produce PAHs, the hazardous food contaminant. The aromatic rings in PAHs are condensed either linearly or peri-condensed. Linear PAHs can be further divided into branched and nonbranched systems. The branched ones are much more stable and less chemically reactive under thermodynamical conditions than the nonbranched PAHs, while the peri-condensed PAHs including alternant and nonalternant depend on the presence of five- or six-membered rings in the molecule. About 660 different compounds belong to the PAH group, and some of them have been shown to be carcinogenic (Jira, 2010; Westman et al., 2013). Benzo[*a*]pyrene has been used as a marker of carcinogenic PAHs and concentrations were measured in some barbequed meat products: 1.5 µg/kg in hamburgers, up to 1.8 µg/kg in pork, up to 4.9 µg/kg in steaks, and 9.2 µg/kg in duck with skin (Aaslyng et al., 2013). Dietary PAHs, such as those from smoked meat products, bind covalently to cellular DNA, causing errors in DNA replication and mutations (Phillips, 1999).

Acrylamide is a carcinogenic compound produced from some foods after heating, frying, baking, and roasting under high temperature and low humidity conditions. Acrylamide is formed via Maillard reaction between amino acids and reducing sugars. Therefore, processed foods containing high carbohydrate and amino acid contents, particularly glucose/fructose and asparagine, are prone to acrylamide contamination. As a rodent carcinogen and a human neurotoxin, acrylamide has been classified as a probable carcinogen to humans by IARC (Lineback et al., 2012). The World Health Organization (WHO) has urged more research into acrylamide in foods after the Swedish scientists found that acrylamide is present in many foodstuffs (FAO/WHO, 2002). Meat products are very low in acrylamide content, lacking the precursors required for its formation, therefore, acrylamide is not a significant safety issue for meat products.

Furan has been reported in foods that underwent heat treatment, particularly in canned and jarred meat products. It is considered a potentially hazardous chemical and is classified as a possible human carcinogen. Furan is formed from ascorbic acid, erythorbic acid, fatty acids, sugars, and mixtures of sugars and amino acids upon thermal processing. There are several distinct pathways responsible for its formation: (i) thermal degradation of reducing sugars, alone or in the presence of amino acids; (ii) thermal oxidation of lipids, especially polyunsaturated fatty acids (e.g., linoleic, linolenic acids) via peroxidation and ring closure; and (iii) decomposition of ascorbic acid or its derivatives, particularly dehydro-ascorbic acid and iso-ascorbic acid (Yaylayan, 2006). High intake of furan (more than 40 mg/kg per day) has been reported to cause hepatocellular necrosis, inflammation, and increased serum transaminases, alkaline phosphatase, cholesterol, triglycerides, and total bilirubin, along with increased blood urea nitrogen and serum creatinine indicative of impaired hepatic and renal functions. Improving the cooking method, such as cooking with cans or jars, could reduce the content of furan compared to cooking in a microwave oven (Kim et al., 2009).

Meat and meat products containing significant amounts of unsaturated fats are prone to oxidation during storage. Lipid peroxidation of unsaturated fats generates breakdown products such as aldehydes, ketones, alcohols, and hydrocarbons, giving off-flavors to the meat products. Still, one of the PITs in meat products that are often neglected is perhaps the trans fatty acids (TFA) induced during processing or storage. While TFA are mostly found in foods rich in partially hydrogenated vegetable oils, processed meat has been found to contain varied amounts of TFA. Apart from the occurrence of health beneficial conjugated linoleic acids found in ruminants, process-induced TFAs in meats may contribute, at least accumulatively, to the health risks of dietary TFA. In both observational cohort studies and randomized clinical trials, TFA adversely affects lipid profiles (including raising LDL and triglyceride levels, and reducing HDL levels), systemic inflammation, and endothelial function

(Micha and Mozaffarian, 2008). More limited but growing evidence suggests that TFA also exacerbates visceral adiposity and insulin resistance. These potential effects of TFA on a multitude of cardiovascular risk factors are consistent with the strong associations seen in prospective cohort studies between TFA consumption and the risk of myocardial infarction and coronary heart disease death (Micha and Mozaffarian, 2008). For details about TFA and other thermal processing-induced chemical contaminants could be found in the related chapters of this monograph.

12.6 TOXINS ENTERING MEAT PRODUCTS FROM OTHER SOURCES

Contaminants migrating from packaging materials to food such as bisphenol A, or heavy metals such as cadmium, lead, mercury, and thallium from contaminated water and soils or animal feed can also end up in processed meat products. These toxic metals can accumulate in the muscle, liver, and kidney leading to possible cardiovascular, nervous, and bone diseases (Rudy, 2009).

Environmental contaminants such as dioxins, polychlorinated dibenzodioxins and polychlorinated dibenzofurans, and coplanar polychlorinated biphenyls (dioxin-like PCBs) have long half-life and can also find their way to edible tissues of exposed animals polluted by municipal, hospital, and industrial hazardous wastes. These toxins are known to cause human immune-toxicity, carcinogenicity, and adverse effects related to reproduction, development, and endocrine functions (Nakamoto et al., 2012). Agrochemical residues including fungicides, insecticides, herbicides, and veterinary drugs have also been found in meat and processed meat products. These toxins may be inhibitors of important enzymes, and may react with life important biomolecules, causing disturbances to the chemical signaling systems in key metabolic pathways, and may alter the physical properties of membranes and cells. They can also mimic hormones, causing reproductive problems and cancer (Jurewicz and Hanke, 2008).

Similar to agrochemical residues, mycotoxins can enter meat products indirectly from contaminated animal feed made from low-grade grains infested with mycotoxin-producing fungi. The most threatening mycotoxins, such as aflatoxins, ochratoxin A, patulin, deoxynivalenol, zearalenone, fumonisins, and T-2/HT-2 toxins, are regulated in foods including meat products (Desmarchelier et al., 2010). For the most recent studies related to mycotoxin contamination in meat products, readers are referred to Bailly and Guerre (2009).

12.7 IMPACT ON HUMAN FOOD SAFETY

As discussed above, chemical contaminants found in meat products can come from environmental sources, or as a form of feed and food additives, or be produced during processing and storage. Once they enter the food system, all these contaminants pose various threats to human safety and health.

Long-term exposure to antibiotics through meat consumption has been found to cause changes in intestinal microflora population and elimination of useful bacterial strains in the human body (Tlaskalová-Hogenová et al., 2011). Addition of antibiotics in animal production has led to the development of antibiotics resistance and super-bugs. Antibiotics-resistant pathogenic bacteria such as *Escherichia coli, Salmonella* spp., *Campylobacter* spp., and *Enterococci* are especially problematic and can pose even life-threatening risks to humans when infected via contact (occupational expo-sure) or through the food chain (Dancer, 2004).

Different from the acute safety risks indirectly caused by overuse or misuse of antibiotics in animal feed, the majority of all other contaminants, additives, or PITs in meat products are of rather chronic nature. Only long-term exposure to low dose or short-term exposure to high dose in animal tests has raised safety concerns. Synthetic hormones, such as 17α-estradiol or 17β-estradiol, were found to cause no adverse effect even at a single oral dose as high as 50 mg followed by an oral dose of 2 mg administered daily for as long as 12 weeks on humans (Moos et al., 2009).

A serious concern about nitrite on the formation of carcinogenic nitrosamines, childhood leukemia, and brain cancer has been raised. Adverse health effects in ani-mals and humans, resulting from acute and sub-acute exposure to excessive nitrite, are typically due to the formation of methemoglobin in the blood (Hord et al., 2009). The consequence of chronic exposure to nitrite, however, is controversial with equiv-ocal evidence of gastric carcinogenicity in female mice. Nitrites are known oxidative stressors that induce methemoglobinemia in a dose-independent manner (Cockburn et al., 2013).

12.8 REGULATORY STATUS

While animal feed and food additives play a key role in the safety and quality and contribute to sustained and low-cost production of meat products, they may impart many negative effects on human health and safety. These substances have been regulated and controlled by agencies at all levels of governments. Various programs have been developed for the control of these and other contaminants, including environmental and process PITs. Same as in other foods, the acceptable daily intake (ADI) levels and tolerable concentrations of contaminants in meat and meat products are determined on the basis of the no observed adverse effect level (NOAEL), which is obtained from animal experiments, and then multiplied by a safety factor (usually 100). The maximum concentrations of chemical contaminants allowed by legisla-tion are monitored and achieved through using good manufacturing practice and other quality control systems such as the hazard analysis and critical control point (HACCP) system and ISO 22000, a new international generic food safety manage-ment system (FSMS) (Mortimore, 2001). Meat and poultry HACCP implementation was completed in January 2000 by The Food Safety and Inspection Service, an agency of the United States Department of Agriculture (USDA) (Arvanitoyannis and Kassaveti, 2009). The antibiotics added in feed are the primary source of con-taminants in meat products; therefore, regulation and control of antibiotics in agri-cultural animals becomes a worldwide food safety issue. However, differences exist in the use and licensing of antibiotics in individual countries. Even among devel-oped countries, there is disunity in banning certain antibiotics (Bampidis et al., 2005). In developing countries where 25% of the world's meat production is pro-duced, antibiotics banned in the EU or the United States, such as chloramphenicol, are still in use (Sánchez-Vargas et al., 2011). Therefore, a consorted international effort must be made to better monitor and manage the misuse and abuse of antibiotics in animal production.

The acceptable limits of seven main growth hormones in animals, according to the European Union and National Residue Control Plan, are as follows: tetracycline, 100.00 ppb/kg; streptomycin, 500.00 ppb/kg; sulfamethazine, 100.00 ppb/kg; trenbolon, 10.00 ppb/kg; estradiol 17β, 0.10 ppb/kg; chloramphenicol, 100.00 ppb/kg; and testosterone 0.10 ppb/kg (Kadim et al., 2010). The ADI for total residues of zeranol is 0.00125 mg/kg of body weight per day according to the FDA Regulations (2013).

A total of 15 PAHs including benzo[a]pyrene (BaP) have been strongly recommended for further investigation on their health risks by the European Commission (European, 2005). Among these 15 EU priority PAHs, 12 are the same compounds identified to be reasonably anticipated by the IARC to be human carcinogens based on sufficient evidence of carcinogenicity in experimental animals conducted from 1973 to 1987 (Wenzl et al., 2006). The maximum limit for BaP in smoked meat products is 5 μg/kg and starting September 2012 an additional maximum limit for four-ring PAHs (PAH4) was included with a level of 30 μg/kg (Table 12.1) (European, 2006, 2011). The sum of PAH4 was highest in beef (average 17.3 μg/kg) compared with pork (average 2.6 μg/kg) and chicken (average 1.1 μg/kg). European exposure estimates suggest that mean dietary exposure to furan may be as high as 1.23 and 1.01 μg/kg bw/day from all foods for adults and 3–12-month-old infants, respectively (Moro et al., 2012).

Current regulations on usage dose of nitrites and nitrates were restricted under 1718 ppm for ingoing nitrate, and 200 and 700 ppm for ingoing sodium and potassium nitrite, respectively, in the United States (USDA, 2005).

12.9 ANALYTICAL METHODS FOR DETECTION

Chemical contaminants in raw and processed meat are generally at very low concentrations. Detection of these toxins, whether as environmental contaminants such as pesticide residues, or feed additives such as antibiotics or growth hormones, or food additives (antioxidants, preservatives, flavor enhancers, colorants), or PITs such as HCAs, PAHs, and furan require highly sensitive and accurate analytical tools. Reviewing all analytical methods available for the diverse groups of the food contaminants is beyond the scope of this chapter. Therefore, only chromatographic methods for most important food chemical toxins are discussed. Environmental pollutants such as pesticides or heavy metals are not included here.

Most of the food/feed additives and PITs are of organic nature, and they are most frequently analyzed using liquid (LC) and gas chromatographic (GC) techniques equipped with various detectors. High-performance liquid chromatography coupled with photodiode array (UV/Vis), fluorescence, mass spectrometric (MS), and nuclear magnetic resonance detectors are most widely used for detecting food chemical contaminants. GC is better suited for volatile compounds, and the common detectors include flame ionization detector, electron capture detector, thermal conductivity detector, and MS detector. It must be pointed out that among all the detection methods, MS is the method of choice for most chemical contaminants in

TABLE 12.1
Overview of Maximum Limits for PAHs in Meat and Meat Products, Set in Certain EU Member States before Commission Regulation (EC) 208/2005 Came into Force

Country	Meat or Meat Products	PAHs	Maximum Limits (µg/kg)
Czech Republic	Meat products	BaA, BbF, BkF, DhA, DhP, DiP, IcP, CHR[a]	3.0
Slovak Republic	Smoked meat products and nonsmoked	BaP	1.0
Italy	Smoked meat	BaP	0.05
Estonia	Smoked meat	BaP	0.05
Denmark	Little sausage	BaP	0.08
Germany	Smoked meat and meat products	BaP	1.0
Belgium	Smoked meat and meat preparations	BaP	2.0
United Kingdom	No legal limits, but some countries have recommended maximum levels		
France			
Ireland			
Cyprus			
Luxemburg			
Slovenia			

Source: European (2006, 2011), Jira (2010), Wenzl et al. (2006).

[a] Abbreviations: Benz[*a*]anthracene, BaA; benzo[*b*]fluoranthene, BbF; benzo[*k*]fluoranthene, BkF; dibenz[*a,h*]anthracene, DhA; dibenzo[*a,h*]pyrene, DhP; dibenzo[*a,i*]pyrene, DiP; indeno[*1,2,3-cd*] pyrene, IcP; chrysene, CHR; benzo[*a*]pyrene, BaP.

foods including meat products. It is not an exaggeration to say that the use of MS detection technique has truly revolutionized the analysis of chemical contaminants in foods, not only providing sensitive and selective detections, but more importantly simultaneous quantitation and structural identification of detected analytes. GC/MS has become popular for many pesticide residues, PAHs, PCBs, and other relatively nonpolar contaminants. For relatively polar, thermolabile, and less volatile chemical contaminants such as veterinary drugs, mycotoxins, and acrylamide, LC/MS is the method of choice. Specific methods for the different chemical contaminants in meat products can be found in recent books and reviews (Oracz et al., 2011; Wenzl et al., 2006).

In addition to the instrumental analysis, other sensitive and selective methods have also been developed. Different immunoassays (enzyme-linked immunosorbent assay, ELISA) have been developed and used for the detection and determination of various contaminants in meat and meat products including bacterial toxins, drugs, anabolic hormones, pesticides, and mycotoxins (Sheu et al., 2009). More recently,

PCR-ELISA method has been developed to detect low contents of allergens, micro-bacterial, and some antibiotics in meat products (Chen and Hsieh, 2000).

At present, biosensors and biochips are the latest technologies in the analysis of food chemical and microbial contaminants. Applications in the detection of chemi-cal contaminants are new and rare for meat products, but have been explored with other food products. An immune-biochip method has been recently developed for the detection of ochratoxin A (Sauceda-Friebe et al., 2011).

12.10 SUMMARY

Different from microbial contamination of foods that causes immediate damage and draws attention to the stakeholders, food chemical contaminants are generally the "silent" killers that put humans in danger over time. Understanding the source, occurrence, and the mechanisms of formation of chemical contaminants in meat products will not only help developing mitigation strategies and regulatory policies, but most importantly also reduce the health risks posed to consumers. The impact of chemical contaminants from environmental sources, animal feeds, or those formed during meat processing and storage can be individually or collectively significant.

There is an increasing trend toward harmonization and stringency of regulatory requirements to prevent chemical contaminants from entering and formation in meat products. Food safety management practices such as HACCP and ISO 22000 have been used to monitor the quality and quantity of chemical contaminants set by vari-ous regulatory agents including FAO, WHO, US-EPA and US-FDA, the Food and Veterinary Office of the European Commission, the European Medicines Evaluation Agency (EMEA, for residues), and EFSA (for contaminants).

Concerted effort made to reduce food chemical toxins throughout the meat food chain is important for producing safe foods. Understanding the chemistry and bio-chemistry of these toxins can help develop strategies for efficient monitoring, reduc-tion, and detoxification of these toxic substances in food.

REFERENCES

Aaslyng, M.D., Duedahl-Olesen, L., Jensen, K., Meinert, L., 2013. Content of heterocyclic amines and polycyclic aromatic hydrocarbons in pork, beef and chicken barbecued at home by Danish consumers. *Meat Science* 93, 85–91.

Arvanitoyannis, I.S., Kassaveti, A., 2009. *HACCP and ISO 22000—A Comparison of the Two Systems, HACCP and ISO 22000.* Wiley-Blackwell: Oxford, United Kingdom, pp. 1–45.

Bailly, J.-D., Guerre, P., 2009. Mycotoxins in meat and processed meat products, in: F. Toldrá (Ed.), *Safety of Meat and Processed Meat.* Springer: New York, pp. 83–124.

Balogh, Z., Gray, J.I., Gomaa, E.A., Booren, A.M., 2000. Formation and inhibition of heterocyclic aromatic amines in fried ground beef patties. *Food and Chemical Toxicology* 38, 395–401.

Bampidis, V.A., Christodoulou, V., Florou-Paneri, P., Christaki, E., Spais, A.B., Chatzopoulou, P.S., 2005. Effect of dietary dried oregano leaves supplementation on performance and carcass characteristics of growing lambs. *Animal Feed Science and Technology* 121, 285–295.

Bouayed, J., Bohn, T., 2010. Exogenous antioxidants—Double-edged swords in cellular redox state: Health beneficial effects at physiologic doses versus deleterious effects at high doses. *Oxidative Medicine and Cellular Longevity* 3, 228–237.

Chen, F.-C., Hsieh, Y.H.P., 2000. Detection of pork in heat-processed meat products by monoclonal antibody-based ELISA. *Journal of AOAC International* 83, 79–85.

Chen, X.-B., Wu, Y.-L., Yang, T., 2011. Simultaneous determination of clenbuterol, chloramphenicol and diethylstilbestrol in bovine milk by isotope dilution ultraperformance liquid chromatography–tandem mass spectrometry. *Journal of Chromatography B* 879, 799–803.

Cheng, K.-W., Chen, F., Wang, M., 2006. Heterocyclic amines: Chemistry and health. *Molecular Nutrition and Food Research* 50, 1150–1170.

Cockburn, A., Brambilla, G., Fernández, M.-L., Arcella, D., Bordajandi, L.R., Cottrill, B., van Peteghem, C., Dorne, J.-L., 2013. Nitrite in feed: From animal health to human health. *Toxicology and Applied Pharmacology* 270, 209–217.

Dancer, S.J., 2004. How antibiotics can make us sick: The less obvious adverse effects of antimicrobial chemotherapy. *The Lancet Infectious Diseases* 4, 611–619.

Demain, A.L., Sanchez, S., 2009. Microbial drug discovery: 80 years of progress. *Journal of Antibiotics* 62, 5–16.

Desmarchelier, A., Oberson, J.-M., Tella, P., Gremaud, E., Seefelder, W., Mottier, P., 2010. Development and comparison of two multiresidue methods for the analysis of 17 mycotoxins in cereals by liquid chromatography electrospray ionization tandem mass spectrometry. *Journal of Agricultural and Food Chemistry* 58, 7510–7519.

Dessau, F.I., Sullivan, W.J., 1961. A two-year study of the toxicity of chlortetracycline hydrochloride in rats. *Toxicology and Applied Pharmacology* 3, 654–677.

Dong, A., Lee, J., Shin, H.-S., 2011. Formation of amino-imidazo-azaarenes and carbolines in fried beef patties and chicken breasts under different cooking conditions in Korea. *Food Science and Biotechnology* 20, 735–741.

Dusi, G., Bozzoni, E., Assini, W., Tognoli, N., Gasparini, M., Ferretti, E., 2009. Confirmatory method for the determination of resorcylic acid lactones in urine sample using immunoaffinity cleanup and liquid chromatography-tandem mass spectrometry. *Analytica Chimica Acta* 637, 47–54.

EFSA (European Food Safety Authority), 2008. Nitrate in vegetables: Scientific opinion of the panel on contaminants in the food chain. *The EFSA Journal* 689, 1–79.

EFSA and ECDC (European Food Safety Authority, and European Centre for Disease Prevention and Control ECDC), 2013. The European Union Summary Report on Trends and Sources of Zoonoses, Zoonotic Agents and Food-borne Outbreaks in 2011; *The EFSA Journal*,11(4):3129, 250 pp.

European, C., 2005. Commission Recommendation 2005/108/EC of 4 February 2005 on the further investigation into the levels of polycyclic aromatic hydrocarbons in certain foods. *Official Journal of the European Union L* 34, 43–45.

European, C., 2006. Commission Regulation (EC) No 1881/2006 of 19 December 2006 setting maximum levels for certain contaminants in foodstuffs. *Official Journal of the European Union L* 364, 5–24.

European, C., 2011. Commission Recommendation of 18 October 2011 on the definition of nanomaterial. *Official Journal L* 275, 10.

Eweka, A.O., Igbigbi, P.S., Ucheya, R.E., 2013. Histochemical studies of the effects of monosodium glutamate on the liver of adult Wistar rats. *Annals of Medical and Health Sciences Research* 1, 21–30.

FAO/WHO, 2002. Health implications of acrylamide in food, Report of a joint FAO/WHO consultation, Geneva.

FDA, 2011. Annual Report on Antimicrobials Sold or Distributed for Food-Producing Animals.

Friedman, M., Zhu, L., Feinstein, Y., Ravishankar, S., 2009. Carvacrol facilitates heat-induced inactivation of *Escherichia coli* O157:H7 and inhibits formation of heterocyclic amines in grilled ground beef patties. *Journal of Agricultural and Food Chemistry* 57, 1848–1853.

Fu, Z., Shrubsole, M.J., Smalley, W.E., Wu, H., Chen, Z., Shyr, Y., Ness, R.M., Zheng, W., 2011. Association of Meat intake and meat-derived mutagen exposure with the risk of colorectal polyps by histologic type. *Cancer Prevention Research* 4, 1686–1697.

Hord, N.G., Tang, Y., Bryan, N.S., 2009. Food sources of nitrates and nitrites: The physiologic context for potential health benefits. *The American Journal of Clinical Nutrition* 90, 1–10.

Jira, W., 2010. Polycyclic aromatic hydrocarbons in German smoked meat products. *European Food Research Technology* 230, 447–455.

Jones, F.T., Ricke, S.C., 2003. Observations on the history of the development of antimicrobials and their use in poultry feeds. *Poultry Science* 82, 613–617.

Jukes, T.H., 1975. Antibiotics in meat production. *JAMA* 232, 292–293.

Jurewicz, J., Hanke, W., 2008. Prenatal and childhood exposure to pesticides and neurobehavioral development: Review of epidemiological studies. *International Journal of Occupational Medicine and Environmental Health* 21, 121–132.

Kadim, I.T., Mahgoub, O., Al-Marzooqi, W., Al-Maqbaly, R., Annamali, K., Khalaf, S.K., 2010. Enzyme-linked immunosorbent assay for screening antibiotic and hormone residues in broiler chicken meat in the Sultanate of Oman. *Journal of Muscle Foods* 21, 243–254.

Keeton, J., 2011. History of nitrite and nitrate in food, in: N.S. Bryan, J. Loscalzo (Eds.), *Nitrite and Nitrate in Human Health and Disease.* Humana Press: New York, pp. 69–84.

Kim, T.K., Lee, Y.K., Park, Y.S., Lee, K.G., 2009. Effect of cooking or handling conditions on the furan levels of processed foods. *Food Additives and Contaminants: Part A* 26, 767–775.

Kowalski, P., Plenis, A., Dzka, I.O., 2008. Optimization and validation of capillary electrophoretic method for the analysis of amphenicols in poultry tissues. *Acta Poloniae Pharmaceutica Drug Research* 65, 45–50.

Lineback, D.R., Coughlin, J.R., Stadler, R.H., 2012. Acrylamide in foods: A review of the science and future considerations. *Annual Review of Food Science and Technology* 3, 15–35.

McEwen, S.A., 2006. Antibiotic use in animal agriculture: What have we learned and where are we going? *Animal Biotechnology* 17, 239–250.

Meyer, H.H.D., 2001. Biochemistry and physiology of anabolic hormones used for improvement of meat production. *APMIS* 109, 1–8.

Micha, R., Mozaffarian, D., 2008. Trans fatty acids: Effects on cardiometabolic health and implications for policy. *Prostaglandins Leukotrienes and Essential Fatty Acids* 79, 147–152.

Moos, W.H., Dykens, J.A., Nohynek, D., Rubinchik, E., Howell, N., 2009. Review of the effects of 17α-estradiol in humans: A less feminizing estrogen with neuroprotective potential. *Drug Development Research* 70, 1–21.

Moro, S., Chipman, J.K., Wegener, J.-W., Hamberger, C., Dekant, W., Mally, A., 2012. Furan in heat-treated foods: Formation, exposure, toxicity, and aspects of risk assessment. *Molecular Nutrition and Food Research* 56, 1197–1211.

Mortimore, S., 2001. How to make HACCP really work in practice. *Food Control* 12, 209–215.

Moutinho, I.L.D., Bertges, L.C., Assis, R.V.C., 2007. Prolonged use of the food dye tartrazine (FD&C yellow no. 5) and its effects on the gastric mucosa of Wistar rats. *Brazilian Journal of Biology* 67, 141–145.

Nakamoto, M., Arisawa, K., Uemura, H., Katsuura, S., Takami, H., Sawachika, F., Yamaguchi, M. et al., 2012. Association between blood levels of PCDDs/PCDFs/dioxin-like PCBs and history of allergic and other diseases in the Japanese population. *International Archives of Occupational and Environmental Health*, 86, 849–859.

Oracz, J., Nebesny, E., Żyżelewicz, D., 2011. New trends in quantification of acrylamide in food products. *Talanta* 86, 23–34.

Phillips, D.H., 1999. Polycyclic aromatic hydrocarbons in the diet. *Mutation Research/Genetic Toxicology and Environmental Mutagenesis* 443, 139–147.

Puangsombat, K., Jirapakkul, W., Smith, J.S., 2011. Inhibitory activity of Asian spices on hetero-cyclic amines formation in cooked beef patties. *Journal of Food Science* 76, T174-T180.

Regulations, F.D.A., 2013. Code of federal regulations 21 CFR 556.760.

Rezaei, R., Knabe, D., Tekwe, C., Dahanayaka, S., Ficken, M., Fielder, S., Eide, S., Lovering, S., Wu, G., 2013. Dietary supplementation with monosodium glutamate is safe and improves growth performance in postweaning pigs. *Amino Acids* 44, 911–923.

Rudy, M., 2009. The analysis of correlations between the age and the level of bioaccumulation of heavy metals in tissues and the chemical composition of sheep meat from the region in SE Poland. *Food and Chemical Toxicology* 47, 1117–1122.

Sánchez-Vargas, F.M., Abu-El-Haija, M.A., Gómez-Duarte, O.G., 2011. Salmonella infec-tions: An update on epidemiology, management, and prevention. *Travel Medicine and Infectious Disease* 9, 263–277.

Sauceda-Friebe, J.C., Karsunke, X.Y.Z., Vazac, S., Biselli, S., Niessner, R., Knopp, D., 2011. Regenerable immuno-biochip for screening ochratoxin A in green coffee extract using an automated microarray chip reader with chemiluminescence detection. *Analytica Chimica Acta* 689, 234–242.

Sheu, S.-Y., Lei, Y.-C., Tai, Y.-T., Chang, T.-H., Kuo, T.-F., 2009. Screening of salbutamol resi-dues in swine meat and animal feed by an enzyme immunoassay in Taiwan. *Analytica Chimica Acta* 654, 148–153.

Solhaug, M.J., Bolger, P.M., Jose, P.A., 2004. The developing kidney and environmental tox-ins. *Pediatrics* 113, 1084–1091.

Songsermsakul, P., Böhm, J., Aurich, C., Zentek, J., Razzazi-Fazeli, E., 2013. The levels of zearalenone and its metabolites in plasma, urine and faeces of horses fed with naturally, Fusarium toxin-contaminated oats. *Journal of Animal Physiology and Animal Nutrition* 97, 155–161.

Sullivan, K.M., Erickson, M.A., Sandusky, C.B., Barnard, N.D., 2008. Detection of PhIP in grilled chicken entrées at popular chain restaurants throughout California. *Nutrition and Cancer* 60, 592–602.

Tehranipour, M., Moghimi, A., 2010. Neuroprotective effects of testosterone on regenerating spinal cord motoneurons in rats. *Journal of Motor Behavior* 42, 151–155.

Tlaskalová-Hogenová, H., Štěpánková, R., Kozáková, H., Hudcovic, T., Vannucci, L., Tučková, L., Rossmann, P., Hrnčíř, T., Kverka, M., Zákostelská, Z., 2011. The role of gut micro-biota (commensal bacteria) and the mucosal barrier in the pathogenesis of inflammatory and autoimmune diseases and cancer: Contribution of germ-free and gnotobiotic animal models of human diseases. *Cellular and Molecular Immunology* 8, 110–120.

USDA, 2005. Food Standards and Labeling Policy Book.

Waters DD, Alderman E.L., Hsia J., et al., 2002. Effects of hormone replacement therapy and antioxidant vitamin supplements on coronary atherosclerosis in postmenopausal women: A randomized controlled trial. *JAMA* 288, 2432–2440.

Wei, M., Wanibuchi, H., Morimura, K., Iwai, S., Yoshida, K., Endo, G., Nakae, D., Fukushima, S., 2002. Carcinogenicity of dimethylarsinic acid in male F344 rats and genetic altera-tions in induced urinary bladder tumors. *Carcinogenesis* 23, 1387–1397.

Wenzl, T., Simon, R., Anklam, E., Kleiner, J., 2006. Analytical methods for polycyclic aro-matic hydrocarbons (PAHs) in food and the environment needed for new food legisla-tion in the European Union. *TRAC Trends in Analytical Chemistry* 25, 716–725.

Westman, O., Nordén, M., Larsson, M., Johansson, J., Venizelos, N., Hollert, H., Engwall, M., 2013. Polycyclic aromatic hydrocarbons (PAHs) reduce hepatic β-oxidation of fatty acids in chick embryos. *Environmental Science and Pollution Research* 20, 1881–1888.

Yaylayan, V.A., 2006. Precursors, formation and determination of furan in food. *Journal fur Verbraucherschutz und Lebensmittelsicherheit* 1, 5–9.

13 Chemistry and Safety of Food Additives

Jing Wang and Baoguo Sun

CONTENTS

13.1 INTRODUCTION

Food additives, as an important part of the modern food industry, play a great role in improving food color, flavor, taste and texture, processing behaviors, oxidative stability, suppressing spoilage, and extending the shelf life of food. Without food additives, there would be no food manufacturing and modern food industry. Food additive technology not only provides reliable technical support and safety protection for the food processing and food catering industries, but also promotes their rapid development. As the continuous emergence of new food additives, the food industry can meet the various needs of consumers, which plays an important role in local economy and the people's livelihood. This chapter reviews the function of food additives, the safety evaluation of food additives by different countries and international organizations, risk assessment of food additives, and risk management and control for food additives.

13.2 CONCEPT OF FOOD ADDITIVES

According to the Codex Alimentarius Commission (CAC) and European Commission definitions, food additive means any substance not normally consumed as a food by itself and not normally used as a typical food ingredient, regardless of whether or not it has a nutritive value, the intentional addition of which a food additive to food for a technological (including organoleptic) purpose in the manufacture, processing, preparation, treatment, packing, packaging, transport, or storage of such food results, or may be reasonably expected to result (directly or indirectly), in it or its byproducts becoming a component of or otherwise affecting the characteristics of such foods. The term does not include contaminants or substances added to food for maintaining or improving nutritional qualities (Codex, 2011; EC, 2008). According to the Food and Drug Administration (FDA), a food additive means any substance the intended use of which results or may reasonably be expected to result directly or indirectly in its becoming a component or otherwise affecting the characteristics of any food. According to a Japanese definition, food additives are substances added during the food manufacturing process, that is, the food processing to mix and infiltrate with food for preservation (Federal Food, Drug, and Cosmetic Act, 1958).

As the definition in Chinese "Food Safety Law," food additives are synthetic or natural substances added to food to improve food quality, color, smell and taste, and meet the need for corrosion protection, preservation and processing technology (China, 2009).

13.3 FUNCTION OF FOOD ADDITIVES

13.3.1 CATEGORIES OF FOOD ADDITIVES

According to various sources, food additives can be divided into natural additives and synthetic additives. Natural food additives are natural substances extracted from animals or plants, or microbial metabolites, and so on. Synthetic food additives are substances obtained by chemical synthesis. Currently, most food additives used are synthetic ones.

According to the different functions, food additives can be divided into several groups, such as acids, acidity regulators, anticaking agents, antifoaming agents, antioxidants, bulking agents, food colorants, emulsifiers, flavors, preservatives, stabilizers, and sweeteners. In China, food additives are divided into 23 categories in the "Standards for Use of Food Additives," including acidity regulators, anticaking agents, antifoaming agent, antioxidants, bleaching agents, leavening agents, candy gum base, colorants, color retention agents, emulsifiers, enzymes, flavor enhancers, flour treatment agents, coating agents, water retention agents, nutrition enhancers, preservatives, stabilizers, coagulants, sweeteners, thickening agents, food spices, food processing aids, and others. In different countries, food additives are divided into different categories, some of which are summarized in Table 13.1.

13.3.2 FUNCTION OF FOOD ADDITIVES

The appropriate food additives are used in food processing to ensure its quality. Food additives play an important role in food industry.

13.3.2.1 Improving and Enhancing Food Color, Aroma, Taste, and Mouthfeel Properties

Food color, smell, taste, texture, and mouthfeel are important indicators for food sensory quality. Food processing procedures may include milling, crushing, heating, pressure, and other physical processes, in which the color and flavor might be altered. In addition, a single process is difficult to meet the needs of the soft, hard, brittle, and tough taste qualities. Therefore, the appropriate use of coloring agent, color stabilizers, edible flavors and aromas, thickeners, emulsifiers, and other quality enhancers can significantly improve the sensory quality of food, and meet the consumers' needs of food flavor and taste.

13.3.2.2 Maintaining and Improving the Nutritional Value of Food

Food preservatives and antioxidants can prevent oxidative deterioration of food, and have an important role in maintaining food nutritional value. For instance, antioxidants are important for ensuring oxidative stability of vegetable oils. An important class of antioxidants used in vegetable oils consists of the phenolic compounds butylhydroxyanisole (BHA), butylhydroxytoluene (BHT), propyl gallate, and tert-butyl hydroquinone (TBHQ) (Aluyor and Ori-Jesu, 2008; Rubalya and Neelamegam, 2012). In addition, food nutrition enhancers may improve the nutritional value of food, which have important implications for preventing malnutrition and nutritional

TABLE 13.1
Classification of Food Additives in Different Countries

Ordinal	China	CAC	EU	USA	Japan
1	Acidity regulators	Acid	Colorants	Anticaking agents and free-flow agents	Preservatives
2	Anticaking agents	Acidity regulators	Preservatives	Antimicrobial agents	Bactericide
3	Antifoaming agents	Anticaking agents	Antioxidants	Antioxidants	Fungicide
4	Antioxidants	Antifoaming agents	Emulsifiers	Colors and coloring adjuncts	Antioxidants
5	Bleaches	Antioxidants	Emulsifying salt	Curing and pickling agents	Bleaches
6	Leavening agents	Filling agents	Thickening agents	Dough strengtheners	Flour improver
7	Candy gum base	Colorants	Coagulants	Drying agents	Thickening agents
8	Colorants	Color retention agents	Stabilizers	Emulsifiers and emulsifier salts	Food spices
9	Color retention agents	Emulsifiers	Flavor enhancers	Enzymes	Insect repellent
10	Emulsifiers	Emulsifying salt	Acid	Firming agents	Chromophoric agents
11	Enzymes	Curing agents	Acidity regulators	Flavor enhancers	Color stabilizer
12	Flavor enhancers	Flavor enhancers	Anticaking agents	Flavoring agents and adjuvants	Colorants
13	Flour treatment agents	Flour treatment agents	Modified starch	Flour treating agents	Flavoring agents
14	Coating agents	Foaming agents	Sweetners	Formulation aids	Acidulant
15	Water retention agents	Coagulants	Leavening agents	Fumigants	Sweetner
16	Nutrition enhancers	Glazing agents	Antifoaming agents	Humectants	Emulsifiers and emulsion stabilizer
17	Preservatives	Water retention agents	Polishing agents	Leavening agents	Antifoaming agents
18	Stabilizers	Preservatives	Flour treatment agents	Lubricants and release agents	Water retention agents
19	Coagulants	Propellant	Curing agents	Nonnutritive sweeteners	Solvents and quality retention agents
20	Sweeteners	Leavening agents	Water retention agents	Nutrient supplements	Leavening agents

TABLE 13.1 (Continued)
Classification of Food Additives in Different Countries

Ordinal	China	CAC	EU	USA	Japan
21	Thickening agents	Stabilizers	Chelating agents	Nutritive sweeteners	Candy gum base
22	Food spices	Sweeteners	Enzymes	Oxidizing and reducing agents	Coating agents
23	Food processing aids	Thickening agents	Filling agents	pH control agents	Nutrient
24			Propulsive gas and packing gas	Processing aids	Extraction solvents
25				Propellants, aerating agents	Food processing aids
26				Sequestrants	Filtering adjuvant
27				Solvents and vehicles	Brewing agent
28				Stabilizers and thickeners	Quality improver
29				Surface-active agents	Coagulator for tofu
30				Surface-finishing agents	Antiadhesive agent
31				Synergists	
32				Texturizers	

deficiencies, maintaining nutritional balance, and improving human health. For instance, vitamins and minerals are added to many foods to make up those lacking in human diet or being lost during processing (Caballero, 2003; Flynn et al., 2009).

13.3.2.3 Contributing to Food Preservation and Extending Food Shelf Life

Fresh foods without preservatives may deteriorate easily. In order to ensure food quality within the shelf life period, preservatives and antioxidants are needed. For instance, weak organic acids such as sorbic and benzoic acids are widely used for food preservation; hydrogen peroxide-mediated systems and chelators such as citric acid and EDTA have the potential use for food preservation; and the naturally occurring preservatives with potential antimicrobial activities also have the prospective use in food preservation, all of which have to be permitted by the regulatory authorities for inclusion in foods (Billing and Sherman, 1998; Brula and Cooteb, 1999; Shee et al., 2010).

13.3.2.4 Increasing the Diversity of Food

Food shelves in supermarkets are filled with wide varieties of foods, in which the indispensable part is food additives in addition to the main raw material of grain and oil, fruits and vegetables, meat, eggs, and milk. The appropriate food additives are added according to the need of processing, formulation, and the food sensory properties. Small quantity of different food additives are in high demand for different food products.

13.3.2.5 Advantageous for Food Processing Operation

Lubrication, antifoaming, leaching, stability, and solidification are required in food processing. Without food additives, most foods cannot be processed. For instance, the machinery used to prepare and process foodstuffs needs lubricants, grease and oil for lubrication, heat transfer, power transmission and corrosion protection of machinery, machine parts, equipment, and instruments (Malone et al., 2003; Moon, 2007).

13.3.2.6 Meeting the Needs of Different Consumers

Sugars are not appropriate for diabetic patients. Sweeteners may be good sugar substitutes for diabetic patients (Busteed et al., 2004; Ceulemans et al., 1997; Jastad et al., 2001). A variety of nutrients such as Ca are needed for infant growth, resulting in the development of formula milk powder fortified with minerals and vitamins.

13.4 PRINCIPLES OF FOOD ADDITIVES

FAO and WHO set the principles for the use of food additives (Codex, 1995).

13.4.1 Food Additive Safety

Only those food additives that can be judged on the evidence presently available from the Joint Expert Committee on Food Additives (JECFA) and present no appreciable health risk to consumers at the levels proposed for use could be used in foods. Where the food additive is to be used in foods for any special groups of consumers (e.g., diabetics, those on special medical diets, sick individuals on formulated liquid diets), the probable daily intake of the food additive by those consumers should be taken into account. The quantity of a selected additive should be at or below the maximum allowed level, and the lowest level necessary to achieve the intended technical effect is preferred. For the same food additive, the maximum allowed levels may differ in different countries. Table 13.2 summarizes the maximum limits of sweeteners used in the water-based drinks in different countries.

13.4.2 Justification for the Use of Food Additives

The use of food additives is justified only when such use has an advantage, and does not present an appreciable health risk to consumers. Their use is justified when they serve one or more of the technological functions set out by Codex and the needs set out below, and only where these objectives cannot be achieved by other means that are economically and technologically practicable. Food additives can be helpful to preserve the nutritional quality of the food, to provide necessary ingredients or constituents for foods manufactured for consumers with special dietary needs, to enhance the quality or stability of a food, or to improve its organoleptic properties, provided that these additives do not change the nature, substance, or quality of the food so as to deceive the consumer, and to provide aids in the manufacture,

TABLE 13.2

Maximum Limits of Sweeteners Used in Water-Based Drinks

	Maximum Limits in Water-Based Drinks				
	CAC	EU	USA	Japan	China
Acesulfame potassium	600 mg/kg	350 mg/L	500 mg/kg	500 mg/kg	300 mg/kg
Aspartame	600 mg/kg	600 mg/L	Not specified	Not specified	Not specified
Aspartame-acesulfame salt	—	350 mg/L	—	—	—
Cyclamates	350 mg/kg	350 mg/L	—	—	650 mg/kg
Erythritol	Not specified	—	—	—	Not specified
Xylitol	Not specified	Not specified	Not specified	Not specified	Not specified
Sorbitol	Not specified	Not specified	Not specified	Not specified	Not specified
Sodium saccharin	300 mg/kg	80 mg/L	300 mg/kg	300 mg/kg	150 mg/kg
Sucralose	300 mg/kg	300 mg/L	400 mg/kg	400 mg/kg	250 mg/kg

processing, preparation, treatment, packing, transport, or storage of food, provided that the additive is not used to disguise the effects of the use of faulty raw materials or of undesirable (including unhygienic) practices or techniques during the course of any of these activities.

13.4.3 GOOD MANUFACTURING PRACTICE

All food additives should be used under conditions of good manufacturing practice (GMP), which include the following: (a) the quantity of the additive added to food shall be limited to the lowest possible level necessary to accomplish its desired effect; (b) the quantity of the additive that becomes a component of food as a result of its use in the manufacturing, processing, or packaging of a food and which is not intended to accomplish any physical, or other technical effect in the food itself, is reduced to the extent reasonably possible amount; and (c) the additive is of appropriate food grade quality, and is prepared and handled in the same way as a food ingredient.

13.4.4 SPECIFICATIONS FOR THE IDENTITY AND PURITY OF FOOD ADDITIVES

Food additives are required to have appropriate food grade quality. In terms of safety, food grade quality is achieved by conformance of additives to their specifications as a whole (not merely with individual criteria) and through their production, storage, transportation, and handling in accordance with GMP.

13.5 SAFETY EVALUATION OF FOOD ADDITIVES

Safety is critical for food additives. Regardless of the differences of the possible toxicity and the dose of food additives, there is a correlation between the dose of food additives and their possible biological effects to human health. The substances have toxicity only above a certain concentration or dose level. As long as food additives are used at or below the maximum allowed level, the food generally shall be safe for consumption.

There are many technical parameters for the safety evaluation of food additives. Accepted daily intake (ADI) is an estimation by the Joint Expert Committee on Food Additives (JECFA) of the amount of a food additive, expressed on a per body weight basis that can be ingested daily over a lifetime without appreciable health risk. Theoretical maximum daily intake (TMDI) is the total daily intake of food based on the international or national maximum residue limit (MRL) value, indicating the theoretical limit of daily intake. For estimated daily intake (EDI), the actual amount of food, the added amount of additives from GMP, and the actual addition procedure during food production are taken into account. Median lethal dose (LD_{50}) is the dose required to kill half the members of a tested population after a specified test duration and is usually expressed as the mass of substance administered per unit mass of test subject, typically as milligrams of substance per kilogram of body mass. Generally recognized as safe (GRAS) exemptions are granted for substances that are generally recognized, among experts qualified by scientific training and experience to evaluate their safety, as having been adequately shown through scientific procedures (or, in the case of a substance used in food prior to January 1, 1958, through either scientific procedures or through experience based on common use in food) to be safe under the conditions of their intended use.

The scope and quantity of additives added to food are strictly regulated. These provisions must be established on the basis of a set of rigorous scientific toxicological evaluations.

13.5.1 Safety Evaluation of Food Additives by JECFA

The Joint Expert Committee on Food Additives (JECFA) is an international expert scientific committee that was administered jointly by the Food and Agriculture Organization of the United Nations (FAO) and the World Health Organization (WHO) in 1956. The Codex Alimentarius Commission (CAC), established by FAO and WHO in 1963, develops harmonized international food standards, guidelines, and codes of practice to protect consumer health and ensures fair practices in the food trade. The General Standard for Food Additives (GSFA) has the goal of being the sole text in Codex dealing with food additives and food colors. The Codex Committee on Food Additives (CCFA) establishes and endorses permitted maximum levels for individual food additives. The CCFA also prepares priority lists of food additives for risk assessment by JECFA on food additives as well as for assigning functional classes to individual food additives.

The safety evaluation of food additives has mainly two aspects: chemical and toxicological information. The general principles of JECFA safety evaluation of food additives include assessment principles and the case-by-case rule. The safety evaluation process is divided into two stages. The first is to collect evaluation data, whereas the second is to evaluate the data. The contents of the evaluation include chemical information evaluation (such as chemical information of petitionary food additives, effects of food ingredients, and quality specifications) and toxicological safety evaluation. Toxicological safety evaluation procedures include endpoints in experimental toxicity studies, the use of metabolic and pharmacokinetic studies in safety, and the influence of age, nutritional status, and health status on the design and interpretation of study results, use of human studies in safety evaluation, and setting the ADI value (WHO, 1987).

According to the results of the safety evaluation, food additives are divided into four categories by JECFA. The first category is GRAS and has no quantity limit. The second category is Class A, being divided into A1 and A2, with Class A1 being evaluated by JECFA, having clear toxicological information, and the formal accepted daily intake (ADI) value; and Class A2 with insufficient toxicological information and a temporary ADI value. The third category is Class B. The safety evaluation has been conducted by JECFA, but ADI value is lacking due to insufficient amount of toxicological data. The fourth category is Class C. The additive is not safe to be used in foods or is used only in specific scopes under strict control.

JECFA has a great authority on the safety assessment of food additives. The evaluation results are the important basis for CCFA to develop utilization standards, specifications, and test methods for food additives (Codex; 2013). The evaluation results are also widely accepted and used by many countries, industrial enterprises, and research centers, and serves as the basis for the management of food additives in many countries.

13.5.2 SAFETY EVALUATION OF FOOD ADDITIVES BY DIFFERENT COUNTRIES

Different countries have similar but slightly different regulation policies for food additives to ensure their safe uses (Magnuson, Munro, Abbot, et al., 2013). In the United States, it is regulated by the Food Additives Supplemental Bill approved by the Federal Food, Drug, and Cosmetic Act (FD&C Act,) in 1958 (FDA, 1993). In EU, EU Council directive 89/107/EEC is the framework for food additive regulation (European Parliament and Council, 1988). In Canada, food additives are divided into 15 categories in part B Division 16 of Food and Drug Regulations, and the maximum level of use, specifications, sales, import, and labeling requirements of colorants are also provided in Division 16 (Health Canada, 1997). Australia-New Zealand Food Standards Code is authorized for the implementation of the common food standards in Australia and New Zealand, in which food additives are distinguishable from processing aids (Standard 1.3.2) and vitamins and minerals (Standard 1.3.3) added to food for nutritional purposes. The Standard 1.3.1 about food additives regulates the use of food additives in the production and processing of food in Australia and New Zealand (Food Standards Australia New Zealand, 2003). In China, the safe use of food additives is regulated by the "Standards for Use of Food Additives" announced by Chinese Ministry of Health.

Producers were required to assure the safety of their food additives. To evaluate the safety of food additives, the applicants should provide the safety assessment data of food additives which include the following aspects: (1) Identity of substance: for chemically defined single substance, formal chemical name (common names, synonyms, or trade names), chemical abstract service (CAS) registry number, empirical and structural formulae, molecular or formula weights, and composition are needed. For mixtures, identifying as many of the components as feasible is recommended to reasonably define their compositions. In addition, information on the chemical composition and identity for each component in the mixture and a material balance should be provided. For food additives of natural origin, information on the source (e.g., systematic name, genus, species, variability based on climate or other geographical factors) is needed; (2) Data on chemical and physical properties of the food additive, that is, melting point, boiling point, specific gravity, refractive index, optical rotation, pH, solubility, and reactivity of the food additives are needed. Chromatographic, spectroscopic, or spectrometric data (e.g., spectra from nuclear magnetic resonance, infrared, electronic absorption, or mass spectra) should also be provided; (3) Manufacturing process, that is, method of manufacture (e.g., the source, the process by which the raw materials are converted into the finished product), production controls, and quality assurance should be included. For chemically synthesized substances, factors such as reaction sequence, side reactions, purification, and preparation of the product to be commercialized should be provided which may assist in determining likely impurities and their influence on toxicological evaluation. For substances extracted from natural sources, information on extraction procedure(s) is needed; (4) Specifications for identity and purity and analytical methods—a limit for lead should be proposed. In addition, limits for arsenic and heavy metals, such as cadmium and mercury, should be considered when their presence becomes a concern. Limits for any known natural toxicants or for microbial contaminants in a food additive derived from a natural source should be proposed when necessary. Limits for residual reactants, reaction by-products, and residual solvents should be proposed. In order to demonstrate the conformance with the proposed specifications, at least five batches of the food additive should be analyzed. Analytical methods for the determination of the substance and its degradation products (where relevant) in the foodstuff should be provided. If the analytical method is a common standard test (e.g., FCC or AOAC International method), only the reference needs to be provided. If the method is not common, or if a common method is applied to a new food additive, or if a modified standard method is used, a detailed description of the method and validation data for the method should be provided; (5) Reaction and fate in food—the stability and any degradation products or reaction products appearing as a result of processing, storage, and preparation of foods and any possible effect on nutrients should be provided; (6) Case of need and proposed uses—technological need, intended use, and benefit to consumer, the quantity to be added to specific foods (intended use levels or maximum use level) and the residues in food should be provided. The quantity of the food additive shall not exceed the amount reasonably required to accomplish its intended technical effect(s) in the food; (7) Exposure assessment—information should be provided on known or anticipated human exposure to the proposed additive from food, including amount (e.g., maximum and average intake or exposure), frequency, and other factors influencing exposure. Information should also be given on any other sources of human exposure to the same substance (e.g., from drinking water, consumer products, etc.). The above exposure calculations should be explained, including any assumptions made. Where possible, information should be provided on consumption of the foods in which the additives are used or intended to be used, including variations affecting particular sections of the population (e.g., by age, sex, disease, etc); (8) Toxicological information—toxicological data of different food additives are obtained by toxicological testing which is to determine whether the substance, when used in the manner and in the quantities proposed, would pose any appreciable risk to the consumer health. Such testing should provide not only information relevant to the ordinary consumer, but also relevant to those population groups whose pattern of food consumption, physiological or health status may make them vulnerable, e.g., young age, pregnancy, diabetes, and so on. In addition to laboratory tests, it may be possible to use human data derived from medical use, occupational epidemiology, or specific studies on volunteers (e.g., on absorption and metabolism) or on critically exposed groups. In the United States, toxicology test methods are the recommended test methods in FDA food additive toxicological safety assessment principles and the "Redbook 2000." In Europe, it is recommended that Organization for Economic Cooperation and Development (OECD) protocols be used. To ensure mutual recognition by Member States of the data submitted, studies should be carried out according to the principles of Good Laboratory Practice (GLP) described in Council Directive 87/18/EEC10 and accompanied by a statement of GLP compliance. In recent years, there has been considerable development of new toxicological methods not based on the use of animals, which have become known as "alternative methods" (Scientific Committee on Food, 2001). The toxicological evaluation test report in China includes the experimental results in four stages: acute toxicity test—LD50, joint acute toxicity, genotoxicity tests, traditional teratogenicity test, and short-term feeding study; subchronic toxicity test—90-day feeding study, reproduction experiments, metabolic testing, and chronic toxicity test (including carcinogenicity test).

China's Food Safety and Toxicology Evaluation Procedures under different conditions may include selected several or all the four stages of the test. The principle of selective toxicity test for food additives includes the following: (1) Spices: the spices that have been recommended for use or the tolerable amount that has been approved by World Health Organization (WHO), or have been approved for use in food by two of the four international organizations, that is, the Flavor and Extract Manufacturers Association (FEMA), the Council of Europe (COE), and International Organization of Fragrance Industry (IOFI), can be evaluated according to the foreign information or requirements. The spices that have been approved by only one international organization should be evaluated by the acute toxicity test or mutation test, and whether the further evaluation tests are needed is determined according to the results of the first evaluation. The spices with no information available, and were not approved for food utilization by any international organizations mentioned above should undergo the first and second stages of toxicity tests. The spices extracted from the edible parts of plants or animals without any information indicating the unsafety are generally not required to undergo toxicity testing. (2) Other food additives: the food additives with the ADI from WHO are required to undergo the acute toxicity tests and two mutagenicity test, whereas the Ames and bone marrow cell micronucleus tests are recommended. The food additives produced by different processes with different purity should undergo the first and second stages of toxicity evaluation, and whether the further testing is needed is determined according to the test results. For the food additives extracted from plants, animals, or microorganisms, the new varieties are required to undergo the first, second, and third stage of toxicity tests; the varieties approved by one of the international organizations or countries are required to undergo the first and second stage of toxicity tests, and whether the further testing is needed is determined by the evaluation results. (3) The toxicological information of the imported food additives should be provided by the importing country, and is examined by the unit designated by the Sanitary Administrative Organization for further toxicity testing.

In addition, food additives should meet three conditions: should be used to achieve a technical effect in food, not misleading consumers, and not adversely affecting the health conditions of users.

13.6   RISK ASSESSMENT OF FOOD ADDITIVES

Risk analysis principles have been widely applied in regulating food additive uses. "Risk analysis principles applied by the Committee on Food Additives and Food Codex Committee on Contaminants" were formulated by the Codex Committee on Food Additives (CCFA) and Codex Committee on Contaminants in Foods (CCCF). JECFA is in charge of the risk assessment of food additives, and has conducted risk assessment for more than 1500 kinds of food additives. The principles and methods of risk assessment of food additives are formulated in the EU, the United States, Japan, Canada, and other countries, and are taken charge of by authorized agencies.

Risk assessment should be based on scientific data most relevant to the national context. It should use available quantitative information to the greatest extent possible and may also take into account qualitative information. Risk assessment should have the four essential components, that is, hazard identification, hazard characterization, intake assessment, and risk characterization (Codex, 2007).

13.6.1   Hazard Identification

Hazard identification of food additives is carried out to determine the potential adverse effects of food additives on human health. The common methods of hazard identification include epidemiological studies, animal toxicology studies, in vitro tests, final quantitative structure–response relationship, and so on. The most common method used in hazard identification of food additives is animal toxicological studies, which can identify potential adverse effects of food additives, determine the necessary exposure conditions and the dose–effect relationship behind the effects, and determine the No Observed Adverse Effect Level (NOAEL). NOAEL is an exposure level at which there is no statistically or biologically significant increases in the frequency or severity of adverse effects between the exposed population and its appropriate control. Animal toxicology studies may provide experimental data for risk assessment. The data may be extrapolated to the subpopulations and have a very important role in the identification of the hazards of food additives.

13.6.2   Hazard Characterization

Hazard characterization of food additives is an essentially qualitative and quantitative analysis of the adverse effects of food additives. The common method of hazard characterization is the threshold method, which is the ratio of NOAEL from the toxicology tests and the safety level or ADI.

When ADI is calculated based on the results of animal tests, a factor of 100 is traditionally used as a safety coefficient by JECFA. ADI value of food additives is obtained through dividing NOAEL by 100. The safety coefficient of 100 is based on the assumption that humans were 10 times more sensitive than experimental animals and that the sensitivity difference between races was 10 times. However, the safety coefficient of 100 is not fixed, and when setting ADI value, different tests should be considered. In the case of insufficient data, the safety coefficient should be greater. If the evaluated food additives are similar to traditional food and metabolized into normal body compositions and have no toxicity, a lower safety coefficient may be used. For the nutrients that can meet normal nutritional needs and maintain human health, safety coefficient of 100 is not enough to provide an adequate quantity to meet the normal need.

ADI value is commonly indicated in the form of interval from 0 to the upper limit, which means the acceptable range of the evaluated substances, emphasizing the upper limit and encouraging the use of the lowest value under the premise of reaching technological feasibility. When the estimated intake of food additives is far lower than its assigned numerical ADI, ADI cannot be prescribed by JECFA. Based on available information, the total daily intake of the evaluated food additives including necessary doses for the purpose of the expected technology and the background content in food should not cause health hazards. Therefore, there is no need to establish a numerical ADI, and food additives satisfying this condition must be used under conditions of GMP.

13.6.3 INTAKE ASSESSMENT

The intake assessment of food additives is a combined qualitative and quantitative evaluation according to the possible intake of food additives from food and other sources of exposure. The intake assessment of food additives is the qualitative and quantitative assessment mainly based on dietary surveys and qualitative levels of additives in food, which can be carried out based on the comprehensive analysis of food consumption index and the contents of food additives.

The intake assessment of food additives can, generally, be divided into assessment before approval and assessment after approval for food applications. For the first case, the content data are mainly from the maximum doses suggested by applicants or food producers. For the latter case, the content data can be obtained from the report of producers, food industry surveys, supervision and monitoring, total diet studies, and scientific literatures. The data of food consumption are mainly from the national food consumption surveys.

At present, a stepwise intake assessment framework is used when conducting intake assessment on chemicals including food additives in food with a number of assessment methods. The initial step of the assessment framework is often to choose substances for evaluation by conservative screening methods. If the conservatively estimated intake amount of the substance exceeds its ADI, the more accurate deterministic assessment and the more accurate model assessment shall be used.

13.6.4 RISK CHARACTERIZATION

As the fourth step of risk assessment, the characterization is to provide scientific foundation based on the comprehensive review of hazard identification, hazard characterization, and intake assessment, and to qualitatively or quantitatively assess the possibility and severity of adverse health effects of a selected food additive under certain conditions, including the uncertainty of every step in the risk assessment. In the risk characterization of food additives, human intake of certain food additives for the intake assessment is compared with the ADI value obtained from the hazard characterization. If the calculated dietary exposure amount was higher than the ADI value of the food additive, specific analysis will be required.

13.7 MANAGEMENT AND RISK CONTROL FOR FOOD ADDITIVES

At present, food additives are used across the world. Supervision, administration, and safety evaluation have been established to standardize the production, business operation, and the use of food additives. Magnuson et al. (2013) compared the regulations, definitions, and approval processes for substances intentionally added to or unintentionally present in human food in various countries including but not being limited to Argentina, Australia, Brazil, Canada, China, the European Union, Japan, Mexico, New Zealand, and the United States.

13.7.1 UNITED STATES OF AMERICA

The United States is a major producer and user of food additives and the output value and types of food additives are at the top of the list in the world. For production, sale, and use of food additives, the United States has established a strict regulation system.

The FDA, affiliated with the American Ministry of Health, is responsible for the regulation of food additives according to the Federal Food, Drug and Cosmetic Act (FD&C). As meat products are regulated by the United States Department of Agriculture (USDA), food additives for meat and poultry products should be certified by both FDA and USDA. As wine and tobacco are regulated by Taxes on Alcohol and Tobacco and Trade Bureau (TTB), food additives used in wine and tobacco products are also regulated by these two agencies. The United States divides the pigment from food additives using a separated regulation mechanism. An Act about pigments (FD&C pigments complement Act) passed by the U.S. Congress in 1960 requires that the premarket pigments used in food and other fields must be approved by the FDA. The basic work of legislation on food additives is often assumed by the associations. For example, the legislation of food flavor is undertaken by the FEMA. If its safety evaluation results are approved by the FDA, it will be graded as GRAS (FEMA, 2011). With the progress of science and technology, accumulation of toxicological data, as well as the improvement of modern analytical techniques, the safety of food additives are re-evaluated and made aware to the public every few years. Paragraph 402 of the U.S. Food and Drug Administration Law requires that food additives can be produced and used in food products only after evaluation and approval; otherwise, it will be affirmed to be unsafe. The food containing unsafe food additives cannot be used for human consumption, and it is prohibited for marketing.

Chemical substances added to food are divided into four categories by FDA. Category (1) food additives: it needs two or more kinds of animal experiments to verify that there is no toxicity and adverse effect on fertility, and does not cause cancer. The quantity for food use is over 1% maximum no-effect level; Category (2) GRAS, substances, such as sugar, salt and spices, and so on can get into the GRAS list recognized by FDA without animal tests. The substances will be removed from the GRAS list if adverse effects are found; Category (3) substances are needed to be reapproved, as long as it is proved unsafe by new test data, the substance is required to confirm the safety again; and Category (4) food colorant, comprehensive safety tests are needed for premarket food colorants. The United States requires that food quality must comply with the provisions of the U.S. Food Chemical Codex (FCC). The Codex in the United States has a "quasi-legal" status, which is an important basis for the FDA to evaluate whether the quality of food additives meets the standard. The first edition of FCC was published in 1966. Before this, the FDA announced safety requirements for food additives by regulations and informal statements. Since the advent of the FCC, it has 5 versions after additions and amendments, and the latest version (VI) was officially launched in 2012. As an authoritative standard in food additive industry, FCC has been widely recognized worldwide. Many food chemical manufacturers, vendors, and users have regarded it as their basis for sale or purchase contracts.

In addition to the FD&C, the administrative regulations on the food additives are incorporated in the Volume 21 of the U.S. Code of Federal Regulations (CFR). The United States revises each volume in the CFR every year and the 21st volume is published on the April 1 of each year. The parts 70–74 and 80–82 in the latest version CFR (April, 2005) are about regulations on pigment. The parts 170–186 are about other food additives, including the terms about general principles, packaging, labeling, and safety assessment.

13.7.2 European Union

The European Union has specialized agencies and special regulations to monitor food additives. DGSANGO is responsible for the regulation of food additives in the EU and mainly for the approval of food additives. EU Scientific Committee on Food (SCF) is mainly responsible for the safety assessment of food additives. If a food additive can pass the assessment, the Commission will start the regulatory amendment process to add it to the appropriate instruction and allow its food utilization. EU legislation on food additives adopts a "hybrid system." Based on the scientific evaluation and consultation, the EU develops food additive regulations that can be accepted by all member nations, and finally announces a list of food additives allowed to be used, the specific requirements, and the quantity limit of usage.

EU Council directive 89/107/EC is the regulation framework for food additives, and it requires that the use of food additives undergo a safety assessment of the SCF. Specific implementation measures of the framework include directive 94/35/EC about the use of sweetener, directive 94/36/EC about the pigment, directive 95/2/

EC on all other additives, and an amending directive. The principle of food additives used in the EU is that food should contain only food additives allowed by the EU and flavors allowed by the member nations. In other words, the use of food additives must comply with the requirements of the relevant provisions of the EU and the general hygiene regulations.

With the development of food industry and indepth research, the EU continues to revise and update the safety standards or regulations on food additives. On January 28, 2002, the European Parliament and the Council Regulation 178/2002 was published. "Regulation 178/2002" means Regulation (EC) No. 178/2002 of the European Parliament and of the Council of January 28, 2002 as amended by Regulation (EC) No. 1642/2003 of the European Parliament and of the Council of July 22, 2003, Commission Regulation (EC) No. 575/2006 of April 7, 2006, Commission Regulation (EC) No. 202/2008 of March 4, 2008, Regulation (EC) No. 596/2009 of the European Parliament and of the Council of June 18, 2009, and Commission Implementing Regulation (EU) No. 931/2011 of September 19, 2011. New food additive regulation EC 1333/2008 was enacted on January 20, 2009. This regulation harmonizes the use of food additives in the community. This includes the use of food additives covered by the Council Directive 89/398/EEC of May 3, 1989 on the approximation of the laws of the Member States relating to foodstuffs intended for particular nutritional uses and the utilization of certain food colors for the health marking of meat and the decoration and stamping of eggs. It also harmonizes the use of food additives in food additive formulations and food enzymes, thus ensuring their safety and quality and storage stability. This has not previously been regulated at the community level (EU, 2008). It is the most important food law in EU and food additive is one of the key areas of its focus. The new law provides important guidance principle for the protection of the quality and safety of food additives in the EU.

In recent years, the European Commission has strengthened the safety of food additives. The new law clearly classifies food additives in the European Common Market and that food additives should be stipulated in the positive list. The new positive list of food additives was made available to the public on April 22, 2013, and food additives not in the positive list will be completely banned in 18 months. The new positive list contains 2100 legitimate additives and the other 400 ones would continue to be available in the market before the review of the European Food Safety Authority.

13.7.3 JAPAN

The Ministry of Health and Welfare in Japan announced the Health Act in 1947, which was a recognition system for chemicals in food, and the food additive regulations were published in 1957. In Japan, food additives are substances used during food processing and preservation by adding, mixing, infiltration, or other methods. Here, food additives are divided into four categories: designated additives, existing additives, natural flavors, and common additives (Japan, 2011). The first official edition of food additives published in 1966 is the standard document for Japanese food additives, formulating regulations about the categories, quality standards, and quantity limit for food additives. With scientific and technological progress and the development of food industry, it has been amended several times, and the latest Japanese official version (Version 7) was launched in September 2000. Food Sanitation Law was revised in 2004 and provided more stringent regulation for food additives. The enforcement provisions and standards for food additives in Food Sanitation Law were revised in 2006.

13.7.4 China

13.7.4.1 Legal and Regulatory System for Food Additives in China

The new "Food Safety Law" and its implementation require improving the regulatory system for food additives actively according to "Notice on strengthening the supervision and regulation of food additives" (Health Authority [2009] No. 89) and "Emergency notice on enhancing the food seasonings and food additives supervision and regulation" (Health Authority [2011] No. 5). In addition, "Procedures and methods on food toxicology safety evaluation" (GB 15193), "Standards for Use of Food Additives" (GB2760), and "Standard for Use of Food Nutrition Fortifiers" (GB 14880) have been developed for safety evaluation and standards.

On November 25, 2011, the Chinese Ministry of Health issued the notice on regulating new food additive licenses, which provides specific requirements, including the necessity of technical materials, requirements of quality specifications of food additives, onsite audit, verifying the product standards of the new food additives, and the contents of this announcement on new varieties of food additives.

"Food safety supervision and management approach on catering," "Food safety operation specification on catering," "Food safety supervision and sampling norms on catering," and "Food safety responsibility interview system" have been established by Chinese government for the food catering industries. In addition, "Food additives management specification on catering" and the catalog of food additives allowed to be used in the catering industry are being developed by the Chinese Ministry of Health and the State Food and Drug Administration to strictly regulate the use of food additives in catering foods.

13.7.4.2 Regulatory System in Food Additives Industry in China

China began to regulate food additives from the 1950s, and strengthened production regulation and quality supervision of food additives in the 1960s. The Food Safety Law implemented on June 1, 2009 established the regulation system for supervision and regulation of food additives and set clear statement about responsibilities for related national departments. The Ministry of Health is responsible for the safety evaluation of food additives and the development of national food safety standards. The General Administration of Quality Supervision is responsible for the production of food additives and the supervision of food additives used by food enterprises; Business Sector is responsible for strengthening quality control of food additives in the circulation process. The Chinese Food and Drug Administration is responsible for the regulation of food additives in the catering industry. The Department of Agriculture is responsible for the supervision of production processes of agricultural products. The Department of Business is responsible for the supervision of pig slaughtering. The Department of Industry and Information is responsible for regulating food additive industry, and the development of industrial policies and guidelines to establish the production enterprise credit system.

13.7.4.3 Standards for Food Additive Industry in China

The primary standards about food additives in China include standards for use, product standards, and testing methods. The Standards for Use of Food Additives (GB2760) and the Standards for Use of Food Nutrition Fortifiers (GB14880) are the basic standards for food additive use in China. The Standards for Food Additive Use (GB2760) provides the definition, scope, category allowed to be used, as well as quantities and principles for food additive use. GB2760 also requires that food additives should not be used to cover the quality defects of the food itself or food produced in food processing, or for the purpose of adulterating, counterfeiting, and faking. GB2760-2011 includes 2310 food additives, among which there are 59 processing aids, 1826 food-flavoring agents, 35 food gums, 51 food enzyme preparations, and other 339 food additives. The Standards for Use of Food Nutrition Fortifier (GB14880) provides definition, scope, the amount of food nutrition fortifier, and so on. Currently, there are about 200 food nutritional fortifiers allowed for food use.

The food additive standards may include, but are not limited to, characteristics, specifications, test methods, inspection rules, mark, label, package, storage, transportation, and so on. According to the preliminary statistics, there are totally 272 food additive standards (not including flavors and nutrition fortifiers) in China, including 84 national food safety standards, 91 national standards, 37 industrial standards, and 60 specified standards. There are totally 39 nutritional fortifier standards, including 12 national food safety standards, 8 national standards, and 19 specified criteria; there are a total of 160 flavoring standards (including the flavoring agent standard and fragrance standard), including 2 national food safety standards, 27 national standards, 97 industry standards, 34 specified standards, and 9 standards for other flavors and fragrance.

The detection method standards of food additives include the dedicated food additive testing standards and other food safety testing methods. According to the incomplete statistics, there are 81 nationally recommended standards (GB/T), 19 inspection and quarantine industry standards (SN/T), 5 Ministry of Agriculture recommendation standards (NY/T), and 1 commercial recommendation standards (SB/T). The standards for food additive testing methods are insufficiency, which are mainly related to preservatives, coloring agents, sweetening agents, and antioxidants.

13.8   CONCLUSION

Food additives are substances added intentionally to foodstuffs to perform certain functions, for example, to color, to sweeten, to better preserve foods, or to improve nutrition quality. The research and application of food additives are one of the important indices indicating the development level of national food science and technology, as well as economy. There would be no modern food industry without food additives. The scope and quantity of the additives added to food are strictly regulated on the basis of rigorous scientific toxicological evaluations. Following the guidelines is necessary for food safety.

REFERENCES

Aluyor, E.O., Ori-Jesu, M. 2008. The use of antioxidants in vegetable oils—A review. *African Journal of Biotechnology* 7, 4836–4842.

Billing, J., Sherman, P.W. 1998. Antimicrobial functions of spices: Why some like it hot. *Quarterly Review of Biology* 73, 3–49.

Brula, S., Cooteb, P. 1999. Preservative agents in foods—Mode of action and microbial resistance mechanisms. *International Journal of Food Microbiology* 50, 1–17.

Busteed, L.K., Vogel, E.J., Weiss, K.E., Borja, M.E. 2004. Sensory acceptance of sugar substitutes by patients with diabetes mellitus. *Journal of the American Dietetic Association* 104, Suppl. 2, 20.

Caballero, B. 2003. Fortification, supplementation, and nutrient balance. *European Journal of Clinical Nutrition* 57, S76–S78.

Ceulemans, G., Aerschot, A.V., Rozenski, J., Herdewijn, P. 1997. Oligonucleotides with 3-hydroxy-N-acetylprolinol as sugar substitute. *Tetrahedron* 53, 14957–14974.

Codex Alimentarius Commission Procedural Manual, twenty-first edition; 2013.

Codex Alimentarius Commission Procedural Manual, 2011.

Codex Alimentarius Commission. Working principles for risk analysis for food safety for application by governments cac/gl 62-2007, 2007.

Codex General Standard for Food Additives, Codex Stan 192–1995.

Environmental Health Criteria 70: Principles for the Safety Assessment of Food Additives and Contaminants in Food. Geneva: World Health Organization, 1987.

European Parliament and Council. Council directive on the approximation of the laws of the member states concerning food additives authorized for use in foodstuffs intended for human consumption, 1988.

European Union: Regulation (EC) no. 1333/2008 of the European Parliament and of the Council of 16 December 2008 on Food Additive, 2008.

FDA. Recommendations for submission of chemical and technical data for direct food additive and GRAS food ingredient petitions. FDA, 1993.

Federal Food, Drug, and Cosmetic Act. 1958. Section 201(s). U.S. Government Printing Office, Washington, DC.

FEMA. FFEMA GRAS Flavoring Substance [EB/OL], 2011.

Flynn, A., Hirvonen, T., Mensink, G.B.M., Ocké, M. C., Serra-Majem, L., Stos, K., Szponar, L., Tetens, I., Turrini, A., Fletcher, R., Wildemann, T. 2009. Intake of selected nutrients from foods, from fortification and from supplements in various European countries. *Food and Nutrition Research* 53, Suppl. 1, 1–51.

Food Safety Law of the People's Republic of China, 2009.

Food Standards Australia New Zealand. Food standard code, 2003.

Health Authority of the People's Republic of China. Regulation no 89/2009. Notice on strengthening the supervision and regulation of food additives. 2009.

Health Authority of the People's Republic of China. Regulation no 5/2011. Emergency notice on enhancing the food seasonings and food additives supervision and regulation. 2011.

Health Canada. Food and Drug Regulations, 1997.

Jastad, L.J., Wheatley, C.J., Gee, D.L., Bennett, V.A. 2001. Two different sugar substitutes produce varied acceptability in white cake. *Journal of the American Dietetic Association* 101, Suppl. 1, A-23.

Magnuson, B., Munro, I., Abbot, P., Baldwin, N., Lopez-Garcia, R., Ly, K., McGirr, L., Roberts, A., Socolovsky, S. 2013. Review of the regulation and safety assessment of food substances in various countries and jurisdictions. *Food Additives and Contaminants: Part A* 30, 1147–1220.

Malone, M.E., Appelqvist, I.A.M., Norton, I.T. 2003. Oral behaviour of food hydrocolloids and emulsions. Part 1. Lubrication and deposition considerations. *Food Hydrocolloids* 17, 763–773.

Moon, M. 2007. How clean are your lubricants? *Trends in Food Science and Technology* 18, S74–S88.

Regulation (EC) No. 1333/2008 of the European Parliament and of the Council of 16 December 2008 on food additives, 2008.

Rubalya, V.S., Neelamegam, P. 2012. Antioxidant potential in vegetable oil. *Research Journal of Chemistry and Environment* 16, 87–94.

Scientific Committee on Food. Guidance on submissions for food additive evaluations by the scientific committee on food evaluations, 2001.

Shee, A.K., Raja, R.B., Sethi, D., Kunhambu, A., Arunachalam, K.D. 2010. Studies on the antibacterial activity potential of commonly used food preservatives. *International Journal of Engineering Science and Technology* 2, 264–269.

The Japan Food Chemical Research Foundation: Food additives http://www.ffcr or jp/zaidan/ ffcrhome nsf/pages/e-foodadditives, 2011.

14 Chemistry and Safety of Melamine and Its Analogs

Jeffrey C. Moore

CONTENTS

14.1 INTRODUCTION

Melamine is a simple, nitrogen-rich triazine compound with no approved uses for direct addition to foods. It became a major global food safety concern following the tandem 2007 and 2008 melamine adulteration scandals involving the intentional, economic adulteration of animal and human food supply chains with melamine and its analogs. Intentional adulteration of food ingredients (often referred to as economically motivated adulteration) is the "the fraudulent addition of nonauthentic substances or removal or replacement of authentic substances without the purchaser's knowledge for economic gain of the seller" (DeVries 2009).

The 2007 incident involved the addition of low-purity melamine-contaminated melamine analogs such as cyanuric acid to animal feed ingredients by Chinese ingredient processors. The addition was carried out intentionally to artificially inflate the total protein contents of the ingredients when measured by the Kjeldahl or Dumas methods for economic gain. The adulteration incident resulted in more than 39,000 companion animals (cats and dogs) developing acute renal failure in North America with an estimated 2000–7000 deaths in North America and South Africa combined (Dalal and Goldfarb 2011).

The later 2008 incident in China involved intentional economic adulteration of milk used to produce infant formula with melamine. This tragic event caused illness in more than 300,000 Chinese infants and young children and six reported deaths, involving kidney stones and renal failure (Gossner et al. 2009). The combination of events also significantly impacted international trade and agribusiness with the importation of many protein-rich ingredients and foods banned by national authorities around the world and the collapse of the Chinese dairy industry.

Although not considered as acute toxins, melamine and its analogs exhibit unique chemical properties that can lead to significant toxicological effects to the urinary system of animals and humans. These effects stem from the ability of melamine to form supramolecular complexes through hydrogen-bonded networks that self-assemble stones or crystals in the kidney. The tragic melamine events of 2007 and 2008 led to an international response and collaboration among food safety authorities worldwide to conduct appropriate risk assessment and risk management steps. This began with a better characterization of the food safety chemistry of melamine and its related compounds including an understanding of its chemistry and toxicological effects. This research has resulted in hundreds of scientific studies and published peer-reviewed manuscripts in support of a growing body of knowledge on the food safety risks and analytical detection methods for melamine.

This chapter provides an overview of risk assessments that have been carried out to date, based on an accumulating body of knowledge on the chemistry, toxicology, exposure, and occurrence of melamine and its analogs in the food supply. It also reports on the latest recognized maximum limits for melamine and its analogs in food and feed ingredients necessary to protect human health, including a detailed explanation for the derivation of the limits. Finally, it reports on the current state of established methods to detect melamine and its related compounds in food and feed, and future research directions aimed at developing new analytical technologies to detect and prevent future "melamine-like" compounds from entering human and animal food supplies.

14.2   CHEMISTRY AND TOXICOLOGY

14.2.1   Chemistry of Melamine and Its Related Compounds

Melamine (Figure 14.1) is a simple, synthetic, nitrogen-rich trizaine compound with no approved uses for direct addition to foods. It is a highly produced industrial

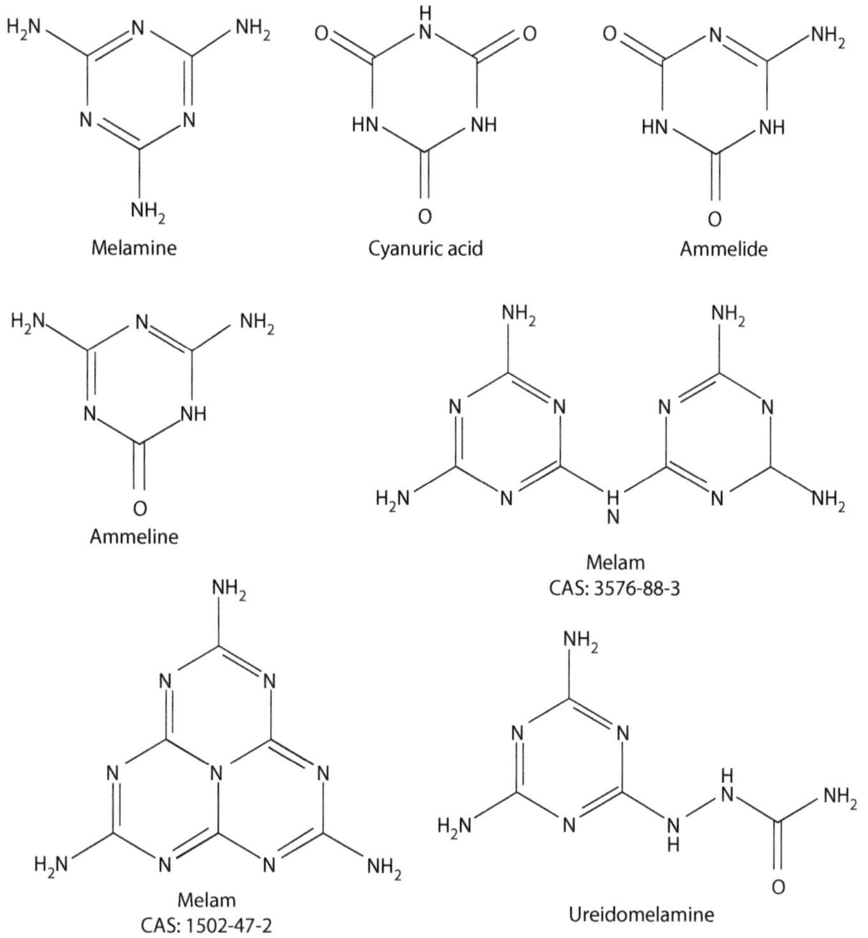

FIGURE 14.1 Structures for melamine and its related compounds and impurities.

chemical with more than 1.2 million tons produced in 2007 (OECD 2004; WHO 2009). It is primarily used in the synthesis of melamine–formaldehyde resins for the production of tableware, plastic coatings, glues, adhesives, and laminates. It also has minor uses as a fertilizer and colorant (WHO 2009). The indirect addition of melamine to foods by migration from food-contact materials made of melamine–formaldehyde plastics has been regulated in some countries since 1986. The European Union, for example, established a 30 mg/kg specific migration limit for melamine in the early 1980s (SCF 1986).

A number of related compounds and analogs of melamine are important to understand the food-safety chemistry of this compound. Cyanuric acid (Figure 14.1) is an oxytriazine melamine analog used primarily as a reactant in the production of trichlortriazine, a swimming pool chlorination chemical, and is also used to stabilize chlorine in swimming pools. As a disinfecting agent, derivatives of cyanuric acid are approved for use in the United States on food-contact surfaces (21 CFR Section 178.1010). Although it is also not approved for direct addition to foods, it is an allowed impurity found in feed-grade biuret, a feed additive for ruminants. Two other related compounds, ammelide and ammeline (Figure 14.1), are monoamino- and diaminooxytriazine analogs, respectively, of melamine and are by-products of melamine synthesis. Ammeline has a reported use in lubricating greases, while ammelide has no reported commercial uses (Tolleson et al. 2009).

The manufacturing chemistry for melamine has been recently reviewed by Tolleson et al. (2009). Melamine can be manufactured from at least three differ-ent starting materials: urea, dicyandiamide, or hydrogen cyanide. Production from urea is reported as the primary production route for commercial materials. Several reaction conditions have been reported for melamine production, all using heat in combination with either high pressure or a catalyst such as modified aluminum oxide or aluminosilicate. The net reaction for both production routes from urea is shown in Figure 14.2. It involves several different intermediates depending on the process used, including isocyanuric acid, cyanuric acid (high-pressure route), and carbodi-imide or cyanamide (catalytic route). The products of the reactions are quenched with water or mother liquor and the reaction yields greater than 95% when using recycling of by-products. Typical purification steps (filtration, crystallization, or cen-trifugation) are used to increase purity further.

Reported impurities in commercial melamine include melam, melem, ammeline, ammelide, cyanuric acid, and ureidomelamine, each with reported levels less than 0.2% (see Figure 14.1). Analysis of melamine-contaminated raw materials used to adulterate infant formula in the 2008 Chinese incident detected median levels of melamine and related compounds as follows: melamine (188,000 mg/kg), cyanuric acid (3.2 mg/kg), ammeline (14.9 mg/kg), and ammelide (293 mg/kg) (WHO 2009). This indicates that a relatively pure, commercial-grade melamine was likely used in the 2008 infant formula incident (WHO 2009). In contrast, analysis of melamine-adulterated pet food samples from the 2007 incident in North America suggests that a lower purity grade of melamine, "melamine scrap," was used to adulterate raw materials in these incidents (WHO 2009). Reports by Dobson et al. (2008) support this conclusion, with concentrations of melamine and related compounds in contami-nated wheat gluten as follows: melamine (84,000 mg/kg), cyanuric acid (53,000 mg/ kg), ammeline (17,000 mg/kg), and ammelide (23,000 mg/kg). It has been hypoth-esized that the "melamine scrap" used in the pet food incident was a recovery prod-uct of melamine production obtained from the mother liquor water stream after the

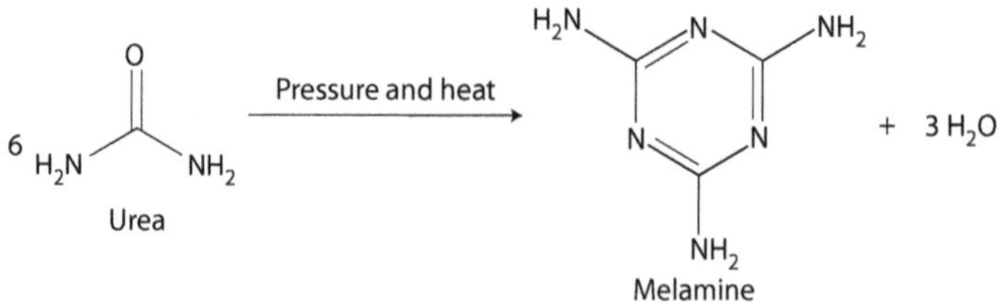

FIGURE 14.2 Net reaction for production of melmaine from urea.

TABLE 14.1

Chemical and Physical Properties of Melamine and Related Compounds

	Melamine	Cyanuric Acid	Ammelide	Ammeline
Chemical formula	$C_3H_6N_6$	$C_3H_3N_3O_3$	$C_3H_4N_4O_2$	$C_3H_5N_5O$
%Nitrogen	66.6	32.6	43.7	55.1
Molecular weights (g/mol)	126.12	129.07	128.09	127.10
Melting point (°C)	345–347[a]	360[a]	Decomposes[a]	Decomposes[a]
Water solubility (mg/L)	3924 at 25°C[a]	2000[b]	76.9[b]	75[b]
pK_a	5.0[b]	6.5[b]		9.65 (40°C)[c]

[a] Chapman et al. (1943).
[b] Ma and Bong (2011).
[c] Tolleson et al. (2009).

recrystallization step was used to increase the purity of melamine (Tolleson et al. 2009). This is supported by a previous 1998 study from Ono and others who analyzed such melamine recovery products and found them to contain significant levels of melamine impurities listed in Figure 14.1, including ammeline, ammelide, and cyanuric acid (Ono et al. 1998; WHO 2009).

An overview of the physical and chemical properties of melamine and its related compounds is listed in Table 14.1. Melamine and its s-triazine ring are very stable, with the ring cleaving only under extreme conditions, such as heating above 600°C. Melamine and cyanuric acid can be hydrolyzed in aqueous solutions to ammonia and carbon dioxide, but only under strongly acidic or basic conditions, or in the presence of alumina catalysts (Zhan et al. 1996). It can also be oxidized stepwise to ammeline, ammelide, and finally to cyanuric acid in the presence of UV light, H_2O_2, and TiO_2 (Bozzi et al. 2004). Both melamine and cyanuric acid are slightly soluble in water, while ammeline and ammelide are practically insoluble. An equation for calculating the aqueous solubility of melamine was reported as

$$\log L = 5.101 - 1642/T$$

with L expressed as mg melamine/100 g water, and T expressed in K (Chapman et al. 1943). Melamine levels in infant formula as high as 4700 mg/kg were reported by Chinese authorities from surveillance studies (Wu et al. 2009), and other studies reported levels as high as 6196 mg/kg in powdered milk products (Gossner et al. 2009). Given the low aqueous solubility of melamine at ambient temperatures and the likely ambient temperature of raw milk when melamine was added to it in adulteration scenarios, it is unlikely that the high levels reported in finished products would be possible without the addition of other additives or thermal processing steps to enhance the solubility of melamine. News reports have suggested that maltodextrin or other emulsifiers were used to enhance the poor solubility of melamine in milk during the 2008 China incident, but no verification of these claims is reported in scientific literatures (Fairclough 2008; Tolleson et al. 2009; Xin and Stone 2008).

Other news reports have suggested that a combination of heat and citric acid was used to overcome the solubility issue (Jia 2008).

Tolleson et al. (2009) recently reviewed the capabilities of triazine compounds to form high-molecular weight, self-assembling complexes through hydrogen bonds and π–π bond aromatic ring stacking. For melamine, this property has been utilized to create molecular scaffolding in the supramolecular chemistry field to produce a variety of micro- and nanoscaled complexes. Melamine is subjected to tautomeric equilibration, but under acidic to neutral conditions expected in biological solutions, the preferred resonance structure is the enamine form. Under biological pH conditions where melamine is in an unionized form, each monomer of melamine has three pairs of unshared electrons that can act as hydrogen bond acceptors (A) from sp2-hybridized nitrogen atoms, and each exocyclic primary amine can provide a pair of hydrogen bond donors (D, Figure 14.3). The donor–acceptor–donor (D–A–D) arrangement for melamine is capable of forming complementary intermolecular bonds with other compounds containing hydrogen bond acceptor–donor–acceptor (A–D–A) arrangements. Cyanuric acid under acidic to neutral conditions prefers the keto form providing a possible A–D–A arrangement. Within the cyanuric acid triazine ring, sp3-hybridized nitrogen atoms act as donors and the six pairs of unshared electrons from the carbonyl oxygen atoms act as hydrogen bond acceptors (Figure 14.3). With similar molecular spacing as melamine, cyanuric acid can interact with melamine to form a melamine–cyanurate hydrogen-bonded complex (Figure 14.4). A recent study (Prior et al. 2013) has characterized the crystalline structure of a melamine–cyanurate complex and reported that melamine and cyanuric acid are arranged in an ordered array with a much larger unit volume than previously reported. Another recent investigation on the assembly of melamine and cyanuric acid in aqueous solutions has suggested that it occurs as an exothermic, proton-transfer-coupled process, as illustrated in Figure 14.5 (Ma and Bong 2011).

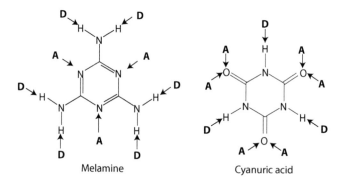

Melamine   Cyanuric acid

FIGURE 14.3 Melamine and cyanuric acid hydrogen bonding under acidic to neutral conditions. **A** stands for hydrogen bond acceptor. **D** stands for hydrogen bond donor. (Redrawn from Tolleson et al. 2009. Background paper on the chemistry of melamine alone and in combination with related compounds. *Prepared for the WHO Meeting on Toxicological and Health Aspects of Melamine and Cyanuric Acid, in collaboration with FAO and supported by Health Canada, 1–4 December 2008.*)

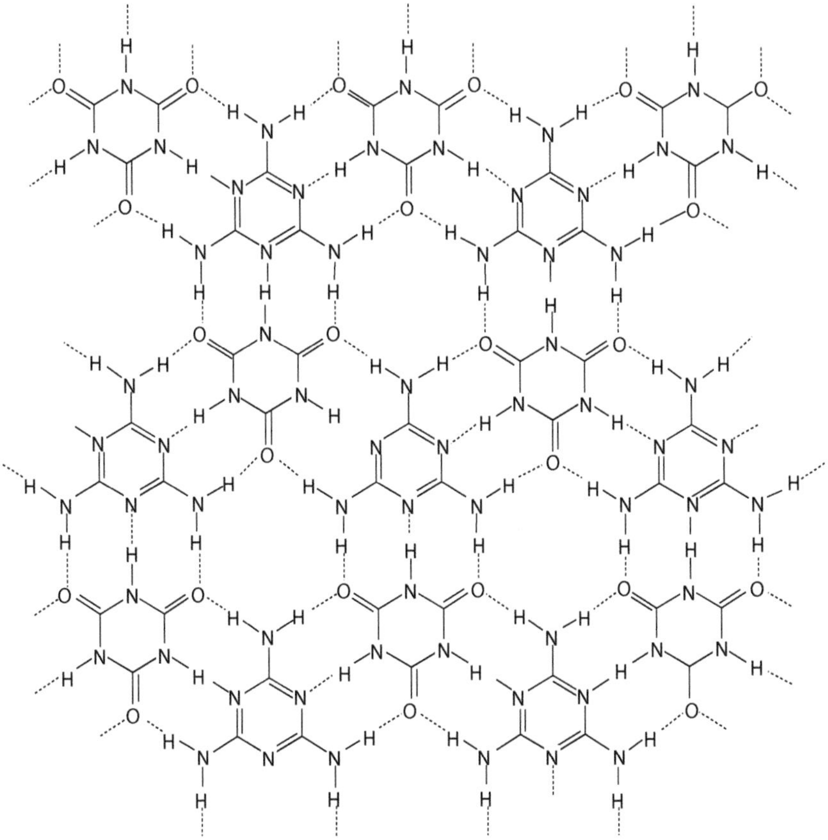

FIGURE 14.4 Proposed structure for the melamine–cyanurate complex. (Redrawn from Sherrington et al. 2001. *Chemical Society Reviews* 30 (2):83–93.)

Solubility minima for complexes of triazines such as melamine and cyanuric acid will occur under conditions that favor the formation of a strong hydrogen bond network, and hence stabilize the complex. Triazine hydrogen bond network can be disrupted by high temperatures or acid–base equilibriums (WHO 2009). The influence of pH on the solubility of the melamine–cyanurate complex under biological conditions has been reported in two studies (Dominguez-Estevez et al. 2010; Tolleson et al. 2009). The reported minimum solubility of the complex under biological conditions occurs at a pH between 5 and 5.5.

Uric acid is an endogenous biomolecule of interest (product of purine metabolism excreted in urine) that is capable of complexing with melamine due to complementary hydrogen bond acceptor/donor features (WHO 2009). A proposed structure for a melamine–urate complex is shown in Figure 14.6. The affinity of melamine for uric acid at pH 7 has been reported to be 29 times weaker than that for cyanuric acid, but the affinity for uric acid increases significantly with decreasing pH conditions (Tolleson et al. 2009). A variety of other cyclic imide biochemicals endogenous to

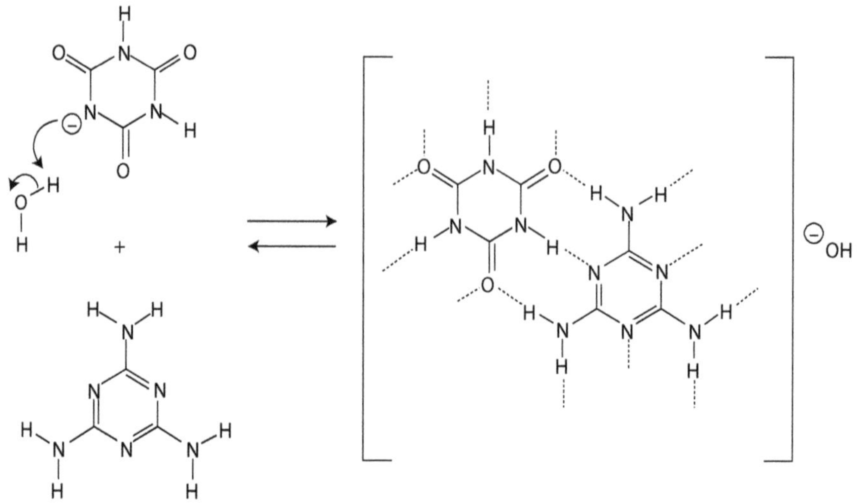

FIGURE 14.5 Proposed proton transfer-coupled assembly of melamine and cyanuric acid at pH 6.8.

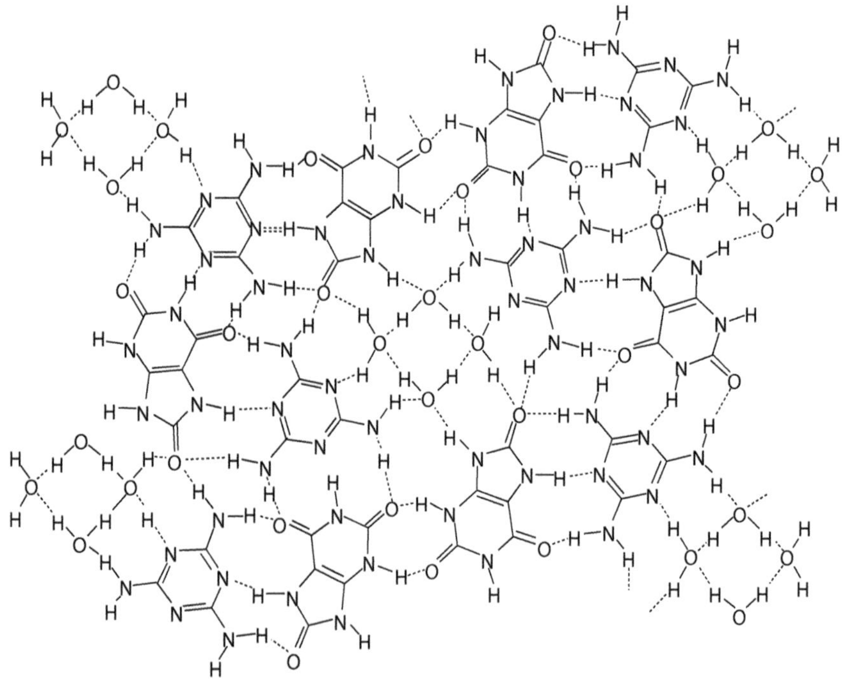

FIGURE 14.6 Proposed crystal structure for melamine–uric acid complex. Dotted lines indicate hydrogen bonds. (Redrawn from Dalal et al. 2011. *Nature Reviews Nephrology* 7 (5):267–274.)

humans and animals such as xanthine, uracil, and allantoin may also be capable of complexing with melamine (Tolleson et al. 2009).

14.2.2　Absorption, Distribution, Metabolism, and Toxicokinetics of Melamine and Cyanuric Acid

Most information available on the absorption, distribution, and toxicokinetics of melamine comes from animal studies. Following oral administration, numerous studies report that melamine is absorbed rapidly in monogastric mammals (Wu and Zhang 2013). The plasma half-life for melamine has been reported to range from 2.7 h in rats to 4 and 4.4 h in pigs and rhesus monkeys, respectively (Baynes et al. 2008; Liu et al. 2010; Mast et al. 1983). Analysis of toxicokinetic results from rat studies suggests that the distribution of melamine after oral administration is primarily limited to the blood or extracellular fluid with insignificant distribution in tissues (Dorne et al. 2013). Elimination of melamine in monogastric animals is primarily by the kidney, and a half-life for rats was reported to be 3 h with a clearance rate of 2.5 mg/min (Mast et al. 1983; Reimschuessel and Gu 2009).

Studies with ruminant mammals and nonmammals have reported some difference in the toxicokinetics, distribution, and excretion of melamine. Research on rainbow trout and goats, for example, has reported significantly greater plasma half-life and mean steady-state distribution volumes for melamine compared to results from monogastric mammals, suggesting much slower elimination in these species (Baynes et al. 2010; Dorne et al. 2013; Xue et al. 2011). Sun et al. (2012) reported significant distribution of melamine in the liver and also in kidney and bladder tissues after oral administration to cows. Studies have also shown that ruminant animals excrete a significant fraction of administered melamine in the feces (Cruywagen and Stander 2011; Sun et al. 2012).

The possibility of triazine compounds being metabolized by gut microorganisms has been studied. An investigation by Jutzi et al. (1982) confirmed that the strains of *Pseudomonas* sp. and *Klebsiella pneumoniae* are capable of metabolizing various triazine compounds. Jutzi's investigation and more recent studies have suggested that the main mode of metabolism is hydrolysis, with melamine being degraded to cyanuric acid through successive deamination reactions. Recent investigations using animal models have supported the possibility for metabolism of melamine by gut microbiota. A study by Sun et al. (2012) found that a significant amount of melamine fed to cows was slowly bioconverted into cyanuric acid in the rumen. Zheng et al. (2013) reported evidence for the bioconversion of melamine into cyanuric acid by gut microbiota in rats. They identified *Klebsiella terrigena* as the possible microorganism responsible for the conversion in rats. Another study by Stine et al. (2011) reported that pigs exposed to melamine developed crystals composed of melamine and cyanuric acid, localized to the kidneys. These data suggest that pigs are capable of bioconverting melamine to cyanuric acid (Stine et al. 2011). Clear causation for metabolism of melamine into cyanuric acid by gut microbiota in pigs, however, is not currently available. Cyanuric acid has also been reported to be metabolized by various bacteria and fungi to urea (Cheng et al. 2005). No evidence is available, however, that demonstrate the metabolism of cyanuric acid by gut microbiota in animals.

Together, these studies suggest a strong possibility that melamine may be metabolized to other triazines in animals, potentially by gut microbiota in some species. If significant bioconversion does take place, this may have important toxicological consequences, given the known stronger toxicity of melamine when coadministered to animals with cyanuric acid (see Section 14.4).

Less information is available on the absorption, distribution, and toxicokinetics of cyanuric acid but it is generally thought to behave similar to melamine (Skinner et al. 2010). A recent review by Dorne et al. (2013) reported that cyanuric acid is eliminated through urinary excretion in rats more rapidly than for melamine. The plasma half-life for oral and intravenous administration of cyanuric acid to rats ranged from 0.5–1 h to 2.5 h for 5 mg/kg b.w. and 500 mg/kg b.w. doses, respectively. A study on dogs reported a plasma half-life of 1.5–2 h for both 5 mg/kg b.w. and 500 mg/kg b.w. doses (Dorne et al. 2013). Data from one human study with swimmers aged 9–17 showed that more than 98% of cyanuric acid was recovered unmetabolized in the urine after 24 h and reported an excretion half-life for cyanuric acid of approximately 3 h (Allen and Pfaffenberger 1982).

14.2.3 Toxicology of Melamine and Cyanuric Acid

Melamine and cyanuric acid by themselves are not considered to be acutely toxic to animals or humans. They have reported lethal doses (LD_{50}) ranging from 3.1 to 7.7 mg/kg in mice and rats when administered individually, approaching the LD_{50} for table salt (NaCl) (Skinner et al. 2010). The target organ for melamine and cyanuric acid toxicological effects is the urinary system including the kidney and bladder. The formation of crystals or stones as a result of melamine and/or cyanuric acid exposure can result in nephrosis, renal failure, kidney inflammation, and hyperplasia and carcinomas of the bladder transitional epithelium (Liu et al. 2012).

The main reported toxicological effects of melamine to numerous animal species and humans are in the kidneys due to their rapid clearance after ingestion in this organ. Dorne et al. (2013) recently reviewed animal studies and reported that a limited number of crystals have been reported in the kidneys of fish, pigs, and rats exposed to melamine. Limited information is available from human studies, but clinical observations and test results of infants who had ingested the melamine-contaminated infant formula in the 2008 China incident have been widely reported. The toxic effects in infants are reported to be as similar to those in young children with reflux nephropathy (formerly known as chronic atrophic pyelonephritis) (Wu et al. 2009). By the end of 2008, after the Chinese incident, more than 294,000 infants and children suspected of melamine exposure were diagnosed with urinary tract stones, some with acute obstructive renal failure that resulted in 50,000 hospitalizations and six confirmed deaths (Chen 2009).

Histopathological and ultrasound investigations of affected infants revealed the formation of kidney stones, which comprised uric acid and melamine at molar ratios ranging from 1.2:1 to 2.1:1, with approximately 19% unidentified material, and no presence of cyanuric acid, in the renal collecting system (He et al. 2009; Sun et al. 2010). This led to the hypothesis that the obstructive nephropathy from melamine observed in infants originates from its precipitation in the lower urinary tract to form melamine–uric acid complexes, as shown in Figure 14.6 (Liu et al. 2012; WHO 2009). It is believed that humans, particularly infants, are significantly more susceptible to melamine–uric acid stone formation than other animals due to the higher concentrations of uric acid in urine (Dorne et al. 2013). Humans lack the urate oxidase enzyme system present in other animals, which converts uric acid into allantoin, resulting in 2–3-fold higher concentrations of uric acid in adult humans compared to rats, and human infants have an additional 4-fold higher level than human adults (Dorne et al. 2013). High uric acid concentrations along with acidic pH conditions in urine, which increase the affinity of melamine to uric acid, are thought to be some of the critical conditions that favored the formation of melamine–uric acid complexes in human infants in the 2008 Chinese melamine incident.

Proteins are known to play an important role in the formation mechanism for kidney stones, and likely play an important role for melamine crystal and stone formations in humans and animals. A recent study by Liu et al. (2012) used a rat model to examine the proteome of melamine urinary bladder stones to better understand the mechanism for stone formation. Stones formed in their studies were composed predominately of melamine and protein. Uric acid was not found in the stones, as expected, since rats have urate oxidase enzyme system. Proteomic analysis of the stones using LC-MS/MS and Western blotting revealed that the majority of proteins were binding proteins with isoelectric points below 7.0, suggesting that the binding proteins had more anionic side groups such as carboxylate groups, than cationic side groups (Liu et al. 2012). Categorization of the identified proteins by both cellular component and molecular and biological functions allowed the authors to suggest a more detailed mechanism for the formation of melamine stones involving endogenous proteins. They hypothesized the first step to be supersaturation of melamine assisted by urinary proteins to form melamine crystals. Formed melamine crystals then mechanically damage renal tubular cells and bladder transitional epithelial cells, causing attachment of crystals to epithelial cells through membrane molecules including proteins. The next step is thought to involve an inflammatory response induced by the damage to the epithelial cells, which increases the retention of crystals. The last steps involve aggregation of crystals anchored to cell membranes, assisted by proteins from membranes or injured cells, and eventually the formation of stones (Liu et al. 2012).

An additional toxicological effect of melamine reported in numerous animal studies is the formation of urinary bladder stones, including microcrystalluria (Wu and Zhang 2013). Melamine studies with rats have reported weight loss in treatment groups compared to control animals, and dose-related urinary bladder stone formation and epithelial hyperplasia (WHO 2009). Melamine studies with mice have also reported body weight reduction for treatment groups, and dose-related ulceration of bladder epithelium and observed kidney stones (WHO 2009). Both rat and mouse studies have reported gender differences with greater rates of bladder stone formation in males compared to females. One study with dogs fed melamine at 1200 mg/ kg body weight per day for 1 year reported persistent crystalluria, large bladder stones, and evidence of chronic cystitis.

One weakness of early studies on melamine toxicity, before 2007, is that formalin fixative was often used to preserve tissues for histopathology investigations. Later studies determined that melamine crystals often dissolve rapidly in formalin, and could have resulted in the underestimation of melamine crystals and stones in studies on melamine toxicity prior to 2007 (WHO 2009).

Reviewing rat studies on the carcinogenicity of melamine, Skinner et al. (2010) concluded that melamine might be carcinogenic to rats when administered in large doses, with tumors forming exclusively in the bladders of animals that developed stones. Based on these results, Skinner et al. (2010) concluded that the carcinogenic effects of melamine are likely mediated by the bladder epithelial irritation and hyperplasia, and not by the genotoxicity of the chemical itself. The International Agency for Research on Cancer (IARC) reviewed studies on the potential carcinogenicity of melamine, and reported that the non-DNA-reactive mechanism by which melamine may produce urinary bladder tumors in male rats occurred only under conditions in which calculi are produced (IARC 1999). Since some studies reported proliferative epithelial lesions in the absence of observable calculi, melamine has not been classifiable as to its carcinogenicity by IARC. A 2009 WHO report has suggested that observable calculi could have been underreported in studies considered by IARC in 1999 due to the common use of formalin known to dissolve melamine-based complexes. Little information is available on the genotoxicity of melamine, but a 2009 WHO report reviewed available studies and concluded that melamine is not a mutagenic agent based on bacterial mutagenicity tests, along with in-vitro and in-vivo assays (WHO 2009).

Information on the toxicological effects of cyanuric acid and other melamine analogs following their individual ingestions in isolation are limited but appear to be similar to that for melamine. Dalal and Goldfarb (2011) reviewed available studies and reported that sodium cyanurate doses of 500–700 mg/kg in rat subchronic studies have resulted in bladder calculi, bladder epithelial hyperplasia, and uremia. Studies with dogs fed the same compound at 8% of their diet for 16–21 months resulted in kidney fibrosis (Dalal and Goldfarb 2011). Another report indicated that cyanuric acid is not considered to be genotoxic orteratogenic, nor is it a reproductive toxicant (Gossner et al. 2009). Limited information on the toxicological properties of ammeline and ammelide are available, but WHO (2009) reported that it is unlikely that these chemicals would be more toxic than melamine.

14.2.4 Metabolism, Toxicokinetics, and Toxicology of Coadministered Melamine and Cyanuric Acid

The combined toxicology of melamine with cyanuric acid represents an illustrative example of the complexity in assessing the toxicological effects of chemical mixtures. As reported in Section 14.2.2, melamine and cyanuric acid present low acute toxicities in a range of animal studies when individually administered. It was not until the 2007 animal feed adulteration incident that the increased toxicity of melamine as a result of coingestion with cyanuric acid was investigated and completely understood.

The tragic 2007 incident involved economically adulterated animal feed ingredients imported from China including wheat gluten, corn gluten, and rice protein tainted with a mixture of melamine and cyanuric acid. The ingredients were likely adulterated with "scrap" grade of melamine contaminated with high levels of cyanuric acid, as described in Section 14.2.1. One reported analysis of a tainted wheat gluten sample in the United States found concentrations of melamine and related compounds as follows: melamine (84,000 mg/kg), cyanuric acid (53,000 mg/kg), ammeline (17,000 mg/kg), and ammelide (23,000 mg/kg) (Dobson et al. 2008). Others have indicated that concentration of melamine in wheat gluten and rice protein ranged from 2000 to 80,000 mg/kg (Dorne et al. 2013). Animals ingesting the tainted feed exhibited symptoms and signs consistent with renal failure including vomiting, anorexia, lethargy, and polyuria. Clinical testing confirmed azotemia and hyperphosphatemia. Histopathology of animals that succumbed to renal failure reported yellowish-brown crystals present in the renal tubules. The incident resulted in more than 39,000 cats and dogs developing acute renal failure in North America, with an estimated 2000–7000 deaths in North America and South Africa combined (Dalal and Goldfarb 2011). Related incidents were reported in several other countries in 2007 including China, the Philippines, Spain, and Italy (Dorne et al. 2013). A report by Luengyosluechakal (2007) indicated that thousands of pigs in Thailand died in 2007 from renal failure after being fed rice protein-based feed economically adulterated with melamine at 3026 mg/kg, ammeline at 958 mg/kg, and cyanuric acid at 69,031 mg/kg.

Laboratory investigations into the causation of the 2007 pet food incident reported by Dobson et al. (2008) began without any prior knowledge that melamine and related compounds were capable of the observed toxicological effects. Other chemicals known at that time, such as heavy metals, pesticides, and mycotoxins, to cause renal failure were first investigated. Toxicologically insignificant levels of these contaminants led to additional analysis and eventually the identification of melamine and related triazine compounds as the likely contaminants. Later in-vivo studies with rodents confirmed one of many working hypotheses to the cause of renal failure. Coingested melamine and cyanuric acid were determined by Dobson et al. (2008) to form an insoluble precipitate in kidney tubules causing renal failure by physical blockage. The insoluble precipitate containing melamine and cyanuric acid was likely the hydrogen-bonded complex, as shown in Figure 14.4. The exact mechanism for formation of this complex exclusively in the kidneys and not in other organs is not yet entirely clear. It has been hypothesized that concentration of urine filtrate in nephrons may lead to nucleation of the complex and crystal growth (Dobson et al. 2008). New insights into the potential mechanism for stone formation as a result of exposure to triazine were reported by Liu et al. (2012), discussed above in Section 14.2.3.

Since 2007, numerous experiments have been conducted using animal models to better characterize the toxicological properties of melamine and cyanuric acid coadministration. Investigations with melamine and cyanuric acid fed at a 1:1 ratio have clearly demonstrated that concomitant ingestion of these compounds cause crystal formation and renal failure in numerous animal species (Stine et al. 2011). Dorne et al. (2013) and Stine et al. (2011) recently reviewed these studies conducted on laboratory animals, pets, and farm animals including rats, pigs, chickens, catfish, tilapia, trout, and salmon. All studies reported the formation of melamine–cyanuric acid crystals as golden brown birefringent crystalline spherulites arranged in clusters and presenting radial symmetry. No observed adverse effect levels (NOAEL) from 0.1 to 10 mg/kg b.w./day have been reported depending on animal species (rat, catfish, pigs, or trout) and study length (1–28 days) (Dorne et al. 2013; Stine et al. 2011). Bench mark dose modeling (BMDL) studies (an alternative to NOAEL studies) have reported $BMDL_{10}$s (95% lower bound of the 10% BMDL) from 1.5 to 10.9 mg/kg b.w./day for pigs in 7 day studies (Dorne et al. 2013; Stine et al. 2011).

Several animal studies have reported important toxicological differences between melamine exposure alone and melamine exposure concomitant with cyanuric acid, suggesting that the latter is significantly more toxic. A study by Jacob et al. (2011) compared the nephrotoxicity for coadministration of melamine and cyanuric acid to that of melamine or cyanuric acid individually administered. They found that coadministration of melamine and cyanuric acid significantly increases toxicity to the kidneys at a comparable concentration of melamine alone. Additional rat studies have supported this conclusion, reporting that combination of melamine and cyanuric acid produced similar toxicity to a 12–20-fold higher dose of melamine administered alone (EFSA 2010). These results provide evidence that the toxicity data derived from toxicological studies conducted with melamine alone may underestimate the toxicity from coexposures to melamine and cyanuric acid. Additional discussion on this and other related studies can be found in Section 14.2.3.

A recent study by Jacob et al. (2012) found that individual and coadministration of melamine and cyanuric resulted in similar pharmacokinetics, while administration of the preformed complex significantly reduced bioavailability of both compounds including lower observed maximum serum concentrations, delayed peak concentrations, and prolonged elimination half-life. They hypothesized that the acidic environment in the stomach delays disassociation of the melamine–cyanuric acid complex into individual compounds, thereby reducing their bioavailability. For coadministered melamine and cyanuric acid, they hypothesized that stomach hydrochloric acid prevents the complexation of melamine and cyanuric acid by protonating the amine groups in melamine and preventing hydrogen bond formation.

The true historical extent of harm to animals and pets from melamine and cyanuric acid contamination is likely larger than currently reported. Once the causative agent and evidence for renal failure in the 2007 pet incident were reported by scientists, evidence from an earlier 2004 outbreak of renal failure in pets in several Asian countries was reinvestigated by Brown et al. (2007). This 2004 incident was originally attributed to mycotoxin contamination, but strikingly similar symptoms and evidence now suggest that melamine and cyanuric acid were likely the causative contaminants in that incident as well (Brown et al. 2007). Another study has reported evidence suggesting that exposure to melamine and its derivatives may have caused the nephrotoxicosis observed in Iberian piglets in Spain from 2003 to 2006 (Gonzalez et al. 2009). The authors of this report indicate that between 300 and 400 postweaning piglets were involved in the incident with morbidity of 40–60% and mortality of 20–40%. Given the significantly greater toxicological effects in animals of melamine combined with cyanuric acid exposure compared to melamine alone, it is fortunate that the former was not used to adulterate the human food supply in 2008.

14.3 ESTABLISHED TOLERABLE INTAKES AND MAXIMUM LIMITS FOR MELAMINE AND RELATED COMPOUNDS IN FOOD AND FEED

Following the 2007 and 2008 melamine adulteration incidents, risk assessments were carried out by several national food-safety authorities to establish tolerable daily intakes (TDI) and corresponding maximum limits for melamine in human foods and infant formula, and animal feed. In response to the potential carryover of melamine from tainted animal feed into animal-based foods (e.g., meat and eggs), the United States Food and Drug Administration (US FDA) carried out a risk assessment using as a basis TDI of 0.63 mg/kg b.w./day for melamine and its analogs (FDA 2007). The TDI was based on a NOAEL of 63 mg/kg b.w./day for bladder stones reported from a 13-week rat study carried out by the National Toxicology Program (NTP) of the United States Department of Health and Human Services in 1983, and also included a 100-fold safety factor (FDA 2007). It has been reported that the NOAEL of 63 mg/kg b.w./day from the NTP study was derived from the dietary concentration of 750 mg/kg and the assumption of a daily consumption of 10 g dry laboratory chow by young rats (EFSA 2010). The 0.63 mg/kg b.w./day TDI reported by US FDA in 2007 was used in 2008 as a starting point for its risk assessment of melamine and its analogs in food for humans in response to the infant formula incident in China (FDA 2008). In light of evidence suggesting an increased toxicity of melamine when combined with cyanuric acid, they added an additional 10-fold safety factor to the already 100-fold factor used originally to derive the TDI, resulting in a new TDI of 0.063 mg/kg b.w./day (FDA 2008). The US FDA also established a maximum limit of 2.5 mg/kg for melamine and its analogs in human foods, as a level that did not raise public health concerns (FDA 2008). They were unable to determine a maximum limit for melamine and its analogs in infant formula at that time but did so later that year. In their updated 2008 risk assessment to address infant formula, they reported a maximum limit of 1.0 mg/kg for melamine and its analogs at the level that did not raise public health concerns. This 1.0 ppm limit was calculated using a worst-case scenario in which the infants' entire total daily diet was contaminated with melamine (FDA 2008).

By 2009, the WHO carried out an independent risk assessment for melamine alone using a TDI of 0.2 mg/kg b.w./day, but did not establish one for melamine analogs (WHO 2009). WHO first evaluated available epidemiological data on kidney damage from the 2008 infant formula adulteration incident, and concluded that due to uncertainties, the data could not be used to derive a conclusive dose–response relationship. WHO also evaluated use of the reported NOAEL of 63 mg/kg b.w./ day, but indicated that a dose response could be established from available rat study data which could be modeled to almost fully capture the general pattern of the dose–response relationship (WHO 2009). WHO, therefore, used this approach in place of the NOAEL and calculated a benchmark dose lower confidence limit for a 10% increased incidence of stone development in the bladder ($BMDL_{10}$). The calculated $BMDL_{10}$ of 415 mg/kg diet was converted into a dose of 35 mg/kg b.w./day by applying conversion factors to adjust from young rats to the general human population. The first was a typical dietary conversion factor of 0.10 for young rats (i.e., assuming

10 g food consumed per day) and the second factor of 14% accounted for an additional feed reduction adjustment based on an additional subchronic toxicity study in rats (WHO 2009). The resultant TDI for melamine was rounded to a single significant figure of 0.2 mg/kg b.w./day and was considered applicable to the whole population, including infants. For risk management purposes, the WHO concluded that a 1 mg/kg limit for melamine in powdered infant formula (1 mg/kg) and 2.5 mg/kg for other human foods would provide a sufficient margin of safety for dietary exposure relative to their established TDI (WHO 2009).

The European Food Safety Authority (EFSA) following the pet and animal food incidents reported a TDI for melamine of 0.5 mg/kg b.w./day based on a 1986 report from the Scientific Committee of Foods (SCF) in Europe, which did not include a derivation for this TDI (EFSA2007). They also proposed a TDI of 1.5 mg/kg b.w./day for sodium cyanurate, based on a NOAEL of 154 mg/kg b.w./day derived from a 2-year rat study and a 100-fold safety factor (EFSA 2007). After the 2009 WHO establishment of a TDI, EFSA undertook another risk assessment for melamine using BMDL from the NTP 1983 study instead of the NOAEL approach. They calculated a $BMDL_{10}$ of 19 mg/kg b.w./day and derived a TDI using a 100-fold safety factor. EFSA carefully considered the appropriateness of the 100-fold factor, a product of 10 to account for interspecies differences and 10 to account for intraspecies differences. They concluded that the factor was appropriate, including for infants, and rounded the TDI to 0.2 mg/kg b.w./day. EFSA did conduct BMDL of the available epidemiological data from the Chinese infant formula incident based on a report from Li et al. (2010) despite its uncertainties. They concluded that the epidemiological data supported the proposed TDI of 0.2 mg/kg b.w./day (EFSA 2010).

It is not surprising that the risk assessments carried out by the US FDA, WHO, and EFSA produced similar results. While each risk assessment derived TDIs using different uncertainty factors (safety factors) and points of departure, all three used the same base data from the NTP 13-week rat study conducted with exposure to melamine alone.

The establishment of risk assessments for melamine led to the establishment of internationally recognized maximum limits for melamine in foods and feed by the United Nation's Codex Alimentarius. A maximum limit of 2.5 mg/kg for food and feed (other than infant formula) and 1 mg/kg for powdered infant formula proposed at the 2010 fourth meeting of the Codex Committee on Contaminants in Foods was adopted by the Codex Alimentarius Commission at their 33rd session later in 2010 (CAC 2010; CCCF 2010). A maximum limit of 0.15 mg/kg for liquid infant formula proposed at the 2012 sixth meeting of the Codex Committee on Contaminants in Foods was adopted by the Codex Alimentarius Commission at their 35th session later that year (CCCF 2012; CAC 2012). These limits are similar to those independently established by numerous national authorities based on their own risk assessments.

Internationally recognized maximum limits for the combination of melamine and other melamine analogs are not currently available based on toxicological data from animal studies related to coexposure of animals to melamine and cyanuric acid. This indicated a research gap—the need to develop a TDI based on dose–response data for the induction of kidney crystals and nephrotoxicity following coexposure to melamine and cyanuric acid. Jacob et al. (2011) carried out an investigation to begin addressing this research gap using an acute rat study. They reported a NOAEL of 8.6 mg/kg b.w./day for crystal formation in rats exposed to both melamine and cyanuric acid for 7 days (Jacob et al. 2011). In contrast, rats exposed to melamine or cyanuric acid alone for the same timelength and substantially higher levels did not show any significant evidence of nephrotoxicity. They calculated TDIs of 8–10 µg/kg b.w./day when applying a 1000-fold safety factor, 40–50 µg/kg b.w./day when applying a 200-fold safety factor, and 80–100 µg/kg b.w./day for a 100-fold safety factor and reported that these TDIs were two to seven times lower than those reported by WHO, US FDA, and EFSA in their risk assessments for melamine only. The same group reported a longer follow-up study in 2012 using a 28-day rat model study (Gamboa da Costa et al. 2012). They reported NOAELs of 2.1 mg/kg b.w./day and <2.6 mg/kg b.w./day for male and female rats, respectively, and BMDL values as low as 1.6 mg/kg b.w./day for both sexes. These lower thresholds compared to their earlier 7-day study indicate that length of exposure is a critical factor when establishing a threshold for toxicity of melamine and cyanuric acid coexposures. Agreeing with their previous study, they concluded that TDIs for the combination of melamine and cyanuric acid based on data from melamine studies alone may underestimate the risk of the combination by a factor of at least 10 (Jacob et al. 2011; Gamboa da Costa et al. 2012). Results from Jacob et al. are also supported by a report from Choi et al. (2010). They carried out a 7-day study with rats coexposed to melamine and cyanuric acid and reported a NOAEL of 3.15 mg/kg b.w./day, suggesting the need to reassess TDI for the combination of melamine and cyanuric acid.

Another study by Stine et al. (2011) used a pig model with coexposure to melamine and cyanuric acid for 28 days. In this investigation, the authors reported a NOAEL of 1 mg/kg b.w./day including a 10-fold safety factor and concluded that their findings supported risk assessments for animal feed and human food based on melamine alone carried out by US FDA, WHO, and EFSA, suggesting that established maximum for melamine alone can be extended to the combination of melamine and cyanuric acid (Stine et al. 2011). However, it is unclear how this conclusion was drawn given differences in the use of safety factors compared to other risk assessments.

In summary, there is a need to reassess the risk for coexposure to melamine and cyanuric acid in light of new data. While strong evidence from rat studies suggest that the risk assessments based solely on data from exposure to melamine may underestimate the risk of coexposure, more studies may be necessary to inform risk assessors.

14.4 EXPOSURE TO MELAMINE AND RELATED COMPOUNDS IN FOOD SUPPLY

Characterizing the potential exposure of humans to melamine and melamine analogs from all sources is complicated by the motivations and entry points for melamine into supply chains for food and feed. Intentional adulteration is one reason for the presence of melamine or related compounds in supply chains that are driven by criminal and economical motivations. It can result in melamine entering food supply chains through animal feed or human food ingredients as evidenced by the numerous product types and levels of melamine reported in the tandem 2007 and 2008 adulteration incidents (Gossner et al. 2009). Intentional adulteration typically involves higher levels of adulteration, enough to make it economically viable. It has been reported in a recent study by Abernethy and Higgs (2013) that melamine levels above 90 mg/kg in raw fluid milk are necessary to make adulteration economically viable. Unintentional contamination is another possible reason for exposure to melamine if food compounds come in contact with materials contaminated by or environmental sources of melamine, but is expected to produce much lower exposures than those from adulteration (WHO 2009). The assessment of the Codex Alimentarius suggests that levels of melamine below 1 mg/kg in foods are most likely the result of unintentional contamination sources (CCCF 2010).

Characterizing human exposure to melamine is also complicated by the potential for carryover between nodes in a production or supply chain (e.g., transfer from adulterated animal feed to foods derived from those animals eating the feed, or from adulterated fluid milk to processed milk products). It is also complicated by potential dilution effects in supply chains. For example, adulterant concentrations will typically be the greatest at the point where adulteration occurs in the supply chain, and may decrease through the supply chain as it is mixed with nonadulterated materials. Figure 14.7 provides a generalized illustration of the potential pathways for melamine exposure in the human food supply. It will be discussed in the below three sections that attempt to characterize melamine exposure from three potential sources, unintentional baseline contamination, intentional adulteration of feed, and intentional adulteration of food ingredients.

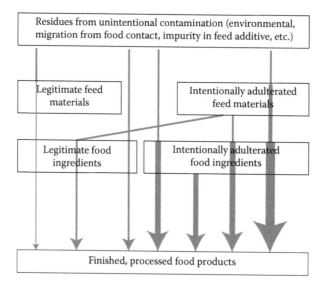

FIGURE 14.7   Potential pathways for melamine exposure in the human food supply chains. Arrow thicknesses are intended to indicate the potential magnitude of exposure, but are not necessarily to scale.

14.4.1   Unintentional Sources of Melamine and Cyanuric Acid

One potential source of melamine in food supplies is the migration of melamine from melamine-containing food-contact materials. Melamine-containing tableware is one possible food-contact exposure source. Studies conducted with a variety of food types and conditions have reported migration levels from 0.05 to 5 mg/kg, the latter being maximum levels under harsh conditions composed of 3% acetic acid under high-temperature conditions (EFSA 2010). WHO, in their assessment, reported a 13 μg/kg b.w./day estimated exposure based on a 1 mg/kg migration level and an assumption that the migrant would be present in 25% of the diet (WHO 2009). Melamine is also a component of some adhesives, and paper and paperboard used in food-contact materials. Estimated human exposures of <0.35 and 0.0019 μg/kg b.w./day have been reported for adhesives and paper/paperboard (WHO 2009).

Another potential unintentional source of exposure to melamine and cyanuric acid for humans is from chlorine-containing disinfectants. Trichloromelamine, which decomposes to melamine, is allowed for use in some countries as a disinfectant on food-packaging materials, food-processing equipment, and utensils. The US FDA has estimated a melamine concentration of 0.14 mg/kg in food as a result of tricholormelamine use, and WHO estimated the exposure for humans to be 7 μg/kg b.w./day (CCCF2010; WHO 2009). Dichloroisocyanurate is a potential source of cyanuric acid residues in food. It is used as a disinfectant for drinking water, and WHO has estimated a dietary exposure of 70 μg/kg b.w./day (WHO 2009).

Cyromazine, an agricultural pesticide and veterinary medical product, is another potential source of melamine in foods. Cyromazine is metabolized by both plants and animals, and melamine is one of the known metabolites. For plants, cyromazine is used as a systemic insect growth regulator on vegetables such as potatoes, lettuce, mushrooms, celery, and melons. It has been reported that for most plants, approximately 10% of cyromazine is converted into melamine, with the exception of mushrooms where approximately 50% is metabolized to melamine. Expected maximum melamine residue levels (MRL) in plants have been reported to be 1 mg/kg, with the exception of mushrooms where MRLs can exceed 2 mg/kg. Estimated adult human exposures to melamine from cyromazine have been estimated to be 0.04–0.27 μg/kg b.w./day (WHO 2009). As a veterinary medical product, cyromazine is used as a feed additive for poultry to control diptera larvae in chicken manure, and on sheep to prevent blow-fly strikes. For sheep meat, an MRL of 4 μg/kg and corresponding human exposure of 0.020 μg/kg b.w./day have been reported with poultry meat at an MRL of 30–36 μg/kg and corresponding human exposure of 0.2 μg/kg b.w./day (Dorne et al. 2013).

Other potential unintentional sources for exposure to melamine and its related compounds in food have been reported. Biuret, a feed additive allowed in some countries, typically contains cyanuric acid as an impurity. In the United States, biuret can contain up to 30% cyanuric acid (EFSA 2010). Levels of cyanuric acid in animal feed are likely very low based on surveillance study data reporting levels below the detection limit of 1 mg/kg (EFSA 2010). Other feed additives including guanidine acetic acid and urea have been reported to contain melamine as an impurity with levels up to 15 and 25 mg/kg, respectively, and cyanuric acid up to 50 and

200 mg/kg, respectively (CCCF 2010). Data to estimate carryover from animal feed to the human food supply are not available, but likely would be very low based on biuret's expected low exposure to animals since it is a feed additive, and the expected low carryover rate from feed to animal tissues (see Section 14.4.2). Other potential sources reported include fertilizers and paints, but no data are available to estimate exposures from these sources (WHO 2009).

Together, the data available for unintentional sources of melamine in the human food supply suggest that numerous sources are possible, but typically at very low levels, which when combined are orders of magnitude less than the established TDI for melamine. This is represented in Figure 14.7, showing that compared to other potential sources of melamine, the risks associated with unintentional sources alone are insignificant from a public health perspective, and do not warrant risk management.

14.4.2 CARRYOVER FROM INTENTIONALLY ADULTERATED ANIMAL FEED INTO FOOD OF ANIMAL ORIGIN

After the 2007 pet and animal food incident involving melamine and cyanuric acid, risk assessors became interested in the potential transfer of melamine and related compounds from adulterated feed to foods of animal origin (e.g., meats, fish, and eggs) as a potential exposure source for melamine and related compounds in human foods. Later reports including an unpublished report by Nestlé cited by the Codex Committee on Contaminants in Foods concluded that melamine contaminated cow feed was the likely source of low levels of melamine reported in some milk and milk products, and that intentional adulteration was not the likely source for those specific products (CCCF 2010; Cruywagen et al. 2009). These reports emphasized the need to further investigate the risks and mitigation strategies to address potential melamine carryover from feed to animal-derived foods.

Several studies investigating melamine carryover were reviewed in the 2010 EFSA risk assessment, and numerous studies have been published since that time. Reported tissue concentrations and transfer rates for melamine into animal-derived foods vary widely, depending on animal, dose, exposure time, and tissue type. Reports for chicken tissues and eggs, duck eggs, pig tissues, cow milk and tissues, and lamb tissues suggested the maximum tissue concentrations from 0.37 to 35 mg/kg and maximum transfer rates ranging from 0.37% to 9.0% (expressed as mg/kg in animal tissue per mg/kg in feed) (Battaglia et al. 2010; Dorne et al. 2013; EFSA 2010; Gao et al. 2010; Lü et al. 2009; Lv et al. 2009; Sunet al. 2011). The animal tissue exhibiting the highest concentrations and transfer rates in feeding studies is generally the kidney. This is expected since it is the primary excretion route in many animals. In a poultry study, for example, broilers that were fed a high concentration of melamine (1000 mg/kg feed) exhibited maximum concentrations in a 28-day study of 8.04, 9.07, and 29.52 mg/kg for breast meat, liver, and kidney, respectively, and average transfer rates of 0.6%, 1.3%, and 2.6%, respectively (Lü et al. 2009). Data from these studies are generally supported by available surveillance data intended to characterize the baseline levels of melamine in animal-derived foods that are likely from animal feed carryover. For example, data for fresh eggs in a study with hens exposed to 5–100 mg/kg melamine in feed were 0.16–1.48% and are comparable to a surveillance of 0.1–4.6 mg/kg reported by WHO from several countries when using a factor of 4 to convert from liquid into dry eggs (Chen et al. 2010; WHO 2009).

The potential for bioaccumulation of melamine in animal tissues has also been investigated. Animal feeding studies have been conducted to estimate bioaccumulation for chicken tissues and eggs, duck eggs, pig tissues, cow milk and tissues, fish tissues, and lamb tissues (Battaglia et al. 2010; CCCF 2010; Chen et al. 2010; EFSA 2010; Gao et al. 2010; Lü et al. 2009; Lv et al. 2009; Shen et al. 2010; Sun et al. 2011). These reports have indicated that very little to no melamine was detected several days after withdrawal of melamine from the diet, including low, if any, levels of bioaccumulation.

Shen et al. (2010) estimated the melamine exposure per day for lactating cows that would result in melamine carryover levels in milk and milk products, which would be concerning to human health. They used the internationally recognized maximum limits for melamine in powdered infant formula and all other foods of 1.0 and 2.5 mg/kg, respectively. They determined that dairy cow exposure to melamine of greater than 312.7 mg would result in powdered milk exceeding the 1.0 mg/kg limit for infant formula, and 715.1 mg melamine would result in milk exceeding the 2.5 mg/kg limit (Shen et al. 2010). These figures can be converted into an estimated melamine concentration in finished cow feed of 20.6 mg/kg for 1.0 mg/kg maximum limit and 47.0 mg/kg for the 2.5 mg/kg maximum limit. This is done using the 609 kg animal weight reported in their study and a typically assumed daily feed amount for one animal of 2.5% of body weight.

Data available on melamine concentrations in finished animal feed likely adulterated by their suppliers were reported by the Codex Committee on Contaminants in Foods. These data, from 85 samples, indicated that 10% of tested samples contained more than 100 mg/kg (maximum level 700 mg/kg), 30% of the samples contained 5–100 mg/kg, and the remaining 60% levels were below the 5 mg/kg detection limit (CCCF 2010). It should also be noted that melamine levels in wheat gluten reported during the 2007 pet food adulteration incident ranged from 2000 to 80,000 mg/kg (Dorne et al. 2013). These figures, along with published transfer rates from animal models described in this section, indicate that carryover of melamine from adulterated animal feed into food alone could be a significant source of melamine in the human food supply that would be problematic for human safety.

As illustrated in Figure 14.7, melamine carryover from intentionally adulterated feed ingredients into human foods does represent a significant or concerning exposure for human diets. This indicates the need for established maximum levels for melamine in feed, as discussed in Section 14.3, and the need for surveillance and routine analytical testing to manage this risk discussed in Section 14.6.

14.4.3 Intentionally Adulterated Food Ingredients

The intentional adulteration of food ingredients can lead to tragic public health consequences for infants as evidenced by the 2008 infant formula scandal in China. Estimated dietary exposure for children exposed to infant formula that contained milk powder adulterated with melamine was reported at median levels which ranged from 8.6 to 23.4 mg/kg b.w./day (WHO 2009). This exposure, exceeding the established

TDI of 0.2 mg/kg b.w./day by 40–120 times, clearly explains the observed toxicological consequences. In addition to milk-derived ingredients, where melamine levels were reported to be as high as 6196 mg/kg, other food ingredients were reported to contain melamine likely from intentional adulteration. These included ammonium bicarbonate and nondairy creamer with reported levels as high as 508 and 6694 mg/kg, respectively (Gossner et al. 2009). WHO's 2009 report also included a dietary exposure estimate for adults from noninfant formula foods containing melamine at adulterated levels. They estimated the exposure to range from 0.16 to 0.7 mg/kg b.w./day, clearly indicating that melamine exposure from intentionally adulterated food ingredients is a public health concern for adults as well. Together, this evidence suggests, as illustrated in Figure 14.7, that melamine sourced from intentionally adulterated food ingredients has the potential to lead to the greatest exposure, justifying the need to focus risk-management resources, including surveillance and testing for both raw materials and finished products, to protect human health.

14.5 OTHER CONSIDERATIONS FOR MELAMINE EXPOSURE AND TOXICITY

Very few investigations have been carried out to understand how food processing or formulation may modulate the exposure or toxicity of melamine and related compounds in contaminated or adulterated food products. The presence of nucleophilic amine groups in its structure suggests that it may react with carbonyl group of reducing sugars to form glycosoamines. Recent studies support this possibility, reporting that melamine can react with lactose and Strecker aldehydes formed in Maillard reactions to produce new adducts (Liu 2013; Ma et al. 2010). No information is available on the toxicology of the reported melamine glycation products, suggesting the need for future research.

14.6 ANALYTICAL TEST METHODS FOR MELAMINE AND RELATED COMPOUNDS

Following the 2007 and 2008 melamine adulteration incidents, several methods were developed and validated for screening and quantifying melamine and related compounds in numerous food and feed matrices. An report by Tittlemier (2010) reviewed methods available at that time, including both selective quantitative methods and quality screening methods. A number of reports since then have introduced new methods and technologies. Methods can be discriminated as screening methods or confirmatory methods based on their selectivity and basic performance characteristics. In addition, new types of analytical screening and confirmatory technologies have been proposed to address the broader challenges associated with economic adulteration and authenticity of protein-based food ingredients: (1) economic adulteration evolution (new adulterants constantly developed to circumvent new QA methods); (2) the need for more selective total protein methods that cannot be tricked by the addition of nitrogen-rich nonprotein adulterants (CCCF 2010; Moore et al. 2010, 2011, 2012; Tittlemier 2010). Sections 14.6.1 and 14.6.2 are intended to provide a brief overview of established and emerging technologies for analyzing food and feed for melamine by category (screening or confirmatory), while Section 14.5.1 focuses on new technologies reportedly being proposed or in development for adulteration.

14.6.1 HIGHLY SELECTIVE AND QUANTITATIVE CONFIRMATORY METHODS FOR MELAMINE AND RELATED COMPOUNDS

Confirmatory methods for melamine and related compounds generally have better performance characteristics compared to screening methods to ensure accurate and sensitive quantitation, and have a higher degree of selectivity to ensure reliable identity confirmation. Reported methods for selective quantitation of melamine in food and feed matrices are HPLC-MS/MS, GC-MS, and GC-MS/MS. These techniques generally involve a simple liquid extraction step with a polar solvent system, sometimes followed by further cleanup with solid-phase extraction.

The current state of GC- and LC-MS/MS methods for melamine and related compound analysis has been recently reviewed by Tittlemier et al. (2010) and Lutter et al. (2011), and is summarized below. For GC analysis, derivatization is necessary given the polar nature of melamine and related compounds, and trimethylsilyl derivatization is generally done for analysis. Early GC methods for melamine and related compound analysis used GC-MS with identity confirmation carried out by selecting target molecules based on molecular mass (single-ion monitoring). Recently, GC-MS/MS methods have been reported. Use of multiple reaction monitoring in GC and HPLC improves selectivity; not only selecting target molecules based on molecular mass, but also adding known fragment ion masses as another identity confirmation. Accurate quantitation is achieved using internal standards or isotope dilution techniques.

For HPLC-MS/MS, most recently reported methods use hydrophilic interaction liquid chromatography separation conditions as reversed-phase methods have poor retention or separation of melamine from matrix components. Melamine is typically analyzed in positive electrospray ionization mode and cyanuric acid in negative mode, requiring a polarity during each run. Monitoring of at least two ion transfers is recommended so that the calculated ratio of responses from the two transitions can be used to further confirm the identity. Accurate quantitation is achieved using internal standards, or isotope dilution techniques.

One significant advantage of the LC-MS/MS and GC-MS methods is their high degree of selectivity, enabling analyte identity confirmation. Lutter et al. (2011) reported performance characteristics of their LC-MS/MS and GC-MS methods for analysis of melamine in milk-based infant formula. For GC-MS measurements of infant formula, they reported a linear range of 0–0.500 mg/kg; the limit of detection (LOD) was 0.009–0.19 mg/kg, and repeatability was 7.6% for a 0.5 mg/kg spike. For the LC-MS/MS measurements of infant formula, the established linear range was 0–2.000 mg/kg; the LOD was 0.05 mg/kg, and a repeatability of 4% for a 0.5 mg/kg spike was reported (Lutter et al. 2011). Overall, the study by Lutter et al. (2011) suggested that the LC-MS/MS method provided the highest level of selectivity and reliability for melamine measurements of the approaches investigated. This is supported by newly established ISO-IDF guidelines for the measurement of melamine and cyanuric acid in milk, milk products, and infant formula using LC-MS/MS (ISO 2010). The major drawback to most of the confirmatory methods reported is the cost of advanced equipment and isotopically labeled internal standards, as well as the level of skill involved in analyses, which are not always possible in routine quality control laboratories.

14.6.2 SCREENING METHODS FOR MELAMINE AND RELATED COMPOUNDS

Acceptable screening methods are generally rapid qualitative or semiquantitative methods that are sensitive enough to screen materials against established maximum limits for melamine and related compounds (generally 0.1–2.5 mg/kg). Screening methods should be simple and inexpensive enough to be practical in routine quality control laboratories. Enzyme-linked immunoassay methods (ELISA) for melamine were developed and reported as early as 2008 (Tittlemier 2010). These methods are useful for high-throughput analysis demands, capable of analyzing 40 samples with replicates on one ELISA plate with results obtained in one hour. These methods have good detection capabilities. A comparative investigation of melamine detection methods by Lutter et al. (2011) reported the LOD and limit of quantitation (LOQ) for cow's milk of 0.06 and 0.1 mg/kg, respectively, and for milk-based infant formula of 0.56 and 1.0 mg/kg, respectively. The main limitations for ELISA methods are selectivity and precision. The same study by Lutter et al. reported a high degree of cross-reactivity of antibodies used in ELISA tests for some melamine derivatives. Ammeline and cyromazine, for example, had cross-reactivities of 100% and 300%, respectively, suggesting selectivity limitations for ELISA methods (Lutter et al. 2011). Lutter et al. (2011) also reported repeatability results (% relative standard deviations) of 10% and 21.1% for analysis of 0.1 and 2.0 mg/kg melamine spiked samples, respectively, suggesting that ELISA test results should be considered semiquantitative when testing in these concentration ranges. Use of ELISA tests for melamine detection in other matrices, including pet food, milk products, eggs, biscuits, and animal feed ingredients, was reviewed by Tittlemier (2010). Available performance data reported in the review suggest that limits of detection can range from 0.008 to 1 mg/kg, and limits of quantitation from 0.1 to 250 mg/kg depending on the matrix (Tittlemier 2010).

HPLC with UV or diode array detection is another established analytical technique suitable for screening raw materials and finished products for melamine. It has been developed and validated for a variety of matrices including milk and milk derivatives, plant protein ingredients, flours, beans, meats, and beverages, and some reported methods can be used for detection of multiple melamine-related compounds including cyanuric acid, ammeline, ammelide, and cyromazine (Tittlemier 2010). An established HPLC-UV method for melamine was investigated by Lutter et al. compared to other techniques. It used both protein precipitation and strong cation exchange steps to reduce the presence of interfering compounds. An approximately 15 min reversed-phase isocratic separation by ion-pairing HPLC with C_{18} column stationary phase was reported in the method. Analysis of the UV spectrum is used to confirm the identity of melamine. Good figures of merit for linear range, sensitivity, and precision for melamine analysis in cow's milk were reported, including 0.03

and 0.09 mg/kg for LOD and LOQ, and 3.6% and 1.7% CV for repeatability at 0.13 and 0.25 mg/kg spike levels, respectively (Lutter et al. 2011). The major limitation of HPLC with UV detection is selectivity. A variety of other UV-absorbing matrix components such as peptides in infant formula that are not fully removed by the cleanup steps can result in significant overestimation of melamine. This suggests that careful cleanup steps and method validation are necessary to ensure reliable use of HPLC-UV methods. Another limitation is sensitivity for analysis of melamine-related compounds. Cyanuric acid, for example, has a much weaker chromophore compared with melamine, resulting in a higher LOD.

A variety of less-established techniques for melamine and related compound detection in food and feed matrices have been investigated. Direct analysis in real time (DART) coupled to time-of-flight mass spectrometry (TOFMS) has been reported as a potential screening method, given its simple sample preparation, rapid analysis time, and adaptability for high-throughput analysis. A recent study reported the development and validation of a DART-TOFMS method with a reported LOD of 170 and 450 μg/kg for melamine and cyanuric acid, respectively, and comparable repeatabilities to LC-MS/MS (Vaclavik et al. 2010). One limitation of this method is its selectivity for melamine if the resolving power of MS is not high enough to address spectral interferences likely to result from matrix components (Tittlemier 2010). The use of ultrahigh resolving power mass analyzers may be one way to address this limitation. Other techniques such as infrared spectroscopy (IR) and surface-enhanced Raman spectroscopy (SERS) analysis have been reported for melamine detection but suffer from high limits of detection (SERS) and the need for matrix specific calibration (IR and SERS) (Tittlemier 2010). Another recently reported rapid and simple method uses a gold-nanoparticle probe with visual colorimetric determination against color standards. Reported validation of the method indicated limited cross-reactivity from melamine analogs, and limits of detection as low as 1 mg/kg for fluid milk. Limitations include validation for only the milk matrix at this point and the lack of commercially available gold nanoparticles required for the method (Storhoff et al. 2004).

14.6.3 NEW TECHNOLOGIES IN DEVELOPMENT TO ADDRESS ECONOMIC ADULTERATION CHALLENGES

The 2007 and 2008 melamine adulteration incidents highlighted significant analytical challenges associated with detecting and preventing such incidents before they become public health issues. By its nature, economically motivated adulteration is deliberately done to evade detection by the food industry and by regulators, with the ultimate incentive of achieving a fraudulent financial advantage in the marketplace. The age-old analytical problem with uncovering adulteration is the underlying constant struggle between the science of deception and the science of detection—new adulterants are developed to evade the latest detection methods available. In the case of melamine, the analytical methods evaded were the Kjeldahl and Dumas methods developed in the 18th century (and still currently used) for determining total protein contents. These measurements are used to determine the economic value of many protein-rich foods and feed ingredients. Both methods estimate total protein by measuring total nitrogen and converting into total protein using ingredient-specific factors. The addition of a nitrogen-rich adulterant such as melamine proved to be a lucrative way to artificially increase the protein content and hence the economic value of foods (Moore et al. 2010). While melamine was a logical choice of adulterant in the 2007 and 2008 incidents, future criminals may likely choose another new inexpensive nitrogen-rich compound to evade detection.

Several potential strategies for advancing the use of analytical chemistry to prevent future protein adulteration incidents have been proposed (Moore et al. 2011). One is the development of new, more selective analytical techniques for measuring total protein contents. Such methods could replace or augment the nineteenth century Kjeldahl and Dumas methods. A recent review highlighted the opportunities and challenges for new total protein estimation methods (Moore et al. 2010).

A second approach is the development of rapid and inexpensive nontargeted methods for confirming identity of compositionally complex protein-rich food ingredients, and thereby excluding adulterants. Conceptually, such technologies would compare the chemical or spectral signatures or fingerprints of a test article to those in a library of authentic samples and provide an estimate of similarity. This approach, thereby, characterizes the intrinsic analytical signature of an authentic ingredient rather than proving the absence of specific adulterants. Analytical techniques such as IR or nuclear magnetic resonance combined with multivariate chemometrics have shown promise for such applications (DiAnibal et al. 2011). Recent reports have been published showing some promise for the use of near-IR for this purpose in the skim milk powder matrix (Botros et al. 2013; Moore et al. 2012). A number of significant challenges remain before the use of this approach can be implemented for routine analysis in food quality control laboratories. These range from the development of spectral libraries of authentic ingredients to characterize the intrinsic signatures of ingredients, to developing standardized methods capable of achieving comparable performance across different instrument platforms and different laboratories.

A third approach is the development of high-throughput screening methods for detecting a wide range of known and potential new nitrogen-rich adulterants. The development and validation of such a method was recently reported by Abernethy and Higgs using LC-MS/MS with multiple reaction monitoring and electrospray ionization. The method uses a thresholding approach for the semiquantitative detection of more than 19 nitrogen-rich adulterants when present at levels consistent with economic adulteration (Abernethy and Higgs 2013). It employs a simple two-step solvent precipitation process to remove proteins using 96-well plates, and hydrophilic interaction liquid chromatography with ultrahigh-pressure liquid chromatography allowing a rapid, 1.2 min separation. The method is reported to be capable of analyzing 80 samples in triplicate with 2 h of preparation time and 6 h of LCMS time.

Most of the analytical approaches described in this section are in their infancy and will require significant research and development to be practical for use by industry and regulatory laboratories. They have the potential to significantly prevent and deter tragic melamine-like events from occurring in the future, and are, thus, greatly needed.

REFERENCES

Abernethy, G, and K Higgs. 2013. Rapid detection of economic adulterants in fresh milk by liquid chromatography–tandem mass spectrometry. *Journal of Chromatography A* 1288: 10–20.

Allen, LM, TV Briggle, and CD Pfaffenberger. 1982. Absorption and excretion of cyanuric acid in long-distance swimmers. *Drug Metabolism Reviews* 13 (3):499–516.

Battaglia, M, CW Cruywagen, T Bertuzzi, A Gallo, M Moschini, G Piva, and F Masoero. 2010. Transfer of melamine from feed to milk and from milk to cheese and whey in lactating dairy cows fed single oral doses. *Journal of Dairy Science* 93 (11):5338–5347.

Baynes, RE, B Barlow, SE Mason, and JE Riviere. 2010. Disposition of melamine residues in blood and milk from dairy goats exposed to an oral bolus of melamine. *Food and Chemical Toxicology* 48 (8):2542–2546.

Baynes, RE, G Smith, SE Mason, E Barrett, BM Barlow, and JE Riviere. 2008. Pharmacokinetics of melamine in pigs following intravenous administration. *Food and Chemical Toxicology* 46 (3):1196–1200.

Botros, L, J Jablonski, C Chang, M Bergana, P Wehling, J Harnley, G Downy, P Harrington, and JC Moore. 2013. Exploring authentic skim milk powder variance for the development of non-targeted adulterant detection methods using NIR spectroscopy and chemometrics. *Journal of Agricultural and Food Chemistry* 61: 9810–9818.

Bozzi, A, M Dhananjeyan, I Guasaquillo, S Parra, C Pulgarin, C Weins, and J Kiwi. 2004. Evolution of toxicity during melamine photocatalysis with TiO_2 suspensions. *Journal of Photochemistry and Photobiology A: Chemistry* 162 (1):179–185.

Brown, CA, K-S Jeong, RH Poppenga, B Puschner, DM Miller, AE Ellis, K-I Kang, S Sum, AM Cistola, and SA Brown. 2007. Outbreaks of renal failure associated with melamine and cyanuric acid in dogs and cats in 2004 and 2007. *Journal of Veterinary Diagnostic Investigation* 19 (5):525–531.

CAC. 2010. Report of the thirty-third session of the Codex Alimentarius Commission. *International Conference Centre*, Geneva, Switzerland, 5–9 July 2010.

CAC. 2012. Report of the thirty-fifth session of the Codex Alimentarius Commission. *FAO Headquarters*, Rome, Italy, 2–7 July 2012.

CCCF. 2010. Proposed draft maximum levels for melmaine in food and feed. (N13–2009). *Joint FAO/WHO Food Standards Program, Codex Committee on Contaminants in Foods (CCCF). 4th Session* CX/CF 10/4/5.

CCCF. 2010. Report of the fourth session of the Codex Committee on Contaminants in Foods (CCCF). Izmir, Turkey, 26–30 April 2010.

CCCF. 2012. Report of the sixth session of the codex committee on contaminants in foods. Maastricht, The Netherlands, 26–30 March 2012.

Chapman, RP, PR Averell, and RR Harris. 1943. Solubility of melamine in water. *Industrial and Engineering Chemistry* 35 (2):137–138.

Chen, J-S. 2009. What can we learn from the 2008 melamine crisis in China? *Biomedical and Environmental Sciences* 22 (2):109–111.

Chen, Y, W Yang, Z Wang, Y Peng, B Li, L Zhang, and L Gong. 2010. Deposition of melamine in eggs from laying hens exposed to melamine contaminated feed. *Journal of Agricultural and Food Chemistry* 58 (6):3512–3516.

Cheng, G, N Shapir, MJ Sadowsky, and LP Wackett. 2005. Allophanate hydrolase, not urease, functions in bacterial cyanuric acid metabolism. *Applied and Environmental Microbiology* 71 (8):4437–4445.

Choi, L, MY Kwak, EH Kwak, DH Kim, EY Han, T Roh, JY Bae, IY Ahn, JY Jung, and MJ Kwon. 2010. Comparative nephrotoxicitiy induced by melamine, cyanuric acid, or a mixture of both chemicals in either Sprague–Dawley rats or renal cell lines. *Journal of Toxicology and Environmental Health, Part A* 73 (21–22):1407–1419.

Cruywagen, CW, WFJ van de Vyver, and MA Stander. 2011. Quantification of melamine absorption, distribution to tissues, and excretion by sheep. *Journal of Animal Science* 89 (7):2164–2169.

Cruywagen, CW, MA Stander, M Adonis, and T Calitz. 2009. Pathway confirmed for the transmission of melamine from feed to cow's milk. *Journal of Dairy Science* 92 (5):2046–2050.

Dalal, R, and D Goldfarb. 2011. Melamine-related kidney stones and renal toxicity. *Nature Reviews Nephrology* 7 (5):267–274.

DeVries, J. *US Pharmacopeia's Food Ingredients Intentional Adulterants Advisory Panel. Presented at USP's 2009 Food Ingredients Stakeholder Forum.* 2009. Available from http://www.usp.org/pdf/EN/stakeholderForums/foodAdditives/Aug2009/2009-08-04FISFPresentationspdf.pdf (slides 80–91).

Di Anibal, CV, I Ruisánchez, and MP Callao. 2011. High-resolution 1H nuclear magnetic resonance spectrometry combined with chemometric treatment to identify adulteration of culinary spices with Sudan dyes. *Food Chemistry* 124 (3):1139–1145.

Dobson, RLM, S Motlagh, M Quijano, RT Cambron, TR Baker, AM Pullen, BT Regg, AS Bigalow-Kern, T Vennard, and A Fix. 2008. Identification and characterization of toxicity of contaminants in pet food leading to an outbreak of renal toxicity in cats and dogs. *Toxicological Sciences* 106 (1):251–262.

Dominguez-Estevez, M, A Constable, P Mazzatorta, AG Renwick, and B Schilter. 2010. Using urinary solubility data to estimate the level of safety concern of low levels of melamine (MEL) and cyanuric acid (CYA) present simultaneously in infant formulas. *Regulatory Toxicology and Pharmacology* 57 (2):247–255.

Dorne, JL, DR Doerge, M Vandenbroeck, J Fink-Gremmels, W Mennes, HK Knutsen, F Vernazza, L Castle, L Edler, and D Benford. 2013. Recent advances in the risk assessment of melamine and cyanuric acid in animal feed. *Toxicology and Applied Pharmacology* 270: 218–299.

EFSA. 2007. EFSA provisional statement on a request from the European Commission related to melamine and structurally related to compounds such as cyanuric acid in protein-rich ingredients used for food and feed. *EFSA Journal* (June 8). Available from http://www.efsa.europa.eu/en/efsajournal/doc/1047.pdf

EFSA. 2010. European Food Safety Authority. Scientific opinion on melamine in food and feed. *EFSA Journal* 8(4):1573.

Fairclough, G. 2008. Tainting of milk is open secret in China. *The Wall Street Journal* 3. Available from http://online.wsj.com/article/SB122567367498791713.html

FDA. *Interim Melamine and Analogues Safety/Risk Assessment*, May 25, 2007. Available from http://www.fda.gov/Food/FoodborneIllnessContaminants/ChemicalContaminants/ucm164658.htm.

FDA. October 3, 2008. *Interim Safety and Risk Assessment of Melamine and Its Analogues in Food for Humans [a]* 2008 October 3, 2008. Available from http://www.fda.gov/Food/FoodborneIllnessContaminants/ChemicalContaminants/ucm164522.htm.

FDA. *Update: Interim Safety and Risk Assessment of Melamine and its Analogues in Food for Humans.* United States Food and Drug Administration 2008. Available from http://www.fda.gov/Food/FoodborneIllnessContaminants/ChemicalContaminants/ucm164520.htm.

Gamboa da Costa, G, CC Jacob, LS Von Tungeln, NR Hasbrouck, GR Olson, DG Hattan, R Reimschuessel, and FA Beland. 2012. Dose–response assessment of nephrotoxicity from a twenty-eight-day combined-exposure to melamine and cyanuric acid in F344 rats. *Toxicology and Applied Pharmacology* 262 (2):99–106.

Gao, C-Q, S-G Wu, H-Y Yue, F. Ji, H-J Zhang, Q-S Liu, Z-Y Fan, F-Z Liu, and G-H Qi. 2010. Toxicity of dietary melamine to laying ducks: Biochemical and histopathological changes and residue in eggs. *Journal of Agricultural and Food Chemistry* 58 (8):5199–5205.

Gonzalez, J, B Puschner, V Perez, MC Ferreras, L Delgado, M Munoz, C Perez, LE Reyes, J Velasco, and V Fernandez. 2009. Nephrotoxicosis in Iberian piglets subsequent to exposure to melamine and derivatives in Spain between 2003 and 2006. *Journal of Veterinary Diagnostic Investigation* 21 (4):558–563.

Gossner, CM-E, J Schlundt, PB Embarek, S Hird, D Lo-Fo-Wong, JJO Beltran, KN Teoh, and A Tritscher. 2009. The melamine incident: Implications for international food and feed safety. *Environmental Health Perspectives* 117 (12):1803.

He, Y, G-P Jiang, L Zhao, J-J Qian, X-Z Yang, X-Y Li, L-Z Du, and Q Shu. 2009. Ultrasonographic characteristics of urolithiasis in children exposed to melamine-tainted powdered formula. *World Journal of Pediatrics* 5 (2):118–121.

IARC. 1999. Melamine. In *Some Chemicals that Cause Tumours of the Kidney or Urinary Bladder in Rodents, and Some other Substances*. Lyon: International Agency for Research on Cancer (IARC). Monographs, Vol. 73.

ISO. 2010. ISO/TS 15495, IDF/RM 203. Milk, milk products and infant formulae—Guidelines for the quantitative determination of melamine and cyanuric acid by LC-MS/MS. ISO.

Jacob, CC, RReimschuessel, LS Von Tungeln, GR Olson, AR Warbritton, DG Hattan, FA Beland, and G Gamboa da Costa. 2011. Dose–response assessment of nephrotoxicity from a 7-day combined exposure to melamine and cyanuric acid in F344 rats. *Toxicological Sciences* 119 (2):391–397.

Jacob, CC, LS Von Tungeln, M Vanlandingham, FA Beland, and G Gamboa da Costa. 2012. Pharmacokinetics of melamine and cyanuric acid and their combinations in F344 rats. *Toxicological Sciences* 126 (2):317–324.

Jia, H. 2008. Chinese melamine crisis prompts call for better tests. *Chemistry World* 09 October.

Jutzi, K, AM Cook, and R Hütter. 1982. The degradative pathway of the s-triazine melamine. The steps to ring cleavage. *Biochemical Journal* 208 (3):679–684.

Li, G, S Jiao, X Yin, Y Deng, X Pang, and Y Wang. 2010. The risk of melamine-induced nephrolithiasis in young children starts at a lower intake level than recommended by the WHO. *Pediatric Nephrology* 25 (1):135–141.

Liu, G, S Li, J Jia, C Yu, J He, C Yu, and J Zhu. 2010. Pharmacokinetic study of melamine in rhesus monkey after a single oral administration of a tolerable daily intake dose. *Regulatory Toxicology and Pharmacology* 56 (2):193–196.

Liu, J-D, J-J Liu, J-H Yuan, G-H Tao, D-S Wu, X-F Yang, L-Q Yang, H-Y Huang, L Zhou, and X-Y Xu. 2012. Proteome of melamine urinary bladder stones and implication for stone formation. *Toxicology Letters* 212 (3):307–314.

Liu, W. 2013. *Ann In vitro Study on the Non-Enzymatic Glycation of Melamine and Serum Albumin by Reducing Sugars*. Department of Chemistry, University of Rhode Island.

Lü, MB, L Yan, JY Guo, Y Li, GP Li, and V Ravindran. 2009. Melamine residues in tissues of broilers fed diets containing graded levels of melamine. *Poultry Science* 88 (10):2167–2170.

Luengyosluechakul, S. 2007. Evidence of melamine and related substances contamination to animal feed in Thailand. *Thai Journal of Veterinary Medicine* 37 (4):7.

Lutter, P, M-C Savoy-Perroud, E Campos-Gimenez, L Meyer, T Goldmann, M-C Bertholet, P Mottier, A Desmarchelier, F Monard, and C Perrin. 2011. Screening and confirmatory methods for the determination of melamine in cow's milk and milk-based powdered infant formula: Validation and proficiency-tests of ELISA, HPLC-UV, GC-MS and LC-MS/MS. *Food Control* 22 (6):903–913.

Lv, X, J Wang, L Wu, J Qiu, J Li, Z Wu, and Y Qin. 2009. Tissue deposition and residue depletion in lambs exposed to melamine and cyanuric acid-contaminated diets. *Journal of Agricultural and Food Chemistry* 58 (2):943–948.

Ma, J, X Peng, K-W Cheng, R Kong, IK Chu, F Chen, and M Wang. 2010. Effects of melamine on the Maillard reaction between lactose and phenylalanine. *Food Chemistry* 119 (1):1–6.

Ma, M, and D Bong. 2011. Determinants of cyanuric acid and melamine assembly in water. *Langmuir* 27 (14):8841–8853.

Mast, RW, AR Jeffcoat, BM Sadler, RC Kraska, and MA Friedman. 1983. Metabolism, disposition and excretion of ^{14}C melamine in male Fischer 344 rats. *Food and Chemical Toxicology* 21 (6):807–810.

Moore, JC, JW DeVries, M Lipp, JC Griffiths, and DR Abernethy. 2010. Total protein methods and their potential utility to reduce the risk of food protein adulteration. *Comprehensive Reviews in Food Science and Food Safety* 9 (4):330–357.

Moore, JC, A Ganguly, J Smeller, L Botros, and M Mossoba. 2012. Standardisation of non-targeted screening tools to detect adulterations in skim milk powder using NIR spectroscopy and chemometrics. *NIR News* 23 (5):9–11.

Moore, JC, M Lipp, and J Griffiths. 2011. Preventing the adulteration of food protein. *Food Technology* 65 (2):56–50.

OECD. *The 2004 OECD List of High Production Volume Chemicals* 2004. Available from http://www.oecd.org/chemicalsafety/risk-assessment/33883530.pdf.

Ono, S, T Funato, Y Inoue, T Munechika, T Yoshimura, H Morita, S-IRengakuji, and C Shimasaki. 1998. Determination of melamine derivatives, melame, meleme, ammeline and ammelide by high-performance cation-exchange chromatography. *Journal of Chromatography A* 815 (2):197–204.

Prior, TJ, JA Armstrong, D Benoit, and KL Marshall. 2013. The structure of the melamine-cyanuricacid co-crystal. *CrystEngComm* 15: 5838–5843.

Reimschuessel, R, DG Hattan, and Y Gu. 2009. Background Paper on Toxicology of Melamine and Its Analogues. *Prepared for the WHO Meeting on Toxicological and Health Aspects of Melamine and Cyanuric Acid, in Collaboration with FAO and Supported by Health Canada, 1–4 December 2008.*

SCF. 1986. Scientific Committee for Food. Report of the Scientific Committee for Food on certain monomers of other starting substances to be used in the manufacture of plastic materials and articles intended to come into contact with foodstuffs. *Seventheenth series (Opinion expressed on 14 December (1984)).*

Shen, JS, JQ Wang, HY Wei, DP Bu, P Sun, and LY Zhou. 2010. Transfer efficiency of melamine from feed to milk in lactating dairy cows fed with different doses of melamine. *Journal of Dairy Science* 93 (5):2060–2066.

Sherrington, DC, and KA Taskinen. 2001. Self-assembly in synthetic macromolecular systems via multiple hydrogen bonding interactions. *Chemical Society Reviews* 30 (2):83–93.

Skinner, CG, JD Thomas, and JD Osterloh. 2010. Melamine toxicity. *Journal of Medical Toxicology* 6 (1):50–55.

Stine, CB, R Reimschuessel, CM Gieseker, ER Evans, TD Mayer, NR Hasbrouck, E Tall, J Boehmer, G Gamboa da Costa, and JL Ward. 2011. A no observable adverse effects level (NOAEL) for pigs fed melamine and cyanuric acid. *Regulatory Toxicology and Pharmacology* 60 (3):363–372.

Storhoff, JJ, SS Marla, P Bao, S Hagenow, H Mehta, A Lucas, V Garimella, T Patno, W Buckingham, and W Cork. 2004. Gold nanoparticle-based detection of genomic DNA targets on microarrays using a novel optical detection system. *Biosensors and Bioelectronics* 19 (8):875–883.

Sun, N, Y. Shen, and L-J He. 2010. Histopathological features of the kidney after acute renal failure from melamine. *New England Journal of Medicine* 362 (7):662–664.

Sun, P, JQ Wang, JS Shen, and HY Wei. 2011. Residues of melamine and cyanuric acid in milk and tissues of dairy cows fed different doses of melamine. *Journal of Dairy Science* 94 (7):3575–3582.

Sun, P, JQ Wang, JS Shen, and HY Wei. 2012. Pathway for the elimination of melamine in lactating dairy cows. *Journal of Dairy Science* 95 (1):266–271.

Tittlemier, SA. 2010. Methods for the analysis of melamine and related compounds in foods: A review. *Food Additives and Contaminants* 27 (2):129–145.

Tolleson, WH, GW Diachenko, and D Heller. 2009. Background paper on the chemistry of melamine alone and in combination with related compounds. *Prepared for the WHO Meeting on Toxicological and Health Aspects of Melamine and Cyanuric Acid, in collaboration with FAO and supported by Health Canada, 1–4 December 2008.*

Vaclavik, L, J Rosmus, B Popping, and J Hajslova. 2010. Rapid determination of melamine and cyanuric acid in milk powder using direct analysis in real time-time-of-flight mass spectrometry. *Journal of Chromatography A* 1217 (25):4204–4211.

WHO. *Toxicological and Health Aspects of Melamine and Cyanuric Acid* 2009. Available from http://whqlibdoc.who.int/publications/2009/9789241597951_eng.pdf.

Wu, Y-N, Y-F Zhao, and J-G Li. 2009. A survey on occurrence of melamine and its analogues in tainted infant formula in China. *Biomedical and Environmental Sciences* 22 (2):95–99.

Wu, Y, and Y Zhang. 2013. Analytical chemistry, toxicology, epidemiology and health impact assessment of melamine in infant formula: Recent progress and developments. *Food and Chemical Toxicology* 56:325–335.

Xin, H, and R Stone. 2008. Chinese probe unmasks high-tech adulteration with melamine. *Science* 322(5906):1310–1311.

Xue, M, Y Qin, J Wang, J Qiu, X Wu, Y Zheng, and Q Wang. 2011. Plasma pharmacokinetics of melamine and a blend of melamine and cyanuric acid in rainbow trout *Oncorhynchusykiss*. *Regulatory Toxicology and Pharmacology* 61 (1):93–97.

Zhan, Z, M Müllner, and JA Lercher. 1996. Catalytic hydrolysis of s-triazine compounds over Al_2O_3. *Catalysis Today* 27 (1):167–173.

Zheng, X, Z Aihua, X Guoxiang, Y Chi, Z Linjing, L Houkai, C Wang, Y Bao, W Jia, M Luther, M Su, JK Nicholson, and W Jia. 2013. Melamine-induced renal toxicity is mediated by the gut microbiota. *Science Translational Medicine* 5 (172):172ra22.

15 Toxins in Foods of Plant Origin

Bing Zhang, Zeyuan Deng, Yao Tang, and Rong Tsao

CONTENTS

15.1 INTRODUCTION

The importance of plants in the human diet has been recognized since ancient times. Everything we eat comes directly or indirectly from plants or organisms that eat plants. Plants are the only organisms that can transform light energy from the sun into excellent macro- and micronutrients that provide energy and sustenance for animals and humans. However, not all plants or their components can be incorporated into foods or feeds. Many naturally occurring compounds in edible plants may even have adverse effects on human health. Incidents of foodborne illness associated with plant-originated toxins have emerged as a food safety issue of global concern that encompasses developed and developing countries, such as Australia, Canada, China, New Zealand, the United States, and the United Kingdom. For example, lectins, a group of toxic plant proteins found in a variety of fruits and vegetables such as apple, banana, and cucumber, as well as in many types of legumes including soybeans, lentils, and kidney beans, are harmful to human health. Lectins, if consumed in large amounts, can cause digestion and immune distress attributing to the ability to interfere with the repair of damaged epithelial cells, as well as the hemagglutinating activity (Wu and Sun, 2012). Many other proteins found in plants including ricin, abrin, viscumin, volkensin, modeccin, soyatoxin, and canatoxin have also been reported to be toxic to humans (Audi et al., 2005). Allergies triggered by consuming certain foods derived from plant materials are also prevalent foodborne health complications, and results of inflammatory responses to such food can often lead to uncomfortable and dangerous situations. Many studies have indicated that the majority of those allergens are related to the proteins in food that is mediated usually by immunoglobulin E (IgE) or by non-IgE (cellular) mechanisms (Radauer et al., 2008). Apart from the above-mentioned macromolecular nutrients, some plant secondary metabolites (PSMs) can also result in potentially adverse effect in humans, causing food poisoning when ingested in large amounts or if processed inappropriately. As such, a comprehensive review on natural toxins in plant foods becomes highly important, but is beyond the scope of this chapter. Instead, the present chapter focuses on the PSMs in food that are toxic to humans.

PSMs are a diverse group of phytochemicals, and a majority of them have been widely investigated due to the toxicity they pose when ingested by humans, even though some PSMs have been used as nutritional diet or therapeutic herbal medicines. In general, PSMs are produced by plants in response to environmental stress conditions or biotic stimulations to protect themselves from invading bacteria, fungi, viruses, herbivorous insects, and other predators. So far, the number of potentially toxic PSMs that have been described is estimated to exceed 100,000 individual substances, many of which have been isolated and identified for chemical structures (Acamovic and Brooker, 2005). The prominent chemical classes of PSMs existing in human food plants that cause clinical symptoms of intoxication can be grouped into glucosinolates, cyanogenic glycosides, alkaloids, psolarens, juglones, cumarins, and plant endocrine disruptors. Many researchers have investigated the source and biosynthesis of these phytochemicals, and studied their toxicity and the biological mechanisms behind them. However, there is little information available on the occurrence of these compounds in commonly consumed food plants and how they may affect human health. This chapter, therefore, provides a brief review on the chemistry, biochemical mechanisms, and impact of toxic PSMs in common plant foods on human health and safety.

15.2 GLUCOSINOLATES

Glucosinolates are the major PSMs commonly present in the *Brassicaceae* plant family. *Brassica* plants such as rapeseed, mustard, cabbage, kale, broccoli, and brussels sprouts are common vegetables in many cultures and are economically important crops. Glucosinolates are stable water-soluble compounds stored in the vacuoles of most of the plant tissues, and their molecules are usually composed of a sulfur-linked β-D-glucopyranose moiety and a variable side chain (R) derived from methionine, tryptophan, or phenylalanine. Over the past few decades, more than 130 natural glucosinolates have been reported with diverse chemical properties (Agerbirk and Olsen, 2012). The biosynthetic pathway of glucosinolates usually involves three phases: (1) amino acid chain elongation, in which additional methylene groups (CH_2) are inserted into the side chain; (2) conversion of the amino acid moiety into the glucosinolate core structure; and (3) subsequent side chain modifications. More recently, Baskar et al. (2012) described the biosynthesis of glucosinolates in *Arabidopsis* and other *Brassica* crops, and a general scheme of glucosinolate biosynthesis pathway is shown in Figure 15.1. Glucosinolates in plants are always accompanied by the

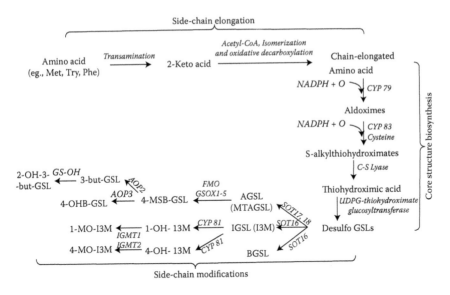

FIGURE 15.1 General scheme of glucosinolate biosynthesis pathway. Abbreviation: GSL, glucosinolate; Met, methionine; Try, tryptophan; Phe, phenylalanine; AGSL, aliphatic GSL; MTAGSL, methyl thioalkyl GSL; 4-MSB-GSL, 4-methyl sulfinyl butyl GSL; 4-OHB-GSL, 4-hydroxy butyl GSL; 3-but-GSL, 3-butenyl GSL; 2-OH-3-but-GSL, 2-hydroxy-3-butenyl GSL; IGSL, indoleglucosinolate; I3M, indol-3-yl-methyl GSL; 1-OH-I3M, 1-hydroxy I3M; 4-OH-I3M, 4-hydroxy I3M; 1-M-I3M, 1-methoxy I3M; 4-M-I3M, 4-methoxy I3M; BGSL, benzoic glucosinolate. (Modified from Baskar, V. et al. 2012. *Appl. Biochem. Biotech.* 168, 1694–1717.)

myrinase (β-thioglucosidase) enzyme (EC 3.2.3.1), which is responsible for the hydrolysis of glucosinolates to release a range of breakdown products including iso-thiocyanates, thiocyanates, thiourea, epithionitriles, oxazolidithione, and nitriles.

The glucosinolates themselves are biologically inactive molecules, while their breakdown products are the bioactive components responsible for the adverse effects in animals including humans. Tanii et al. (2004) reported that nitriles in the diet caused health-degradation in mice. The thyroid function of rats may be influenced by thiocyanate, thiourea, and oxazolidithione intake due to disrupted iodine avail-ability to the thyroid (Wallig et al., 2002). Isothiocyanates are usually responsible for the pungent taste in food or condiments such as mustard or wasabi. Long-term exposure to glucosinolates and their breakdown products has been reported to cause additional potential adverse effects such as goitrogenecity, mutagenecity, hepa-totoxicity, nephrotoxicity, and antinutritive effect (Tanii et al., 2004). Among all the adverse above-mentioned effects, the effects on thyroid metabolism have been well studied, and the goitrogenecity to animals is the major limiting factor in the commercial application of brassica feedstuffs, particularly rapeseed meal, a food safety issue that many researchers have focused on. A comprehensive review on the negative effects of glucosinolate breakdown products in animals was provided by Tripathi and Mishra (2007). While many studies have been conducted on animals, in principal, the glucosinolate breakdown products could also result in undesirable health problems if humans are exposed to the same or similar extent of these compounds. Heating the foods can inactivate myrosinase, thus preventing the formation of these breakdown products. This could be a potential means for detoxification or reducing toxic effect of glucosinolates. Despite the toxicity and negative impacts of glucosinolates and their breakdown products, these compounds have also shown beneficial health effects such as the antimicrobial and anticarcinogenic properties in recent years (Aires et al., 2009). Allyl isothiocyanate, the prevailing breakdown product of sinigrin in brown mustard, inhibited microbial growth at very low concentrations (Pang et al., 2013). Another isothiocyanate sulforaphane from broccoli together with other products such as indole-3-carbinol have been reported to have antioxidant, anticancer, and other beneficial properties, and consumption of broccoli has been widely recommended as a healthy practice (Jeffery and Araya, 2009). These seemingly controversial findings suggest a need for further studies on the balance between the adverse and beneficial effects of food plants containing glucosinolates. Glucosinolates and their breakdown products have been analyzed using conventional chromatographic methods such as HPLC (Tsao et al., 2002), or more advanced techniques using negative ion electrospray ionization LC–MS/MS multiple reaction monitoring (MRM) (Song et al., 2005).

15.3   CYANOGENIC GLYCOSIDES

Cyanogenic glycosides (cyanogens) are considered as one of the largest and most extensively studied classes of PSMs that are found in more than 2600 plant species, although edible food plants containing cyanogenic glycosides are somewhat limited to cassava, flax, bamboo shoots, tree nuts, almond, apricot, and plum. Chemically, cyanogenic glycosides are β-glucosides of α-hydroxynitriles. Although great structural diversity exists among cyanogenic glycosides, almost all of them are derived from the six different amino acids including L-valine, L-isoleucine, L-leucine, L-phenylalanine, L-tyrosine, and a nonprotein amino acid cyclopentenyl-glycine. According to Ganjewala et al., the biosynthetic pathway of cyanogenic glycosides includes the following three steps: (1) a precursor amino acid is converted into aldoxime through two successive N-hydroxylation of the amino group of the parent amino acid by an enzyme of the cytochrome-P450 family; (2) the aldoxime in turn is converted into cyanohydrin, which is catalyzed by another cytochrome-P450 enzyme; and (3) cyanohydrins get glycosylated by a soluble enzyme UDP-glucosyltransferase (Ganjewala et al., 2010).

Cyanogenic glycosides are recognized as toxic food constituents, and have detrimental effects on human health due to their potential to release the toxic hydrogen cyanide (HCN) after contacting with β-glucosidases and α-hydroxynitrilelyases when plant tissues are injured. While cyanide is a known acute poison due to its affinity for the terminal cytochrome oxidase in the mitochondrial respiratory pathway, the amount of cyanogenic glycosides in food is usually expressed as the level of equivalent hydrogen cyanide. The central nervous system is the most susceptible region for acute cyanide intoxication. Recently, a case of cyanide intoxication was reported after a girl consumed a large quantity of apricots, plum, and apricot seeds (Dogan et al.,

2013). Cyanide intoxication may be accompanied by fever, dullness, and convulsion. Cyanogenic glycoside contents in food plants can vary significantly with species, cultivar, climatic conditions, plant part, and the degree of processing. Their typical levels for some commonly consumed food plants are shown in Table 15.1.

Food processing procedures such as crushing, soaking, or fermentation can activate the β-glucosidase for immediate release of hydrogen cyanide, and further heating and storage may reduce the volatile toxin in the finished food products. Ngudi et al. (2003) reported that boiling of cassava leaves in water with added palm oil resulted in a 96–99% reduction in cyanogen levels. Obilie et al. (2004) found that steaming of the cassava product (akyeke) or fermentation of cassava pulp or dough

TABLE 15.1

The Major Cyanogenic Glycoside and General Content Containing in Commonly Consumed Plant Food

Food Plants	Major Cyanogenic Glycoside	Cyanogen Content (mg HCN/kg fresh weight)	References
Cassava (*Manihot esculenta*)—root	Linamarin	10–500	Siritunga and Sayre (2003)
Cassava (*Manihot esculenta*)—leaves	Linamarin	200–1300	Siritunga and Sayre (2003)
Sorghum (*Sorghum vulgare*)—leaves	Dhurrin	750–790	Rezaul Haque and Howard Bradbury (2002)
Flax (*Linum usitatissimum*)—seed meal	Linamarin, linustatin, neolinustatin	360–390	Rezaul Haque and Howard Bradbury (2002)
Giant taro (*Alocasia macrorrhizos*)—leaves	Triglochinin	29–32	Rezaul Haque and Howard Bradbury (2002)
Bamboo (*Bambusa arundinacea*)—Unripe stem	Taxiphyllin	3000	Hunter and Yang (2002)
Lima beans (*Phaseolus lunatus*)		100–4000	Hunter and Yang (2002)
Cherry (*Prunus* spp.)—seed	Amygdalin	1000	Hunter and Yang (2002)
Apple (*Malus* spp.)—seed	Amygdalin	690–790	Rezaul Haque and Howard Bradbury (2002)
Peach (*Prunus persica*)—Kernel	Amygdalin	710–720	Rezaul Haque and Howard Bradbury (2002)
Apricot (*Prunus armeniace*)—seed	Amygdalin	400–4000	Hunter and Yang (2002)
Plum (*Prunus* spp.)—Kernel	Amygdalin	696–764	Rezaul Haque and Howard Bradbury (2002)
Nectarine (*Prunuspersica varnucipersica*)—Kernel	Amygdalin	196–209	Rezaul Haque and Howard Bradbury (2002)
Bitter almond (*Prunus dulcis*)	Amygdalin	2900	Hunter and Yang (2002)

for 4–5 days resulted in a reduction in total cyanogen contents of 74–80% and 52–63%, respectively. Furthermore, Montagnac et al. (2009) reviewed the processing methods used to reduce the toxicity of cassava, and suggested that pounding or crushing is most effective for cyanogenic glucoside removal because it ruptures cell compartments, allowing the activation of enzymes and escape of the toxins.

Given the acute poisoning nature of cyanogenic glycosides to humans, many organizations such as the Joint FAO/WHO Expert Committee on Food Additives (JECFA), Food Standards Australia New Zealand (FSANZ), and the International Programme on Chemical Safety (IPCS) have assessed the safety of consuming foods containing these toxins. However, none of these assessments established a safe level of exposure to cyanogenic glycosides due to lack of quantitative toxicological and epidemiological information. Only the World Health Organisation (WHO) has set a recommended safe limit of 10 ppm total cyanide for cassava flour, which has been adopted by many countries (Kolind-Hansen and Brimer, 2010).

The contents of cyanogenic glycosides in their native form can be determined directly by various chromatographic methods. However, such methods are not widely practical due to lack of cyanogenic glycoside standards or their high cost. Indirect determination via quantification of HCN released after acidic or enzymatic hydrolysis of cyanogenic glycosides has been extensively employed in the quantification of cyanogenic glycosides. A ninhydrin-based spectrophotometric micromethod has been recently adapted to indirectly assess the total cyanogenic compounds (cyanogens) in plants (Surleva and Drochioiu, 2013).

15.4 ALKALOIDS

Alkaloids are nitrogen-containing naturally occurring compounds with very diverse structures. Most plant alkaloids are derived from amino acids and exist in about 20% of plant species. There are by far more than 3000 alkaloids identified, many of which have been exploited as pharmaceuticals, stimulants, narcotics, and poisons due to their potent biological activities (Vina et al., 2012). For the same reason, because of the highly toxic nature of these compounds, only a few alkaloid-rich plant species are consumed as foods. In this chapter, particular attention is paid to the *Solanaceae*, or nightshade family of plants, as they are one of the most economically important and widely consumed vegetable crops, including potato, tomato, eggplant, chili, and bell peppers. Nearly all *Solanaceae* family foods contain alkaloids. Solanaceous steroidal glycoalkaloids are compounds consisting of a steroidal alkaloid backbone and a carbohydrate moiety through ester bond. Generally, these compounds can be categorized into two groups according to the structure of the aglycone: the group possessing an oxo-azaspiro structure including tomatine from *S. lycopersicum* (tomato) and solasonine from *S. melongena* (eggplant), and the other group possessing a cyclic amine structure, such as α-solanine and α-chaconine from *S. tuberosum* (potato) (Ohyama et al., 2013). All of these steroidal glycoalkaloids are believed to be derived from cholesterol. Details on the biosynthesis of the steroidal alkaloids in *Solanaceae* plants can be found in Ohyame et al. (2013).

Steroidal glycoalkaloids, particularly α-solanine and α-chaconine in potato, can lead to serious toxic effects on human health. At the cellular level, these compounds exhibit strong lytic properties and inhibition activity against acetylcholine and butyrylcholinesterases. The symptoms of steroidal glycoalkaloid poisoning usually include gastrointestinal disorders, confusion, hallucinations, partial paralysis, convulsions, coma, and even death (Krits et al., 2007). Al Chami et al. (2003) reported that the adverse effects of exposure to steroidal glycoalkaloids also included lower respiratory activity and blood pressure, interference with sterol and steroid metabolism, and bradycardia or hemolysis resulting primarily from membrane disruption. In addition, consumption of foods containing high steroidal glycoalkaloid content has been considered as a possible environmental factor for schizophrenia.

Typically, the contents of steroidal glycoalkaloids found in commercial potatoes are less than 100 mg/kg fresh weight with the majority in between 25 and 52 mg/kg fresh weight (Mweetwa et al., 2012). Potato tubers taste bitter if the amount increased to 200 mg/kg fresh weight (Ginzberg et al., 2009). The concentration of glycoalkaloids is largely associated with environmental factors such as light, mechanical injury, and storage conditions. In general, production of these glycoalkaloids is raised by the same conditions that promote the development of chlorophylls such as exposure to light. Therefore, the contents of these glycoalkaloids are higher in potato sprouts, green potato skins, tomato vines, and immature green tomatoes. Physical damage and wounding may also activate sterol and steroidal glycoalkaloid synthesis in potato. In order to avoid the intoxication of glycoalkaloids, consumers should avoid consumption of potatoes that show signs of sprouting, greening, physical damage, or rotting. However, as the amount of these glycoalkaloids cannot be significantly affected by cooking or frying under high temperature, commercial potato cultivars meant for human consumption normally contain low concentrations of glycoalkaloids, and content in tubers exceeding 200 mg/kg fresh tuber weight is considered very undesirable (Ginzberg et al., 2009). Furthermore, since glycoalkaloids are usually accumulated in the skin and just below its surface, peeling of tubers prior to preparation for consumption is considered as the most effective step to reduce the risk.

Many analytical methods have been developed for glycoalkaloids in foods such as potato and eggplant, most of which are based on chromatographic detections (Abreu et al., 2007). Other methods such as immunoassays have also been explored; those include enzyme-linked immunosorbent assay (ELISA), fluorescence polarization immunoassay, and solution phase immunoassay with capillary electrophoresis. These methods are particularly useful for the analysis of their aglycone forms (Driedger et al., 2000).

15.5 COUMARINS AND FUROCOUMARINS

Coumarin (benzo-α-pyrone) is a fragrant compound naturally occurring in a variety of plants, although it can be synthetically produced and used as a food additive, or a spice and perfume ingredient due to its distinctive sweet scent. One natural source of coumarin is edible tonka beans, known by the French as *coumarou*, from which coumarin was first isolated. Coumarin and its derivatives such as 7-hydroxycoumarin can also be found in many other commonly consumed food plants such as cassia (bastard cinnamon or Chinese cinnamon), cinnamon, melilot (sweet clover), green tea, peppermint, celery, bilberry, lavender, honey (derived both from sweet clover and lavender), and carrots, as well as beer, tobacco, wine, and other edible materials (Felter et al., 2006). The contents of coumarin in these plants may vary greatly, from less than 1 mg/kg (ppm) in celery to 900–40,600 mg/kg in cinnamons, while cassia was reported for the highest detected coumarin content of 87,000 mg/kg (Felter et al., 2006; Lake, 1999). The amount of human dietary exposure to coumarin has been estimated to be 0.02 mg/kg/day; the dose can be up to 0.06 mg/kg/day if added the exposure from cosmetic products (Lake, 1999).

Addition of synthetic coumarin to foods was banned in the United States in 1954 based on reports of the hepatotoxicity in rats. It was suspected to have genotoxic and carcinogenic effects in the 1980s (Abraham et al., 2010). Based on these potential toxicities, the Codex Alimentarius recommended that specific maximum levels for coumarin in the final product consumed by human can be established (Alimentarius, 1985). In general, the maximum level is 2 mg/kg for foods and beverages, with the exception of special caramels and alcoholic beverages for which the maximum level is 10 mg/kg. The Codex Alimentarius maximum levels were subsequently introduced into European law in 1988. However, more recent evidence has suggested that coumarin may not be a genotoxic agent (Lake, 1999). Effort has then been focused on the carcinogenic and hepatotoxic effects of coumarin. Abraham et al. (2010) reviewed the toxicology and performed a risk assessment of coumarin, and concluded that hepatotoxicity and carcinogenicity were indeed observed in many laboratory animals, but as for the adverse effect on human, they found that a subgroup of the human population may be more susceptible to the hepatotoxic effect than the animal species investigated. However, Felter et al. (2006) suggested that the adverse effects in humans caused by exposure to coumarin are restricted to high doses associated with various oral clinical therapies. The side effects following coumarin treatment were found to include mild dizziness, diarrhea, and vomiting, in addition to the hepatotoxicity in humans administered highdose (50–7000 mg/day) for therapeutic purposes (Loprinzi et al., 1999). Based on the nonobserved-adverse-effect level (NOAEL) for hepatotoxicity, in October 2004, the European Food Safety Authority (EFSA) reviewed coumarin toxicity and established a tolerable daily intake (TDI) of 0.1 mg/kg bw (Sproll et al., 2008). The German Federal Institute for Risk Assessment (BfR) also accepted the same TDI under consideration of human data available from the pharmaceutical use of coumarin (Sproll et al., 2008). Furocoumarins are structurally related to coumarins. Psoralen is one of the furocoumarins often found in plants belonging to the *Rutaceae* and *Umbelliferae* families such as parsnip, parsley, celery, and carrots. Psoralen and its derivatives (psoralens) are formed from coumarins in the shikimate pathway, and this biosynthesis process is usually catalyzed by a cytochrome P450-dependent monooxygenase (psoralen 5-monooxygenase), and requires cofactors (NADPH) and molecular oxygen.

Phototoxicity is the common form of poisoning caused by psoralen exposure. When skin comes in contact with wet foliage of plants containing psoralens and is then exposed to sunlight, it may result in second-degree sunburns due to the absorption of wavelengths of light not normally taken in. Placzek et al. reported that fragrances containing psoralens are phototoxic, particularly to the skin when exposed to sunlight. Psoralen, 5-methoxypsoralen (5-MOP, bergapten), and 8-methoxypsoralen

(8-MOP, xanthotoxin or methoxsalen) are three most active forms of psoralens that can cause photodermatitis. They can be released after exposure to near UV light (320–380 nm) and form adducts with DNA and DNA-cross links, thus leading to cell death, mutations, and chromosome aberrations, and ultimately to cancer development. Dolan et al. (2010) found that 5-MOP and 8-MOP produced skin tumors in experimental animals in the presence of ultraviolet A radiation. Skin cancers have also been reported in patients treated with 8-MOP and long-wave ultraviolet light for psoriasis or *Mycosis fungoides* (Stern et al., 1997).

In commonly consumed food plants, the concentration of psoralens was found to be from 45 to 100 mg/kg in celery, 40 to 145 mg/kg in parsnip, and 112 mg/kg in parsley (Coulombe, 2001; Lombaert et al., 2001). The levels of these natural toxins in plants are influenced by various stress factors such as mechanical damage, attack by insects or fungi, and unfavorable climatic/storage conditions. Schulzová et al. (2007) provided an overview on various factors influencing the levels of psoralens in plants, and indicated that the concentrations of psoralens found in organically grown parsnip and celery were higher than that in the conventionally grown products, and a large increase was triggered by mechanical injury and/or infestation by pests (fungi, insects) during storage. It should be noted that these compounds are heat-stable and are not destroyed by cooking (boiling or microwave) (Ivie et al., 1981). Peeling and blanching are considered as efficient procedures in reducing dietary psoralen exposure.

Dietary exposure to psoralens in humans has been estimated in several countries to be 1.3 mg (the United States), 1.45 mg (Germany), and 1.2 mg (the United Kingdom) per person per day, which is corresponding to 0.020–0.023 mg/kg bw/day for a 60 kg person (Guth et al., 2011). The NOAEL for 8-MOP was determined to be 25 mg/kg bw/day for liver toxicity, and 37.5 mg/kg bw/day for cancer induction in rats (Dolan et al., 2010). The phototoxicity threshold of psoralens is equivalent to about 15 mg 8-MOP per person/day (Dolan et al., 2010). Therefore, regular dietary consumption of psoralens is not considered to be a significant risk factor for humans. There is no FDA regulation or guideline specifically for the presence of psoralens in food.

Coumarin compounds are mainly analyzed by HPLC with spectrophotometric or fluorescent detection (Hroboňová et al., 2013). Other novel methods including two-dimensional thin layer chromatography (2D-TLC), capillary electrophoresis (CE) with laser-induced fluorescence detection, capillary electrochromatography (CEC), micellarelectrokinetic chromatography (MEKC), and gas chromatography (GC) have been employed in the quantitative analysis of coumarins (Cieśla and Waksmundzka-Hajnos, 2009; Sgorbini et al., 2010).

15.6 JUGLONE

Most members of the walnut family (*Juglandaceae*) such as black walnut, pecan, hickory, butternut, and related plants can produce juglone (5-hydroxy-alphanaptho-quinone) in all parts of these plants. In particular, black walnut and butternut contain the highest quantity of juglone, at concentrations sufficient to cause allelopathic phytotoxicity to other adjacent plant species. Juglone inhibits enzymes in the respiration chain of sensitive plants (Dana and Lerner, 2001). Symptoms of plants affected by juglonephytotoxicity range from stunting of growth, to partial or total wilting, to death. Juglone may be toxic to insects and possess antimicrobial activities, but humans are seldom seriously affected by juglone. In general, the highest concentrations of juglone are found in the roots of the tree and the hulls of the nut, hence consumption of walnut kernel does not normally result in juglone toxicity. However, people may experience a symptom if they are allegic to juglone. In addition, some people can also suffer from a rash when handling the fruits and nuts of black walnut. The contents of juglone in plants are usually analyzed using HPLC (Cosmulescu et al., 2011).

15.7 PLANT ENDOCRINE DISRUPTORS

Plant endocrine disruptors can be defined as substances which are naturally produced in plants and may alter normal hormone regulations in animals. The majority of plant endocrine disruptors are also known as phytoestrogens, which embody several groups of nonsteroidal estrogens including isoflavones, coumestans, prenylated flavonoids, and lignans that widely occur in plants including food plants. Almost all of these phytoestrogens are believed to possess weak estrogenic properties, affecting the endocrine system owing to their structural similarity to that of the inherent hormone estradiol; that is, the key structural element of these compounds is the phenolic ring that binds to estrogen receptors in the same fashion as estradiol.

In general, human exposure to phytoestrogens is mainly through daily diets such as cereals, rice, and soybeans. A review on the distribution and contents of a variety of phytoestrogens in dietary plants can be found in Yang et al. (2006). Isoflavonoids are the most studied group among different phytoestrogens. Isoflavones are found almost exclusively in the *Leguminosae* family plants such as soybean, alfalfa, and clover. Soybeans are abundant in isoflavones, particularly genistein, daidzein, and glycitein, although mostly as glycosides or acylglycosides. The typical content of isoflavones in soybeans is approximately 2 g/kg fresh weight. Red clover, although mostly used as a forage plant for animals, contains mostly methylated isoflavones, and its total isoflavone content was found to be several folds greater than those found in soybeans (Tsao et al., 2006). Coumestrol, which is the most common and active form of coumestans, is structurally related to isoflavones. The highest concentration of coumestrol found in clover and fresh alfalfa sprouts was 5.6 and 0.7 mg/g dry weight, respectively (Kurzer and Xu, 1997). As incorporating sprouts into our diet become increasingly popular, attention must be paid to this compound. Lignans, a class of nonflavonoid phytoestrogens, are usually found in fiber-rich foodstuffs such as flaxseed, whole wheat, legumes, cereal, and unhulled soybeans. The highest concentration of lignans was found in dried flaxseeds (80 mg/100 g) and unhulled soybeans (20.5 mg/100 g), while lesser amounts were found in other food plants ranging from 0.08 to 0.9 mg/100 g (Kurzer and Xu, 1997; Yang et al., 2006). It must be emphasized that the phytoestrogen content in plants can vary among varieties and through processing.

While many studies have focused on the potential health benefits of the aforementioned phytoestrogens, particularly the soy isoflavones, it is still a controversial subject about the risks and benefits of phytoestrogens in human. There is an increasing concern about the long-term consequences of exposure to these chemicals due to the potential to disturb the endocrine system and thus cause adverse health effects. Cederroth et al. (2012) pointed out that while consumption of phytoestrogen-containing food has been frequently associated with beneficial health effects, the potentially adverse effects on development, fertility, and the reproductive and endocrine systems are likely underappreciated. Animal experiments have shown that defatted soy causes thyroid carcinoma. Doerge and Sheehan (2002) reviewed the thyroid toxicity of genistein, and concluded that the possibility of widely consuming soy products may cause harm in the human population via either or both estrogenic and goitrogenic activities. In addition, a great deal of publications clearly demonstrated that developmental exposure to genistein can lead to adverse reproductive effects in female rodents such as alterations in ovarian development, the timing of vaginal opening, estrous cyclicity, ovarian function, HPG axis, subfertility, and an increased incidence of uterine adenocarcinoma (Delclos et al., 2009). In humans, there is an apparent scarcity of studies on evaluating the effects of phytoestrogens on fertility and reproductive parameters. Among limited available information, Chavarro et al. (2008) found that dietary intake of soy-based foods and isoflavones was inversely related to sperm counts. Recently, numerous concerns have been raised on infants, a susceptible population to be exposed to phytoestrogens through soy-based infant formula. Therefore, infants exposed to substantial levels of phytoestrogens during the development period, a very sensitive period for endocrine disruption, can have potentially long-term detrimental effects on reproduction, fertility, and behavior. Bernbaum et al. (2008) had conducted a prospective study that showed human infants fed with soy-based formula at times exhibited an estrogenized vaginal epithelium when compared with those fed with cow-based formula or breastmilk. Similarly, women fed with soy formula during their childhood were found at increased risk of developing benign smooth-muscle tumors of the uterus (D'Aloisio et al., 2010). Therefore, it is important to investigate the long-term detrimental effects of exposure to high-level phytoestrogens on humans using epidemiological or clinical studies.

Many analytical methods have been developed for quantitative and qualitative analyses of the different phytoestrogenic compounds in food samples. Among these methods, reversed-phase liquid chromatography (RP-LC) with UV/Vis, fluorescence, and MS detection are the most frequently used ones because of their capability in simultaneous analysis, separation efficiency, and high sensitivity (Wu et al., 2004).

15.8   OTHER TOXINS IN FOOD PLANTS

Apart from the above-mentioned plant toxins, other naturally occurring phytochemicals in food plants, such as some antinutrient saponins, phytic acid, and tannins, have also been found to be capable of eliciting deleterious effects on human health depending on their type and quantity. Saponins are glycosides of the aglycone sapogenins. According to the chemical structures of the sapogenins, saponins can be categorized into two major groups: steroidal and triterpenoid saponins, which are usually found in soybeans, sugar beets, peanuts, spinach, asparagus, broccoli, potatoes, apples, eggplants, alfalfa, and ginseng root. Many saponins provide health beneficial effects such as antioxidant, antitumor, cardiovascular and central nervous system effects, antidiabetes effects, and immunomodulation (Yuan et al., 2010). The primary toxic effects of saponins include their capability to hemolyse red blood cells and to trigger diarrhea and vomiting, which are associated with the reduction of surface tension (Onder and Kahraman, 2009).

Phytic acid, known also as phytate in its salt form, is the principal storage form of phosphorus in many plant tissues, as well as an energy storage and a source of cations and myoinositol (a cell wall precursor), but is generally considered as an antinutrient agent. It constitutes about 1–5% by weight of many cereals and legumes (Cheryan and Rackis, 1980). Phytates are not digestible by humans or nonruminants due to the lack of phytase, an enzyme responsible for generating free phosphoric acid and inositol. Phytic acid is a strong chelator of divalent minerals including copper, calcium, magnesium, zinc, and iron (Reddy and Sathe, 2002), capable of forming insoluble complexes with these minerals, thus decreasing the bioavailability of important minerals. Furthermore, since the phytates are ionic in nature, they can directly or indirectly bind to charged groups of proteins, adversely affecting protein digestion and bioavailability.

Tannins are a complex class of polyphenolic compounds that can be widely found in vegetal foodstuffs, particularly in fruits, cereal grains, and beverages (red wine, tea, and beer). They can be structurally grouped into condensed (polymers of catechin) and hydrolyzable tannins (esters of glucose with gallic acid). While free phenolic acids such as gallic acid and condensed tannins such as oligomeric procyanidins are strong antioxidants and play significant roles in risk reduction for many chronic diseases (Tsao, 2010), tannins, in general, acting similarly to the phytates, can also form complexes with proteins, starch, and digestive enzymes, causing a reduction in nutritional values of foods. Since intake of these antinutrients may result in negative effects on human health, detection of these compounds in various foods becomes important. Many analytical methods are currently available for total estimation and specific quantification of these antinutrients (Doria et al., 2012).

15.9 SUMMARY

There is a growing awareness that some inherent compounds originated from food plants may have adverse effects on human health or may be toxic to humans. Understanding the occurrence, contents, and the chemistry, biochemistry, and toxicology of food plant toxins will help in establishing guidelines and strategies to eliminate and reduce these toxins and antinutritional agents, providing consumers with safe and healthy foods through improved breeding, production, processing, and storage. While only the major and most important groups of plant food toxins were included in this chapter, it is hoped that information on these compounds and related discussions, including the concentrations in commonly consumed foods and the analytical methods used, can serve as a reference for researchers, consumers, food industry, and regulatory agencies. It must also be pointed out that, plant-based foods available in the market, especially in developed countries, are generally safe, and the amount of the toxic compounds or food-nutrients consumed through normal dietary practices will not be of concern to consumers. In-depth reviews on the compounds discussed in this chapter can be found in the literature. A general review such as this chapter also serves as a reminder of the importance of food chemical safety to different stakeholders, particularly in times such as now when consumers are constantly bombarded with conflicting scientific discoveries on phytochemicals.

REFERENCES

Abraham, K., Wöhrlin, F., Lindtner, O., Heinemeyer, G., Lampen, A., 2010. Toxicology and risk assessment of coumarin: Focus on human data. *Mol. Nutr. Food Res.* 54, 228–239.

Abreu, P., Relva, A., Matthew, S., Gomes, Z., Morais, Z., 2007. High-performance liquid chromatographic determination of glycoalkaloids in potatoes from conventional, integrated, and organic crop systems. *Food Control* 18, 40–44.

Acamovic, T., Brooker, J.D., 2005. Biochemistry of plant secondary metabolites and their effects in animals. *Plant Nutr. Soc.* 64, 403–412.

Agerbirk, N., Olsen, C.E., 2012. Glucosinolate structures in evolution. *Phytochemistry* 77, 16–45.

Aires, A., Mota, V.R., Saavedra, M.J., Rosa, E.A.S., Bennett, R.N., 2009. The antimicrobial effects of glucosinolates and their respective enzymatic hydrolysis products on bacteria isolated from the human intestinal tract. *J. Appl. Microbiol.* 106, 2086–2095.

Al Chami, L., Méndez, R., Chataing, B., O'Callaghan, J., Usubillaga, A., LaCruz, L., 2003. Toxicological effects of α-solamargine in experimental animals. *Phytother. Res.* 17, 254–258.

Alimentarius, C., 1985. General requirements for natural flavourings (CAC/GL 29.1987).

Audi, J., Belson, M., Patel, M., Schier, J., Osterloh, J., 2005. Ricin poisoning a comprehensive review. *J. Am. Med. Assoc.* 294, 2342–2351.

Baskar, V., Gururani, M., Yu, J., Park, S., 2012. Engineering glucosinolates in plants: Current knowledge and potential uses. *Appl. Biochem. Biotech.* 168, 1694–1717.

Bernbaum, J.C., Umbach, D.M., Ragan, N.B., Ballard, J.L., Archer, J.I., Schmidt-Davis, H., Rogan, W.J., 2008. Pilot studies of estrogen-related physical findings in infants. *Environ. Health Persp.* 116, 416.

Cederroth, C.R., Zimmermann, C., Nef, S., 2012. Soy, phytoestrogens and their impact on reproductive health. *Mol. Cell. Endocrinol.* 355, 192–200.

Chavarro, J.E., Toth, T.L., Sadio, S.M., Hauser, R., 2008. Soy food and isoflavone intake in relation to semen quality parameters among men from an infertility clinic. *Hum. Reprod.* 23, 2584–2590.

Cheryan, M., Rackis, J.J., 1980. Phytic acid interactions in food systems. *Crit. Rev. Food Sci. Nutr.* 13, 297–335.

Cieśla, Ł., Waksmundzka-Hajnos, M., 2009. Two-dimensional thin-layer chromatography in the analysis of secondary plant metabolites. *J. Chromatogr. A* 1216, 1035–1052.

Cosmulescu, S.N., Trandafir, I., Achim, G., Baciu, A., 2011. Juglone content in leaf and green husk of five walnut (*Juglans regia* L.) cultivars. *Not. Bot. Horti Agrobo. Cluj.* 39, 237–240.

Coulombe, R.A., 2001. *Natural Toxins and Chemopreventives in Plants.* CRC Press: Boca Raton, FL, USA.

D'Aloisio, A.A., Baird, D.D., DeRoo, L.A., Sandler, D.P., 2010. Association of intrauterine and early-life exposures with diagnosis of uterine leiomyomata by 35 years of age in the sister study. *Environ. Health Persp.* 118, 375.

Dana, M.N., Lerner, B.R., 2001. *Black Walnut Toxicity.* HO-193, Department of Horticulture, Purdue University, Cooperative Extension Service, West Lafayette, IN, pp. 1–2.

Delclos, K.B., Weis, C.C., Bucci, T.J., Olson, G., Mellick, P., Sadovova, N., Latendresse, J.R., Thorn, B., Newbold, R.R., 2009. Overlapping but distinct effects of genistein and ethinyl estradiol (EE2) in female Sprague–Dawley rats in multigenerational reproductive and chronic toxicity studies. *Reprod. Toxicol.* 27, 117–132.

Doerge, D.R., Sheehan, D.M., 2002. Goitrogenic and estrogenic activity of soy isoflavones. *Environ. Health Persp.* 110, 349.

Dogan, M., Yilmaz, C., Kaya, A., Caksen, H., Taskin, G., 2013. Cyanide intoxication with encephalitis clinic: A case report. *East. J. Med.* 11, 22–25.

Dolan, L.C., Matulka, R.A., Burdock, G.A., 2010. Naturally occurring food toxins. *Toxins* 2, 2289–2332.

Doria, E., Campion, B., Sparvoli, F., Tava, A., Nielsen, E., 2012. Anti-nutrient components and metabolites with health implications in seeds of ten common bean (*Phaseolus vulgaris* L. and *Phaseolus lunatus* L.) landraces cultivated in southern Italy. *J. Food Compos. Anal.* 26, 72–80.

Driedger, D.R., LeBlanc, R.J., LeBlanc, E.L., Sporns, P., 2000. A capillary electrophoresis laser-induced fluorescence method for analysis of potato glycoalkaloids based on a solution-phase immunoassay. 1. Separation and quantification of immunoassay products. *J. Agr. Food Chem.* 48, 1135–1139.

Felter, S.P., Vassallo, J.D., Carlton, B.D., Daston, G.P., 2006. A safety assessment of coumarin taking into account species-specificity of toxicokinetics. *Food Chem. Toxicol.* 44, 462–475.

Ganjewala, D., Kumar, S., Devi, A., Ambika, K., 2010. Advances in cyanogenic glycosides biosynthesis and analyses in plants: A review. *Acta Biol. Szegediensis* 54, 1–14.

Ginzberg, I., Tokuhisa, J.G., Veilleux, R.E., 2009. Potato steroidal glycoalkaloids: Biosynthesis and genetic manipulation. *Potato Res.* 52, 1–15.

Guth, S., Habermeyer, M., Schrenk, D., Eisenbrand, G., 2011. Update of the toxicological assessment of furanocoumarins in foodstuffs (Update of the SKLM statement of 23/24 September 2004)—Opinion of the Senate Commission on Food Safety (SKLM) of the German Research Foundation (DFG). *Mol. Nutr. Food Res.* 55, 807–810.

Hroboňová, K., Lehotay, J., Čižmárik, J., Sádecká, J., 2013. Comparison HPLC and fluorescence spectrometry methods for determination of coumarin derivatives in propolis. *J. Liq. Chromatogr. R. T.* 36, 486–503.

Hunter, I.R., Yang, F.E., 2002. *Cyanide in Bamboo Shoots*. International Network for Bamboo and Rattan (INBAR), WP39, p. 7. http://www.inbar.int/publications/?did=110

Ivie, G.W., Holt, D.L., Ivey, M.C., 1981. Natural toxicants in human foods: Psoralens in raw and cooked parsnip root. *Science* 213, 909–910.

Jeffery, E.H., Araya, M., 2009. Physiological effects of broccoli consumption. *Phytochem. Rev.* 8, 283–298.

Kolind-Hansen, L., Brimer, L., 2010. The retail market for fresh cassava root tubers in the European Union (EU): The case of Copenhagen, Denmark—A chemical food safety issue? *J. Sci. Food Agr.* 90, 252–256.

Krits, P., Fogelman, E., Ginzberg, I., 2007. Potato steroidal glycoalkaloid levels and the expression of key isoprenoid metabolic genes. *Planta* 227, 143–150.

Kurzer, M.S., Xu, X., 1997. Dietary phytoestrogens. *Annu. Rev. Nutr.* 17, 353–381.

Lake, B.G., 1999. Coumarin metabolism, toxicity and carcinogenicity: Relevance for human risk assessment. *Food Chem. Toxicol.* 37, 423–453.

Lombaert, G.A., Siemens, K.H., Pellaers, P., Mankotia, M., Ng, W., 2001. Furanocoumarins in Celery and Parsnips: Method and multiyear Canadian survey. *J. AOAC Int.* 84, 1135–1143.

Loprinzi, C.L., Kugler, J.W., Sloan, J.A., Rooke, T.W., Quella, S.K., Novotny, P., Mowat, R.B., Michalak, J.C., Stella, P.J., Levitt, R., 1999. Lack of effect of coumarin in women with lymphedema after treatment for breast cancer. *New Engl. J. Med.* 340, 346–350.

Montagnac, J.A., Davis, C.R., Tanumihardjo, S.A., 2009. Processing techniques to reduce toxicity and antinutrients of cassava for use as a staple food. *Compr. Rev. Food Sci. Food Saf.* 8, 17–27.

Mweetwa, A.M., Hunter, D., Poe, R., Harich, K.C., Ginzberg, I., Veilleux, R.E., Tokuhisa, J.G., 2012. Steroidal glycoalkaloids in *Solanum chacoense*. *Phytochemistry* 75, 32–40.

Ngudi, D.D., Kuo, Y.H., Lambein, F., 2003. Cassava cyanogens and free amino acids in raw and cooked leaves. *Food Chem. Toxicol.* 41, 1193–1197.

Obilie, E.M., Tano-Debrah, K., Amoa-Awua, W.K., 2004. Souring and breakdown of cyanogenic glucosides during the processing of cassava into akyeke. *Int. J. Food Microbiol.* 93, 115–121.

Ohyama, K., Okawa, A., Moriuchi, Y., Fujimoto, Y., 2013. Biosynthesis of steroidal alkaloids in Solanaceae plants: Involvement of an aldehyde intermediate during C-26 amination. *Phytochemistry.* 289, 26–31.

Onder, M., Kahraman, A., 2009. Antinutritional Factors in Food Grain Legumes, Conference Papers, 1st International Symposium on Sustainable Development, Sarajevo, June 9–10, 2009.

Pang, Y.H., Sheen, S., Zhou, S., Liu, L., Yam, K.L., 2013. Antimicrobial effects of allyl iso-thiocyanate and modified atmosphere on *Pseduomonas aeruginosa* in fresh catfish fillet under abuse temperatures. *J. Food Sci.* 78, M555–M559.

Radauer, C., Bublin, M., Wagner, S., Mari, A., Breiteneder, H., 2008. Allergens are distributed into few protein families and possess a restricted number of biochemical functions. *J. Allergy Clin. Immunol.* 121, 847–852.e847.

Reddy, N.R., Sathe, S.K., 2002. *Food Phytates*. CRC PressI LLC, Boca Raton, FL.

Rezaul Haque, M., Howard Bradbury, J., 2002. Total cyanide determination of plants and foods using the picrate and acid hydrolysis methods. *Food Chem.* 77, 107–114.

Schulzová, V., Hajšlová, J., Botek, P., Peroutka, R., 2007. Furanocoumarins in vegetables: Influence of farming system and other factors on levels of toxicants. *J. Sci. Food. Agr.* 87, 2763–2767.

Sgorbini, B., Ruosi, M.R., Cordero, C., Liberto, E., Rubiolo, P., Bicchi, C., 2010. Quantitative determination of some volatile suspected allergens in cosmetic creams spread on skin by direct contact sorptive tape extraction–gas chromatography–mass spectrometry. *J. Chromatogr. A* 1217, 2599–2605.

Siritunga, D., Sayre, R.T., 2003. Generation of cyanogen-free transgenic cassava. *Planta* 217, 367–373.

Song, L., Morrison, J.J., Botting, N.P., Thornalley, P.J., 2005. Analysis of glucosinolates, iso-thiocyanates, and amine degradation products in vegetable extracts and blood plasma by LC–MS/MS. *Anal. Biochem.* 347, 234–243.

Sproll, C., Ruge, W., Andlauer, C., Godelmann, R., Lachenmeier, D.W., 2008. HPLC analysis and safety assessment of coumarin in foods. *Food Chem.* 109, 462–469.

Stern, R.S., Nichols, K.T., Väkevä, L.H., 1997. Malignant melanoma in patients treated for psoriasis with methoxsalen (psoralen) and ultraviolet A radiation (PUVA). *New Engl. J. Med.* 336, 1041–1045.

Surleva, A., Drochioiu, G., 2013. A modified ninhydrin micro-assay for determination of total cyanogens in plants. *Food Chem.* 141, 2788–2794.

Tanii, H., Takayasu, T., Higashi, T., Leng, S., Saijoh, K., 2004. Allylnitrile: Generation from cruciferous vegetables and behavioral effects on mice of repeated exposure. *Food Chem. Toxicol.* 42, 453–458.

Tripathi, M.K., Mishra, A.S., 2007. Glucosinolates in animal nutrition: A review. *Anim. Feed Sci. Tech.* 132, 1–27.

Tsao, R., 2010. Chemistry and biochemistry of dietary polyphenols. *Nutrients* 2, 1231–1246.

Tsao, R., Papadopoulos, Y., Yang, R., Young, J.C., McRae, K., 2006. Isoflavone profiles of red clovers and their distribution in different parts harvested at different growing stages. *J. Agr. Food Chem.* 54, 5797–5805.

Tsao, R., Yu, Q., Potter, J., Chiba, M., 2002. Direct and simultaneous analysis of sinigrin and allyl isothiocyanate in mustard samples by high-performance liquid chromatography. *J. Agr. Food Chem.* 50, 4749–4753.

Vina, D., Serra, S., Lamela, M., Delogu, G., 2012. Herbal natural products as a source of monoamine oxidase inhibitors: A review. *Curr. Top. Med. Chem.* 12, 2131–2144.

Wallig, M.A., Belyea, R.L., Tumbleson, M.E., 2002. Effect of pelleting on glucosinolate content of Crambe meal. *Anim. Feed Sci. Tech.* 99, 205–214.

Wu, Q., Wang, M., Simon, J.E., 2004. Analytical methods to determine phytoestrogenic compounds. *J. Chromatogr. B* 812, 325–355.

Wu, W., Sun, R., 2012. Toxicological studies on plant proteins: A review. *J. Appl. Toxicol.* 32, 377–386.

Yang, M., Park, M.S., Lee, H.S., 2006. Endocrine disrupting chemicals: Human exposure and health risks. *J. Environ. Sci. Health C* 24, 183–224.

Yuan, C., Wang, C., Wicks, S., Qi, L., 2010. Chemical and pharmacological studies of saponins with a focus on American ginseng. *J. Gins. Res.* 34, 160.

Index